BUSINESS DATA COMMUNICATIONS

BASIC CONCEPTS, SECURITY, AND DESIGN

Wiley Series in Computers and Information Processing Systems in Business

BUSINESS DATA COMMUNICATIONS
BASIC CONCEPTS, SECURITY, AND DESIGN

Jerry FitzGerald

Jerry FitzGerald & Associates

John Wiley & Sons

New York
Chichester
Brisbane
Toronto
Singapore

Cover Photo   © E. B. Weill/The Image Bank
Cover Design   Sheila Granda

Copyright © 1984, by John Wiley & Sons, Inc.

Library of Congress Cataloging in Publication Data:

FitzGerald, Jerry.
    Business data communications: Basic concepts, security, and design.

    (Wiley series in computers and information processing
systems for business)
    Includes index.
    1. Data transmission systems.   2. Computer networks.
3. Office practice—Automation.   I. Title.   II. Series.

TK5105.F577   1984   621.38   83-14798
ISBN 0-471-89549-0

Printed in the United States of America

10 9 8 7

ABOUT THE AUTHOR

Dr. Jerry FitzGerald is the principal in Jerry FitzGerald and Associates, a management consulting firm. He has extensive experience in computer security, audit and control of computerized systems, data communications security, and systems analysis. In addition to consulting, he also conducts training courses and seminars in these subjects.

Prior to starting his own firm, Dr. FitzGerald was a senior management consultant with an international consulting firm, an associate professor in data processing for a state university system, and a senior systems analyst at both a major medical center and a computer manufacturer.

As a consultant, Dr. FitzGerald has been active in numerous system design projects, EDP audit reviews, new system development control reviews, EDP audit training, internal control reviews of on-line systems, and control/security of data communication networks. This work has included development of the computer security administration function within organizations, redesign of the system development life cycle process, development of data communication networks for organizations, review of the internal EDP audit function on behalf of management, and development of requests for proposals with regard to selection and purchase of computer systems.

Dr. FitzGerald has over 20 years of business and consulting experience with organizations in many fields. These include:

- Aerospace
- Banking
- Computers
- CPAs
- Education
- Government
- Health Care
- Manufacturing
- Pharmaceuticals
- Stock Brokerage

He has conducted many seminars for both private firms and government agencies in the United States and other countries.

In 1980, Dr. FitzGerald was the annual recipient of the Joseph J. Wasserman Award. This award is given by the EDP Auditors Association to the person who made the most outstanding contributions in the areas of EDP auditing, control, and security during the year.

In addition to numerous articles, he is the author of four books, three of which have been translated into Spanish. The books are *Designing Controls into Computerized Systems*, *Fundamentals of Systems Analysis*, *Internal Controls for Computerized Systems*, and *Fundamentals of Data Communications*.

Dr. FitzGerald's education includes a Ph.D. in Business Administration from the Claremont Graduate School, an M.B.A. from the University of Santa Clara, and a Bachelor's Degree in Industrial Engineering from Michigan State University. He is also a Certified Information Systems Auditor (CISA) and has a Certificate in Data Processing (CDP).

PREFACE

Data communications is a dynamic technology that is revolutionizing the manner in which business and government conduct their operations. This book reflects the current state of the art in data communications.

The book is systems oriented in order to integrate the technical concepts of data communications into the automated business office. It incorporates the systems approach for understanding, designing, managing, securing, and implementing data communication networks.

The book is divided into three parts in order to introduce the *fundamentals of data communications* (Part One), to define the *network management/security/ control* aspects (Part Two), and to describe the *network design fundamentals* Part Three).

Your learning experience should be taken in three separate segments as shown on the next page.

Part One (Chapters 1 to 6) is devoted to the basic fundamentals and introductory concepts of modern-day data communication networks.

After the introductory chapter, Chapter 2 begins with a detailed review of the hardware of data communication networks. Chapter 3 is a collection of all the "technical and application details." Chapter 4 describes the basic components and traces the flow of a message as it moves from the terminal, through the data communication circuits, and then to the host computer. Chapter 5 is a detailed

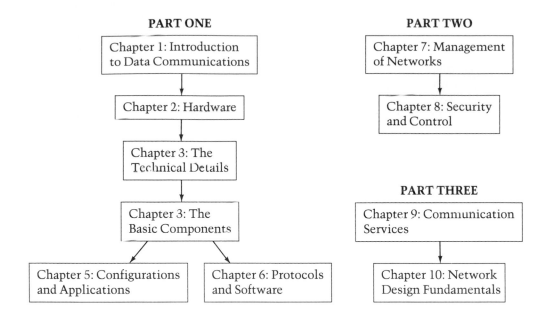

review of the various configurations of data communication networks. It also in-cludes selected applications, such as the automated office, so you can visualize both the network configuration and its use in the business environment. Chapter 6 is a description of protocols and software. It starts with the same message flow as Chapter 4, but this time the message is passing through the different protocol/software packages in the various pieces of hardware. Chapter 6 also describes the seven layers of the ISO standard as well as protocols, software packages, and net-work architectures.

Part Two (Chapters 7 and 8) is devoted to the human and organizational aspects of managing networks as well as the security requirements and the controls that are necessary.

Chapter 7 covers network management, control reports, test equipment, orga-nizational charts, and the management skills required to manage a network. Chapter 8 describes the security and control aspects of data communication net-works. There is an increasing emphasis on security and control that is correlated with our increasing usage and dependence upon on-line real-time data communi-cation networks. This chapter clearly depicts the security/control points for a net-work, specific controls that can be implemented, risk analysis, and a methodol-ogy for identifying/documenting/evaluating the security controls in a network.

Part Three (Chapters 9 and 10) is devoted to designing data communication net-works as well as describing the available communication services.

Chapter 9 describes the communication services such as private leased services, measured use, and other special services. Besides the United States, selected Canadian and worldwide services are included. Chapter 10, the closing chapter, describes thirteen detailed steps to follow during the design of data communication networks. This is a step-by-step systems approach that can be used when one is designing new networks or enhancing current ones.

At the end of the book there is an extensive Bibliography, which includes computerized literature resources, and a comprehensive Glossary. You will notice the index is very detailed and is quite complete in its coverage.

The appendixes contain various network control matrices, lists of specific communication-oriented controls, a methodology for conducting a risk analysis for communication networks, typical costs for data communication channels/circuits, and a methodology for evaluating teleprocessing monitors.

This book is intended for *both* college and university courses, as well as the working professional who needs to learn about the world of data communications/teleprocessing. You do not need any prior data communications knowledge to understand this book, although some basic knowledge of computers or data processing would be helpful.

You probably will be surprised at the amount of knowledge gained by reading this book, and you will feel confident that you have a firm grasp on the world of data communications.

The following reviewers were very helpful with suggestions and comments on the manuscript and their contribution is much appreciated: Professor Carl Taylor of Orange Coast College, Professor Herman Washington of La Guardia Community College, Professor Curtis Rawson of Kirkwood Community College, Jeffrey Held of Network Strategies, Professor Paul Licker of University of Calgary, and Professor John Walter of University of California, Dominguez Hills.

Jerry FitzGerald
Redwood City, California

ACKNOWLEDGMENTS

Several prominent people in the business and academic world of data communications have made contributions to various sections of this textbook. We would like to formally acknowledge their efforts here and thank them for their contributions. Listed below, in alphabetic sequence, are the organizations and the individuals who contributed to this textbook.

Contel Information Systems
130 Steamboat Road
Great Neck, NY 11024

David A. Rubin, Vice-President of Network Management Systems, contributed the section on Modeling Networks.

Mort Fortgang, Vice-President of Operations, contributed the section on Local Area Networks (LAN).

Contel Information Systems provides customized software and turnkey systems for telecommunications, data processing, automated manufacturing applications; computer aided design/analysis for communication networks; and consulting services in data and voice communications/office automation. They also have developed the CONTELNET Local Area Network.

CYLIX™ Communications Network
800 Ridge Lake Boulevard
Memphis, TN 38119

D. Dwight Drinkard, Director of Marketing, contributed the section on Cost Analysis and Figures 10-11 through 10-17.

The CYLIX™ Communications Network Corporation designs and develops custom networks using their own intelligent modem for their clients.

Prime Factors
6529 Telegraph Avenue
Oakland, CA 94609

Michael Schwartz, the owner, contributed the sections on Hardware Encryption and Encryption.

Prime Factors is a company that specializes in the implementation of practical encryption systems for all types of organizations, especially the banking and financial industries.

Software Design Associates, Inc.
71 Fifth Avenue
New York, NY 10003

Jackie Herbst, Manager of Support Services, contributed Appendix 6, on Evaluating Teleprocessing Monitors.

Software Design Associates, Inc., provides data processing services to financial service, insurance, manufacturing, and telecommunications industries. SDA markets a complete range of data processing, education, and training services, and a software product line for large data processing operations.

Telephone Museum
Telephone Pioneers of America
George S. Ladd Chapter No. 27
1145 Larkin Street
San Francisco, CA 94109

Don T. Thrall, Archivist at the Telephone Museum, contributed numerous photographs that appear throughout this book.

University of Calgary
Facility of Management
2500 University Drive, N.W.
Calgary, Alberta T2N 1N4 Canada

Professor Paul Licker contributed the section on Canadian Communication Services. He is a member of the faculty of management and teaches courses on data communications and other related subjects.

Warren Stallings & Associates
22 Admiralty Place
Redwood City, CA 94065

Warren Stallings, the owner, contributed the sections on the Automated Office and UNIX.

Warren Stallings & Associates is a consulting firm that specializes in the development and implementation of microprocessor/microcomputer-based systems.

CONTENTS

PART ONE

FUNDAMENTALS OF DATA COMMUNICATIONS

Part One of this book is devoted to the basic fundamentals and introductory concepts of modern-day data communication networks.

1

INTRODUCTION TO DATA COMMUNICATIONS

The purpose of this chapter is to introduce data communications and to show the progression of systems as they moved toward today's networks. A description of "how to use this book" is included to help adapt it to your specific needs. Generic types of data communication networks will be defined and the layout and structure of a basic data communication system will be discussed. The chapter closes with a discussion on the future of networks.

WHY STUDY DATA COMMUNICATIONS

The reasons for studying data communications can be summed up in the occupational history of the United States. In the 1800s we were an agricultural society dominated by farmers. By the 1900s we had moved into an industrial society dominated by labor and management. Now, as we approach the twenty-first century, we clearly have moved into the information society which is dominated by computers, data communications, and highly skilled individuals who use brain power instead of physical power. The industrial society has reached its zenith and the communication/computer era, which is dubbed the information society, is advancing rapidly.

In an industrial society, the strategic resource is capital. In an information society, the strategic resource is knowledge which creates information that must flow on communication networks. This information society started in the mid-1950s.

Knowledge of data communications is even more important when you realize that satellites are transforming the earth into a global city. In other words, the compression of time that is achieved through satellite communications allows us to be in immediate contact with all other companies or people and to utilize business information in a timely manner.

In an information society, dominated by computers and communications, value is increased by knowledge as well as the speed of movement of that knowledge. This new information economy will completely destroy Karl Marx's "labor theory of value," because in such a society what increases value is not the labor of individuals, but information. Knowledge/information can be created, it can be destroyed, and it is synergetic in that the whole usually is greater than the sum of the parts. In fact, the whole may be many times greater than the sum of the parts if you have the proper communications network to transmit the information. Knowledge that cannot be disseminated (transmitted) may be of zero value.

The main stream of the information age is communications. The value of a high-speed data communication network that transmits knowledge/information is that it brings the message sender and the message receiver closer together in time. The effect of this is to collapse the information lag, which is the time it takes for information to be disseminated throughout the world. For example, in the 1800s it might have taken several weeks for specific information to reach the United States from England. By the 1900s it could be transmitted within the hour. Today, with modern data communication systems, it can be transmitted within seconds. Collapsing the information lag speeds up the incorporation of new technology into our daily lives. In fact, today's problem may be that we are unable to handle the quantities of information that we already have.

Finally, the transition from an industrial to an information society means that you will have to learn many new technologically based skills. Instead of becoming a specialist in a certain subject and planning on working in that area for the rest of your life, you will have to adapt and possibly retrain yourself several times during your lifetime. For that reason, the study of data communications will become a basic tool that can be used throughout your lifetime. You will incorporate your knowledge of data communications into several careers such as circuit designer, programmer, business system application developer, communication specialist, business manager, and so on. Even the basic physical job tasks of our society now require technical knowledge in the use of data communications such as: citizens band radios in trucks, microcomputers in your home that are connected to national or international communication networks, and personal communication devices such as mobile telephones (cellular radio).

In closing let me forecast that the collapsing of the information lag may be the single most important point in your current study of communications. This is because new communication technology is being incorporated into the fabric of the information society as fast as people are able to learn how to maintain and use this technology. Once the basics are learned from this textbook, you will need to "keep up with the communication technology" for the remainder of your life.[1]

ABOUT THIS BOOK

Data communications can be a complex and sometimes confusing subject, although it need be neither complex nor confusing. As you read this book, you will encounter many new terms because the world of data communications has its own jargon. A comprehensive glossary has been included at the end of the book because it is not always possible to interrupt the presentation of seemingly complex subjects with a thorough definition of a new word. The author has evaluated the text carefully to ensure that all the jargon of data communications has been included in the Glossary and also that the miscellaneous technical details are described thoroughly.

This book introduces the world of data communications in enough technical detail so that you can put your new-found knowledge to work. The chapters are:

1. Introduction to Data Communications (1)
2. Data Communications Hardware (2)
3. Technical Details (3)
4. Basic Components (4)
5. Configurations/Applications(5)

---

[1]Many of the ideas in this section have been extrapolated from the first chapter of the book *Megatrends* by John Naisbitt (New York: Warner Books, Inc., 1982).

6. Protocols/Software (6)

7. Network Management (7)

8. Security/Control (8)

9. Communication Services (9)

10. Network Design Fundamentals (10)

There are many good books and periodicals on data communications. A bibliography at the end of this book provides a selected sample of current books, serial publications, and computerized literature resources. At this point you might review both the Glossary and the Bibliography.

HOW TO USE THIS BOOK

This book is divided into three parts. Part One introduces the fundamentals of data communications. Part Two defines the network management/security/control aspects of networks. Part Three describes the network design fundamentals.

Your learning experience should be taken in three separate segments as shown in Figure 1–1. The first four chapters in Part I take you from the introduction on to hardware, the technical details, and the basic componenets of data communications. Chapters 5 and 6 (Configurations and Applications; Protocols and Software) should be read, in either sequence, after the first four chapters are digested. At this point you will have the necessary knowledge to go on to the more advanced topics in Parts Two and Three.

Parts Two and Three can be covered in either sequence, depending upon your needs. It is suggested that Network Management and Security (Part Two) be read first because these two topics should be taken into account during network design.

DATA COMMUNICATIONS

Data communications involves the movement of information by means of communication circuits. *Data communications* is the movement of encoded information from one point to another by means of electrical or optical transmission systems. Such systems are often called *data communication networks.* In general, data communication networks are established to collect data from remote points (called *terminals*) and transmit those data to a central point equipped with a computer or another terminal, or to perform the reverse process, or some combination of the two. Data communication networks facilitate more efficient use of computers. Networks improve the day-to-day controls of a business by providing

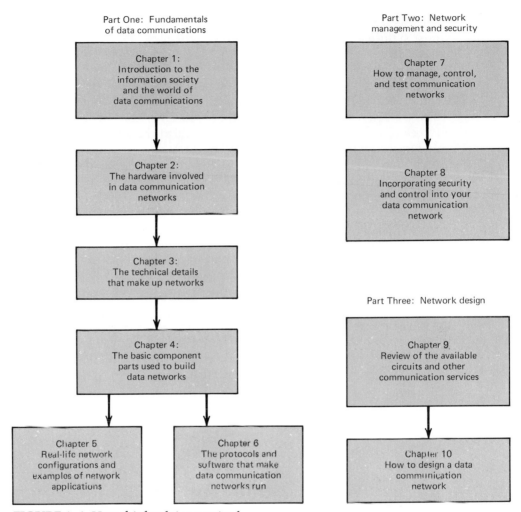

FIGURE 1–1 How this book is organized.

faster information flow. They provide message switching services to allow terminals to talk to one another. In general, they offer better and more timely interchange of data among their users and bring the power of computers closer to more users.

Telecommunications and *teleprocessing* are other terms used to describe data transmission between a computing system and remotely located devices. The terms data communications, telecommunications, and teleprocessing are used interchangeably by many writers in this field. Data communications is the term we will use because of the combination of computing and communications, especially as it relates to the concept of the automated office of the future. In fact, the most significant change in the business office will be the integration of data communications, voice communications, and imaging systems, along with the use of host computers and microprocessors.

Figure 1–2 shows the basic components for a data communications system. You must have a sender, which is usually a terminal. It might be a teleprinter terminal, a video terminal, or some other type. Once the user has entered a message, it goes to the encoder, which is usually called the *modem*. In this example, the modem converts the signal from its direct electrical pulses (baseband) into a series of varying frequency tones (broadband). The purpose of this encoding process is to put the transmission into a mode that is compatible with the various transmission facilities such as copper wires, microwave, satellite, fiber optic, or other facilities.

In our figure these transmission facilities are referred to as a *circuit*. These are the telephone company circuits over which your message will move. Finally, when your message reaches the distant host computer it first passes through the decoder, which is another modem. This modem converts the signal from broadband (frequency tones) back to baseband (electrical voltages). Finally, your signal (in reality your message request) is passed on to the host computer for whatever processing might be required. While Figure 1–2 depicts only a simple data communication connection, it will suffice for now. Later we will present more complex pictures of the various types of configurations that can be utilized when one is building data communication networks.

A BASIC SYSTEM

Figure 1–3 depicts a basic data communication system. This system includes terminals, connector cables, a line-sharing device, modems, local loops, telephone company switching offices, interexchange channel (IXC) facilities, a front end communications processor, and a host computer.

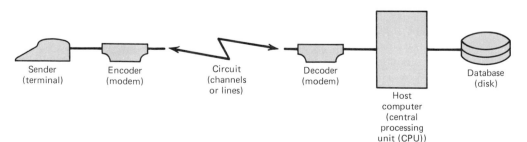

FIGURE 1–2 Basic components for a data communication system.

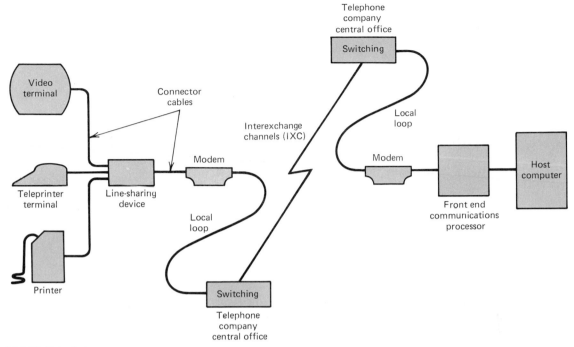

FIGURE 1–3 Basic system.

The terminals involve a human-to-machine interface device where people can enter and receive data or information. This type of device might have a video screen or a teleprinter type of printing mechanism, and a keyboard. In the future this device may be voice actuated.

The *connector cables* in Figure 1–3 are special cables containing many wires that interconnect the terminal to the modem.

The *line-sharing device* allows multiple terminals to share a single modem. Each terminal sequentially has its turn to transmit and receive data/information.

The *modem* is a solid state electronic device that converts direct electrical baseband signals (+ and − voltages of electricity) to modulated broadband signals that can be sent over data communication circuits. The most common form of modulated signal is a frequency modulated signal where the direct electrical voltages are converted to frequency tones. For example, a high-pitched tone might equal a binary 1 and a low-pitched tone equals a binary 0.

The *local loops* in Figure 1–3 are the intracity connections or "last mile" that interconnects your home or office to the telephone company central office (switching office), or to the special common carrier network if you are using a connection other than the telephone company.

The *central office* (sometimes called *end office* or *exchange office*) contains the various switching and control facilities that are operated by the telephone company or other special common carrier. When you utilize the dial-up telephone facilities, your data transmission goes through these switching facilities. When you have a private leased circuit, however, the telephone company wires your circuit path around the switching facilities in order to provide a clear unbroken path from one modem to the other.

The *interexchange channels/circuits* (sometimes called *IXC circuits*) are the long-haul circuits between cities. In reality, they go from one telephone company central office to another telephone company central office. These circuits are usually microwave circuits, but they may be copper wire pairs, coaxial cables, satellite circuits, optical fibers, or other transmission medium.

The *front end communication processor* is a specialized minicomputer with very special software programs (protocols/telecommunication access programs). These software programs, along with the front end hardware, control the entire data communication network. For example, a powerful front end communication processor may have 100 or more modems attached to it through its ports (circuit connect points).

Finally, the *host computer* is the central processing unit (CPU) that processes your request, performs database lookups, and carries out the data processing activities required for the business organization.

SYSTEM PROGRESSION

The natural evolution of business systems, governmental systems, and personal systems has forced the widespread use of data communication networks to interconnect these various systems.

In the 1950s we had batch systems with discrete files, and the users carried their paper documents to the computer for processing. The data communications of that era involved human beings physically carrying paper documents (see Figure 1–4).

During the 1960s we added communication circuits (telephone lines) and gave the users on-line batch terminals. At this point the users entered their own batches of data for processing. The data communication aspect involved the transmission of signals (messages) from these on-line batch terminals to the computer and back to the user.

During the late 1960s and into the 1970s we started developing on-line real-time systems that moved the users from batch processing to single transaction oriented processing where the response back to the user department was required within approximately three seconds or less. It was during this time that data communications really became a necessity.

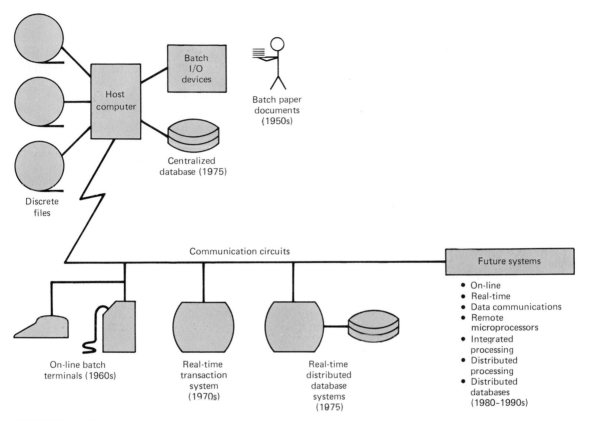

FIGURE 1–4 System progression.

As the 1970s progressed, we added database management systems that replaced the older discrete files. We also started developing integrated systems whereby one business system might automatically create and pass transactions to some other business system. With integrated systems, the entry of an on-line real-time transaction could automatically trigger two or three other transactions. For example, when an on-line terminal user from a purchasing department entered data indicating the purchase of 100 executive desks, the system might initiate three other related transactions. Transaction 1 would go to the Accounts Payable System. It would set up the original matching file where the purchase order would be matched to the invoice, which in turn would be matched to the receiving dock ticket showing that the goods were received. Transaction 2 might go to the receiving dock in order to prenotify the personnel that they should expect 100 executive desks in two months. Transaction 3 might go to the Cash Flow Accounting System so preparations can be made to pay for (cash availability) these executive desks. As you can see, both data communications and data processing are interconnected in on-line real-time systems.

As we enter the 1980s we are fast approaching completely distributed systems where user departments will be given their own computers (probably microprocessors with disks) and the data communication network will have to be even larger and more reliable. The office of the future that interconnects typewriters, word processing machines, facsimile machines, copy machines, teleconferencing equipment, microprocessors, mainframe host computers, and other equipment will put tremendous demands on data communication networks. Also, local area networks (loop circuits within buildings) will offer greater reliability and speed.

Finally, the ultimate in reliability of the network must be achieved before we move to distributed databases. High reliability is necessary because if one distributed site uses another distributed site's database and the communication circuit fails, it might lock up the database so no one could use it until the communication of the transaction is completed.

TYPES OF NETWORKS

Basically there are seven overlapping categories of networks.

- Single application
- Multiple application
- Organization-wide
- Multiorganization
- Value added
- Common carriers
- International

The *single application* network is built within a single corporation or government agency. It is used for one specific purpose. For example, in banking you might have a network for the bank balance inquiries for the automated teller machines, checking accounts, or passbook savings.

The *multiple application* network is designed to handle many different applications which can share the network and the common database and/or processing facilities. A multiple application network might be seen in a manufacturing organization. This type of network might handle business systems involving raw material inventory, production planning, the manufacturing process, finished goods inventory, sales and distribution, general ledger accounting, cash flow, accounts payable, accounts receivable, and so on.

Organization-wide networks are developed by large corporations and government agencies that have many computer centers. The networks are designed to interconnect the multiple computer centers. For example, a large government agency could have multiple computer centers in order to keep track of agricultural data, farming patterns, crops, and acreage records. They might place the computer centers in various locations around the country. This organization-wide network would serve its users by allowing local and remote access to any of the data centers and transmission between these data centers.

Multiorganization networks have been constructed to serve groups of similar corporations such as airlines or universities. When you make airline reservations, if any leg of your trip is to be on another airline, the multiorganization reservations network handles transmission of the data to the other airline so proper reservations can be guaranteed.

A *value added* network provides a network constructed with leased lines (circuits) and it serves many customers in different geographical areas. It is usually a general purpose computer network like the ones developed by public companies such as Telenet, Tymnet, Sprint, or MCI. These value added carriers may transmit either data or voice. Their objective is to allow many different users to use their network for a fee, which is dependent upon the amount of time the user is using the network (voice calls) or the volume of data being transmitted (data calls). In other words, they lease circuits from the telephone company, build a network, and by doing so, add value to the raw communication circuits because these circuits are now reliable functioning systems.

Common carriers, such as telephone companies, also provide nationwide data networks which you can use for a set fee. You also can lease communication circuits from them to build your own network for single or multiple applications that may be either organization-wide or multiorganization. These common carriers are now offering value added networks. There are also *special common carriers* which lease communication circuits in competition with the telephone companies; therefore, your circuits may or may not be provided by the telephone company.

International networks may be single/multiple application or organization-wide/multiorganization and they span the globe. In other words, an international network passes over the borders between countries. Special limitations may be imposed upon these international networks with regard to the flow of information (transborder data flow controls). These controls are enacted by the government of each country.

THE FUTURE OF NETWORKS

By 1990, data communications/teleprocessing will have grown faster and become more important than computer processing itself. Both go hand in hand, but recently we have moved from the computer era to the communications era.

There are many exciting prospects in the next ten years with regard to communications. Some of these prospects will be discussed in the following paragraphs.

Videotex will be connected to private homes. Videotex involves the two-way transmission between a television set in your home and many other organizations outside of your home. It allows you to carry on a two-way dialogue with a doctor, take courses in your home, provide security services for your house, review information retrieval databases, perform teleshopping from your local store, conduct teleconferences (picture and voice) from your home, play video games, have community access to political meetings, view first run movies, interconnect with satellite television programming, utilize packet switching networks for delivery of electronic mail messages, connect with your post office, review the most current news stories, and utilize voice store and forward message systems. Canada already has videotex (called Telidon there), as do England and France.

With regard to voice and data, there is already equipment available that combines both voice transmission and data transmission over a single communication circuit. Combining these would be very cost effective in most governmental and private business organizations because much higher circuit usage could be achieved at reduced costs. Approximately three-quarters of today's communication costs are for voice and one-quarter are for data transmission.

We will have home satellite TV. There might be a satellite dish antenna, located on the roof of a house, which will enable us to communicate directly with each other via the satellite. This satellite dish antenna might lead us to transmit either voice or data directly from our house in the United States to someone else's house in Canada or England (see Figure 1–5).

Such use of satellites by individuals leads to widespread possibilities with regard to the freedom and flow of information and ideas. Because the borders of a country may no longer be able to be closed to the free flow of data, information, and ideas, people around the world may become more politically aware. For this reason transborder data flow will become more significant.

Many countries today restrict the flow or movement of data across their national boundaries. The United States is probably the most open with regard to the flow of information into and out of the country. Even the United States, however, limits data flow through restrictions upon the sale or delivery of some technological equipment or information to countries that are viewed as less than friendly. Canada requires that the initial processing of all bank transactions be done in Canada and that foreign networks cross the border only at one crossing point. France is investigating the possibility of taxing data. Sweden has a data inspection board that must approve the export of data files or the transmission of personal data out of Sweden. In England you must share your secret encryption key with the postal and telegraph service. Belgium and France have up to a $400,000 fine for transmitting data that are defined as sensitive. In Spain you must deposit money in an escrow account before data files can be transmitted out of the coun-

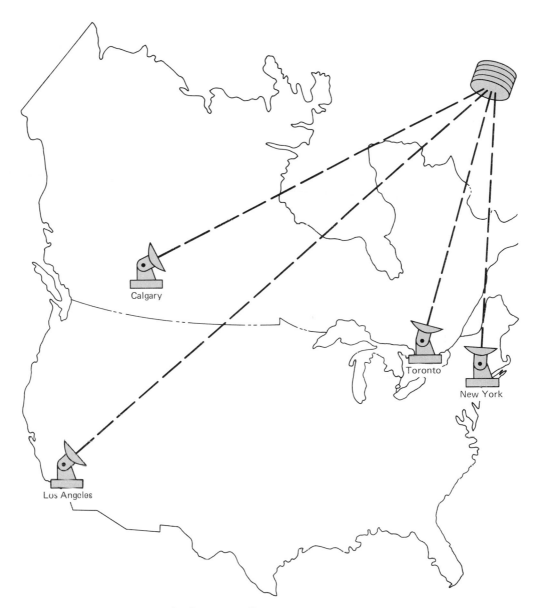

FIGURE 1–5 International information flow.

try. Such data protection laws are a type of tariff or duty on the free flow of information.

As we move into the next decade, you will note that we are moving from the
manufacturing/management era to the information era. In other words, informa-

tion is the single most valuable resource of an enterprise. Information is more important than management structure, manufacturing ability, or financial capabilities. Thus, a country that restricts information will most likely slow down its economic growth, thereby lowering the standard of living for all its citizens.

The normal voice telephone systems will have store and forward capability for voice information. In other words, if you call someone who is not at home, the telephone system will accept your voice message and forward it later to the person when he or she calls in. The system can even try repeatedly to contact the person until their telephone is answered and your stored message is delivered.

There will be more encryption used on data communication circuits. *Encryption* is the method of encoding your data to make it secret during transmission. Encryption on public networks will become standard, as will Forward Error Correction (FEC). *Forward error correction* is the process of automatically correcting most circuit originated transmission errors without retransmitting the message that contains the error.

There will be more public networks utilizing packet switching. These public networks will have "standard interfaces" that will connect almost any terminal to anything. In other words, any terminal or microprocessor will be able to communicate with any other terminal or microprocessor on the public packet switching network.

Along with more satellites to make better use of television/news, we may even have citizens band radio via satellite. The laws will be changed to encourage satellite, and governments (for economic reasons) will create a good data communications environment within their countries.

More companies will purchase their own computer-based private telephone systems for use within their company. They will interconnect with the rest of the world through rooftop antennas that will transmit via satellite.

Another way these companies will transmit to the rest of the world may be through cellular radio local loops. With cellular radio, radio transmission towers are placed in strategic locations throughout a city. The messages from companies or private homes are transmitted over the airways as radio frequency transmissions to these towers. The towers then connect to land-based communication circuits, microwave circuits, or satellite transmission systems for the long-haul (Interexchange Channel-IXC) transmission of voice and data. Cellular radio local loops would replace the copper wire local loops that you see throughout the city. In other words, cellular radio towers would both augment and replace telephone lines/telephone poles or underground copper wire cables.

One of the major requirements for adequate data communications in business systems is a large document storage and retrieval system for information that is contained in business documents and for archival storage of documents, market data, books, television shows, movies, local news, and the like. We are on the verge of getting mass data storage on optical disks. These optical disks are of the same type that you see today for recorded movies. It is forecasted that optical disks in the future will cost about $10.

If you stacked optical disks (similar to the way records are stacked in juke boxes), you could store the entire contents of the U.S. National Archives on only 1000 12-inch disks. These disks are a "write-once" medium so you have security in the knowledge that no one can modify your data. Optical disks have approximately a ten-year life.

The U.S. Postal Service (England, Canada, and France are ahead of the United States) is getting into the data communications area for electronic message transmission. The service is called Electronic Computer Originated Mail (ECOM). Initially the system will allow users to transmit computer-generated messages, via private telecommunication carriers, to computers in major post offices across the country. The messages are printed in hard copy form at the post offices and then delivered within two days by first class mail. This service is similar to one offered by Western Union called Mailgram.

The cable television companies will be increasing their role for two-way communications into and out of each person's home. They are in direct competition with rooftop satellite antennas and the major television networks. These cable TV companies and the common carrier business of data communications may merge into a business cable for the private, commercial business, and government market. This communication pipeline into your home, via both the telephone system and cable TV, is a very critical issue because of privacy and security. As an individual consumer, you will need to impose extra privacy on your data and your life. For example, just by using cable TV, someone might determine which television shows you watch and therefore build a profile of your personality. Also, the TV company could keep track of your purchases, your financial transactions, and anything else that is received or transmitted on the cable to your personal television set. Personal privacy may be a grave concern here.

Did you know that every day 20 million meetings are held in the United States and that more than three-quarters of all meetings last less than 30 minutes? Over one-half of all meetings could be handled by voice communications only, one-third of all meetings are for the exchange of information only, and almost 90 percent of air travel in the United States is business travel. For these reasons you will see a distinct increase in video teleconferencing. With teleconferencing, people who want to attend a business meeting but are in diverse geographic locations can get together in both voice and picture format. In fact, even documents can be shown and copied at any of the remote locations.

Communications will be enhanced further because integrated circuit chips containing 1 million components per chip will be available. These chips will allow us to develop hardware solutions (firmware) for our current software problems. This capability will definitely increase both the speed and reliability of data communication networks. The chips, which currently allow us to have a video screen, will soon give us an entire wall as our picture screen, and in the future an entire holographic wall for our data pictures. Early in the twenty-first century there will be virtually no paper used in business communications. Everything will be stored using microchips and large-scale memory devices such as bubble memory or optical disks.

The competition for Digital Termination Systems (DTS) will become very strong. DTS is a solution to the telephone company local loop bottleneck or "last mile" between a user application and long-haul digital transmission (IXC) facilities. DTS is a set of technology and service options that might include microwave, cable TV, telephone company wire pairs, infrared, digital radio broadcast (cellular radio), optical fiber, and rooftop antenna satellite services.

Both electronic mail and voice mail will grow quite rapidly. Electronic mail will grow primarily in the business sector and between the private homes of individuals who have microprocessors. Voice mail will grow as more special common carriers and telephone companies offer voice store and forward systems.

We are living in an era that is controlled, and soon will be dominated, by data communications. If you think the computer has had an impact upon your life or your lifestyle, then you might be surprised, when you look back in 1990, to try to determine the changes that were brought about because of data communications. The ultimate in data communications has been a standard part of the television series *Star Trek*, where they use a "transporter" to beam people down from space ships to various planets. While today this is science fiction, it might not be science fiction in the twenty-first century.

QUESTIONS

1. Is there a difference between data communications and telecommunications?
2. Define the progression of systems from the 1950s to the present date.
3. Is it possible for a large business organization to have a combination of all seven categories of networks that were described in this chapter?
4. Describe the most recent data communication development that you have read about in the newspaper or other periodical.
5. Why would restrictions on transborder data flow hamper business?

2

THE HARDWARE

This chapter defines each piece of hardware that might be utilized in a data communication network. The hardware's technical capabilities are described along with its purpose within the network. Numerous figures show the specific location of the hardware within the network environment.

HOST COMPUTER

The host computer generally is considered to be the central computer or central processing function for a business or government data communication application processing system. In distributed processing, there may be several host computers tied together by the data communication network. While this host computer is not truly a part of the network, it may perform many network functions. This is because many of the network functions currently are shared between the host computer and the front end communication processor.

In the beginning of data communication networks, during the 1960s, the host computer handled all of the functions. It wasn't until the middle sixties that powerful front end communication processors were developed to remove the network control functions from the host computer. The general trend today is to remove everything that you can from the host computer and move it further out into the network. Figure 2–1 shows the downline movement of some of the functions that can be moved out of the host computer. This movement increases the efficiency of each piece of hardware because it offloads some of its duties to the next piece of hardware.

For example, the data channel (a part of the host computer) handles the movement of all data into the host computer memory and the movement of completed

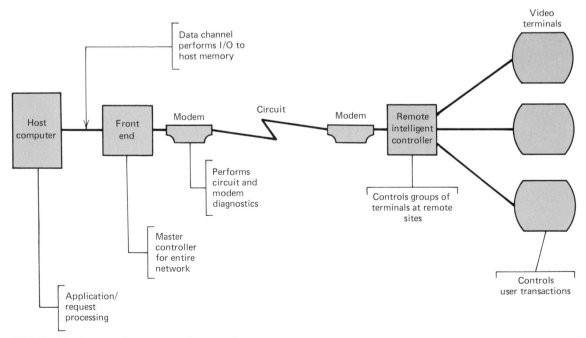

FIGURE 2–1 Downline network control.

processing out of the host computer memory. The front end communication processor is now handling most if not all of the data communication related control functions. Modems perform modulation and specific diagnostic checks. Next, some of these control functions have been passed off to other devices such as switches, statistical multiplexers, and remote intelligent controllers that might be located hundreds or even thousands of miles away from the host computer. This also implies that the massive quantities of software that used to reside solely in the host computer have now been removed and are located downline in other pieces of hardware such as front ends and remote intelligent controllers.

Finally, the ultimate proof of this movement of software and hardware functions out of the host computer is exemplified by distributed data processing where application user departments now have minicomputers or microprocessors to perform their own data processing functions. The result might be that user application departments might be given their own distributed database so they can store their own files. When this time comes, each business or government function might have its own data processing host computer, even though it might be a small microprocessor, and its own distributed database files. At that time, the data communication networks will be the fabric or glue that holds the business organization together; there will be multiple host computers that range in size from a mainframe on down to a minicomputer or microprocessor.

The problem (incompatible architecture and software) of one host computer talking to another host computer is beginning to be solved with the advent of code conversion techniques, protocol converters, local networking interrupt techniques, the new fourth generation PBX/CBX automated switchboards, and standard protocols such as X.25.

Networks can operate in a *central control mode* or an *interrupt mode*. The central control mode involves the centralized polling of each station device (terminal or node) on the network. *Polling* is the process of individually giving each of the terminals permission to send data one at a time.

The interrupt philosophy implies that when a terminal sends data, the incoming data stream interrupts the host computer and stops processing in order to handle the incoming data. Since this mode of operation is wasteful of processing capacity, it usually is not utilized on host computers except for some minicomputer systems that have very few terminals.

The host computer contains software to operate the data communication network. For example, teleprocessing monitors are used to control the routing, scheduling, and movement of data within the host computer. Also, telecommunication access programs might be utilized to handle the routing, scheduling, and polling of terminals out in the network, although in today's marketplace these telecommunication access programs usually are offloaded to a front end communication processor.

Finally, as we move toward the twenty-first century, you might find that the term *host computer* becomes a little blurred. This is because some of the new "microprocessor" chips (Intel 432) are 32-bit chips that when coupled together can operate at 2 million instructions per second. As a result, they have the power of today's mainframe host computers. In other words, the computing power of a host mainframe computer today may be in the microprocessor of the future.

FRONT END COMMUNICATION PROCESSOR

A front end communication processor is a computer that has been specifically programmed to perform many different control and/or processing functions required for the proper operation of a data communication network. This computer might be a mainframe, a minicomputer, or, as is becoming the case more often today, a microprocessor-based digital computer.

As was stated in the section on host computers, the primary purpose of this device is to offload some of the processing and control functions of the data communication network from the host computer system to a specially designed and programmed communication processor. These devices are programmable and they come with extensive software packages. The software is what defines the architecture of the system; it determines which of the various protocols and software programs are used for communicating with this communication processor device.

The primary application for this type of device is the interface between the central data processing system (host computer) and the data communication network with its hundreds or even thousands of input/output terminals or nodes. Many of the newer and more powerful communication processors can perform message processing because they have enough storage capacity, processing power, and disk units. For example, the processor might receive inquiry messages from remote terminals, process the messages to determine the specific information required, retrieve the information from an on-line random access storage unit, and send it back to the inquiring terminals without involving the host computer. In systems of this type, application-oriented processing is of equal importance with message receipt and transmission.

Another processing-related function might be performing message switching. This occurs when the communication processor receives a message, determines that it only needs to be switched to some other terminal or node, and performs the circuit switch or message switch. It also could utilize a store and forward process to hold the message and forward it at a later time.

Some of the basic component parts of a front end communication processor are as follows (see Figure 2–2):

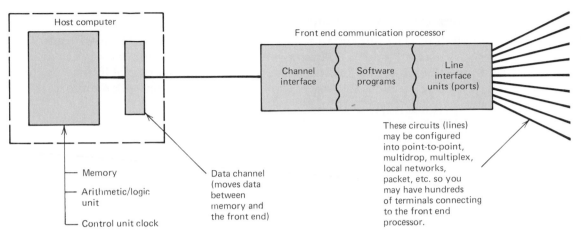

FIGURE 2-2 Front end communication processor.

- Channel interface is the hardware interface that permits the communication processor to connect directly to the standard data channel of a host computer.
- Line interface units (also called *ports*) are hardware devices that link the communication processor with the modems that terminate each communication circuit. Along with the line interface units there might be a communication multiplexer if multiplexing is built into the processor itself. A specific front end may have the capacity for 10 or 200 ports (circuits).
- Software is the set of stored programs that are usually highly specialized and that define the specific architecture/protocols of the front end. The software is what determines which of the standard protocols (such as X.25) are utilized for a given front end. Some of this software is now being built into firmware where the programs themselves are coded into circuit chips instead of being programmed logic. Some sophisticated communication processors might have firmware that can serve as a protocol converter so you can interconnect the X.25 protocol to the protocols of other host mainframe computers.

There are over two dozen vendors of front end communication processors, including each of the major host mainframe manufacturers.

The general functions that might be performed by front end communication processors are discussed below. All communication processors do not perform all of these functions; you must check an individual model to be sure it performs the functions necessary for your network.

- **Communication Line Control**

  Polling/selecting of individual terminals, intelligent terminal controllers, or network nodes. *Polling* involves asking each terminal whether it has a message to send. *Selecting* involves asking each terminal if it is in a condition to receive a message. Both of these imply a network architecture that involves central control.

  Automatic answering, acknowledgment, and dial for outgoing calls.

  Port selection allowing several circuits to share a single port. The port is the plug or connection point where the individual lines enter the front end communication processor. You can control which of a group of terminals can have access to the host computer. This implies that the communication processor might have several incoming circuits, each of which are in contention for a single incoming port. These ports also can be put on a priority basis, with certain incoming communication circuits having priority over others.

  Ability to address messages to specific circuits or terminals. Examples are: a broadcast address that might go to all terminals of the system, a multiple address that might go to a select group of several terminals, or a single address that might go to a specific communication circuit or specific terminal.

  Circuit switching which allows one incoming circuit to be switched directly to another when it is available. This creates a straight through transmission path for the movement of messages from one terminal location to another. Associated with circuit switching is a store and forward capability similar to that in voice mail or electronic mail systems. When the second half of the circuit path is unavailable, the communication processor can record the incoming text or voice message and can transmit it to the other terminal when the circuit is free.

  Automatic routing of messages to a backup terminal when a specific terminal or circuit is out of order. This is a type of switching function, but it is used when there are problems in the network.

  Addition or deletion of communication line control codes. Line control codes (the grammar of data communications), such as end-of-block or beginning-of-block or start-of-message, must be deleted before the message is passed to the host computer or must be added before the message is passed to the outgoing communication circuits.

- **Protocol/Code Conversion**

  Code conversion, that is, the software or hardware conversion from one code format to another, such as ASCII to EBCDIC. Code conversion is available from any code to any code.

Conversion from one protocol to another, which provides the ability to have different machines talk to each other if they all use different protocols such as HDLC, SDLC, X.25, and the like.

- **Assembly of Character/Messages**

    Assembly and disassembly of bits into characters. Bits are transmitted in serial fashion on a communication circuit. The front end assembles these serial bits into parallel characters and possibly into parallel words.

    Assembly/disassembly buffering in order to handle synchronous or asynchronous modes of transmission.

    Handling of transmission speed differences where different communication circuits transmit at different bit per second rates, such as 2400 bits per second versus 56,000 bits per second.

- **Data and Message Editing**

    Control editing, which involves adding items to a message, rerouting messages, rearranging data for further transmission, or looking for nonexistent addresses.

    Message compression or compaction, a methodology for transmitting meaningful data messages but through the transmission of fewer data bits.

    Application editing, using some of the expanded processing and storage capabilities of the newest front ends. It provides for editing of either application type errors or human factors type errors that occur during data entry.

    Triggering special remote alarms if certain parameters are exceeded.

    Signaling abnormal occurrences to the host computer.

    Assigning consecutive serial numbers to each message and, possibly, time stamping and date stamping each individual message.

- **Message Queuing/Buffering**

    Buffering several messages in the main memory of the front end processor before passing them to the host computer or out to the remote terminal station (node).

    Slowing up the flow of messages when the host computer or the remote terminal station (node) is overburdened by traffic.

    Queuing up messages into distinct input queues and output queues between the front end communication processor and the host computer or between the front end communication processor and the outgoing communication circuits.

    Priority systems that give priorities to different communication circuits or automatically assign priorities to various types of messages in order to speed the throughput for various message types.

Handling the "time out" facility, such as when a specific terminal station does not respond or when a circuit ceases to respond. The system will time-it-out and in the future skip it in order to continue its normal operation.

- **Error Control**

   Error detection for parity on single characters.

   Error detection and automatic retransmission for parity checks on message blocks (cyclical redundancy checks and others).

   Forward error correction techniques to reduce the errors flowing through the communication circuits.

- **Message Recording**

   Logging all inbound and outbound messages on magnetic tape for a historical transaction trail.

   Logging the most recent 20 minutes on a magnetic disk for immediate re-start and recovery purposes.

   Monitoring for specific messages in order to identify trouble or as a security check.

- **Statistical Recording**

   Maintaining a continuous record of all data communication traffic such as number of messages processed per circuit, minutes of downtime per circuit, number of errors per circuit per hour or per day, number of errors encountered per program module, terminal stations that appear to have a higher than average error rate, average length of time in the queue for each message, number of busy signals on dial-up circuits, and any other unique statistical data collection required by the organization.

   On-line development of various pictorial graphs and charts showing the efficiency of the network on an hourly or daily basis.

   Keeping track of on-line diagnostics performed by vendors.

- **Other Functions**

   Multiplexing.

   Dynamic allocation of task management queues.

   Automatic switch-over to a backup host computer in the event of a primary host failure.

   Circuit concentration where a number of low-speed communication circuits might be interfaced to a higher-speed communication circuit.

   Performing some of the functions of the Packet Assembly/Disassembly (PAD), if the front end also is serving as a Switching Node (SN) in a packet switching network. In this situation it would contain the multiple databases that are required for the alternate routing and packetizing of messages.

LINE ADAPTERS

Line adapters allow terminals to be interconnected to computers in many different configurations. Terminal users who would like to connect a single terminal to more than one computer at the same time, without the need for time-consuming dialogue, can turn to a *line interface module*. This device enables a terminal user to connect to both a host computer and a local minicomputer, alternately accessing screens from either computer through simple keyboard instructions (see Figure 2–3).

Machine cycles and memory space are tied up when a host computer has to perform protocol conversion, terminal emulation, and polling operations. When these tasks can be performed externally by the line interface module, the computer is free to do what it was originally intended to do. This device requires no additional wires and switches, and the terminal can be connected on either a dial-up or lease-line basis. The line interface module performs some of the functions of a front end communication processor.

A *port-sharing device*, or *bridge*, treats several point-to-point lines as if they were a single multipoint line. In this way, it conserves ports at the expense of al-

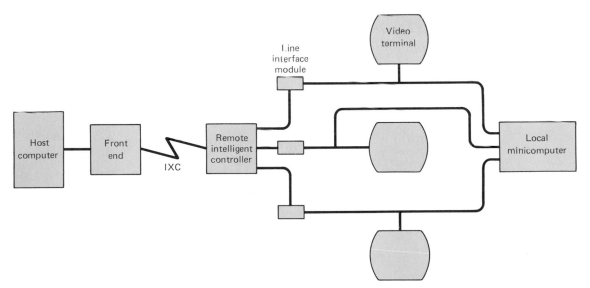

FIGURE 2–3 Line interface module.

lowing only one terminal to transmit to the computer at a time. Figure 2–4 shows a port-sharing device. Port-sharing devices are available that allow connection of up to four low-speed terminals (either local or remote). In this way, one front end communication processor port is expanded to handle four terminals. This type of device can be utilized to avoid installing a second front end communication processor when all ports of the first are occupied. This may not be a long-term solution but it can be a short-term "holding action" until a new network can be configured or new hardware can be purchased.

Another line adapter might be an *intelligent port selector*, which replaces the function of the old-fashioned telephone rotary. It provides the same contention facility as a rotary and will give a response equivalent to a busy signal if no ports are available. This type of device is used when you have many dial-up terminals but not enough ports for all the terminals to connect at the same time. It is configured into the network in the same position as the port-sharing device. Intelligent port selectors will handle different speeds of transmission and different codes; they can interface either dial-up or dedicated ports, can offer any caller receiving a busy signal the opportunity to stay on the line and be placed in a queue for future connection, and will collect statistics on utilization.

A *line splitter* (see Figure 2–5) is similar to a port-sharing device except that it usually is located at the remote end of the communication circuit. It functions as a switch in that it treats several terminals as if they were a multidrop line, but the terminals must be located close to one another. The benefit of a line splitter is that several terminals at a single location can be handled as though they were a

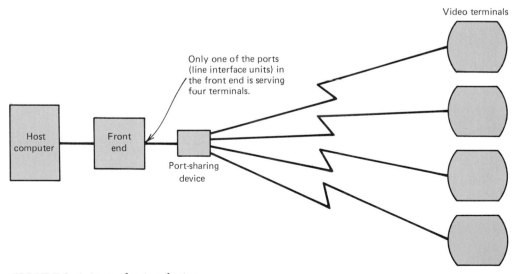

FIGURE 2–4 Port–sharing device.

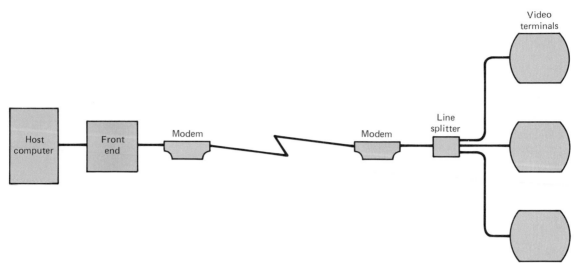

FIGURE 2–5 Line splitter.

multidrop line. With a line splitter, costs are reduced to the cost of one communication circuit and one modem pair between the remote site and the distant front end communication processor.

REMOTE INTELLIGENT CONTROLLERS

Remote intelligent controllers, sometimes called *intelligent terminal controllers*, usually reside at the distant or far end of a communication circuit (see Figure 2–6). A remote intelligent controller is nothing more than a scaled-down front end communication processor. In fact, you could have a remotely located front end processor for an area of the country or world where hundreds of terminals are located. In this case you would connect the remotely located front end to the host computer front end by high-speed data circuits.

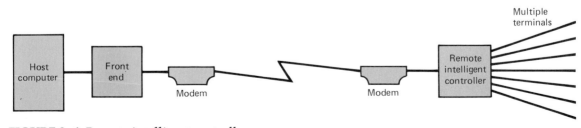

FIGURE 2–6 Remote intelligent controller.

Remote intelligent controllers control four to sixteen local terminals, although some are much more powerful. There is a unique address for each controller and, therefore, a further address or unique memory space for each terminal connected to the controller. One of the major reasons that remote intelligent terminal controllers are used is that they allow an organization to have full duplex transmissions between the two electronic devices (front end and remote local controller).

A remote intelligent controller might perform any or all of the functions performed by a front end, but usually it is scaled down and not as powerful, although each vendor's terminal controller has its own set of functions. These controllers started out as simple devices to control four or eight video terminals. They have been increasing in power ever since so that today one device might control sixteen multifunctional terminals in a branch bank. In fact, the terminals in a branch bank might be entirely different from each other, say, four video terminals, four simple teller inquiry terminals, two high-speed printers, one sophisticated wire transfer data entry terminal, one facsimile machine, and so on. This variety makes a remote intelligent controller especially desirable.

MODEMS

The way data bits are moved about electronic computers and terminals is different from the way they are moved across data communication circuits. The modem is a device that bridges this gap. In other words, modems *mo*dulate the signal

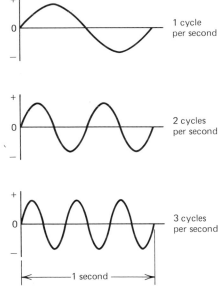

FIGURE 2–7 Sine waves.

at the transmit end of a data communication circuit and *de*modulate the signal at the receiving end. To *modulate* is to adapt a signal so it can be transmitted over the type of transmission media in use.

In other words, the basic function of a modem is to convert electrical signals, such as plus and minus voltages of electricity, to modulated signals that can be transmitted over communication circuits. The most common modulated signal is *frequency modulation*. Modulation is the process of putting intelligence on a carrier wave. The carrier wave is the basic unchanged sine wave. Basically a signal can be shown as a sine wave (see Figure 2–7). See the section on Analog Modulation in Chapter 3 for more on frequency modulation.

We will use frequency modulation or changing the number of cycles per second in Figure 2–8. Notice where it shows how a stream of zeros and ones going across a communication circuit might be represented by changing the frequencies or cycles per second (called *hertz*). To our ears, the change in frequencies results in a change in the pitch of the tone. For example, a high-pitched tone is a higher frequency, and therefore a greater number of cycles per second, whereas a low-pitched tone has a fewer number of cycles per second or hertz.

Figure 2–9 shows the data stream of zeros and ones that might be coming from a terminal, going into a modem, and going on to the communication circuit. Notice how the zeros and ones from the terminal are represented as plus five volts of electricity for a 1, and minus five volts of electricity for a 0. Next, the modem modulates or changes that into two frequency tones which are represented as 2225 cycles per second for ones and 2025 cycles per second for zeros. In other words, when the modem receives a minus five volts of electricity, it converts that to a signal of 2025 cycles per second.

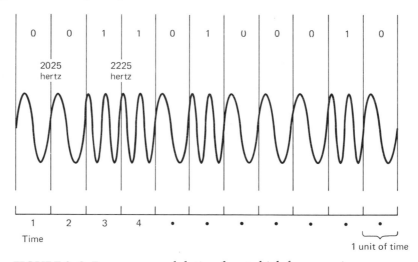

FIGURE 2–8 Frequency modulation (low to high frequency).

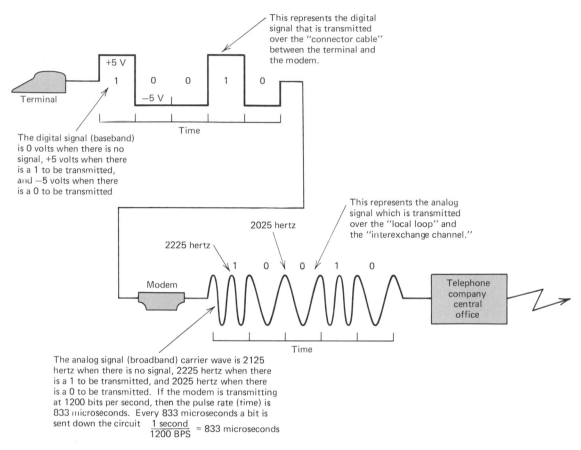

FIGURE 2–9 Operation of a modem.

In reality, a much more complicated sine waveform is sent down the communication circuit; it is a waveform containing the combination of all the frequencies (2025, 2125, and 2225). The oversimplified description above suits our purpose because we are trying to show how a modem converts electrical signals (baseband) to frequency tones (broadband). Frequency tones (called *frequency shift keying*, or FSK) are the most popular method of transmitting data across telephone lines, although at high bit-per-second rates other modulation techniques are used.

Sometimes direct electrical voltages are sent over wire pairs for short distances. When these direct electrical voltages are sent, the signal is called a baseband signal. While it is possible to send baseband signals over short distances without using a modem, you may still need a modem for timing characteristics and other control purposes.

Some of the basic transmission speeds of modems are 300, 1200, 4800, 9600, 14,400, 16,000, 50,000, and 56,000 bits per second.

Digital transmission (discrete on and off pulses) involves sending baseband signals between your premises and the telephone company central office and then transmitting digital pulses over the long-distance links between cities. Even when you go to digital transmission, you use a digital type of modem which shapes the digital pulses. The telephone company calls this modem either a *channel service unit* (CSU) or a *data service unit* (DSU). The channel service unit performs transmit and receive filtering, signal shaping, longitudinal balance, voltage isolation, and equalization, and it supports remote loopback testing. The data service unit is a channel service unit that, in addition, provides bipolar conversion functions, which ensure proper signal shaping and signal strength for transmission that will be almost totally error free. Digital transmission is described more fully in Chapter 3.

Transmitting on an optical circuit requires use of a fiber optic modem. This type of modem converts electrical signals from a terminal to pulses of light that are transmitted down the optical fiber. It is the same as an FSK modem except it outputs two different frequencies of light.

Another type of modem is a short-haul modem in which you use your own cable and transmit direct electrical baseband signals. Sometimes, this is called a 20 milliamp circuit. Transmission is at a maximum of 19,200 bits per second. This type of modem can be used for short haul (several miles) around your facility or between various buildings in the same locale.

Another type of modem is an *acoustic coupler* (see Figure 2–10). This modem is used primarily for dial-up; it will interface with any basic telephone hand set. All you do is dial the computer and place the telephone hand set into the acoustic coupler. The coupler performs the typical modem functions of converting direct electrical signals from the terminal to frequency modulated tones that can be sent over any telephone communication circuit.

Before we discuss the features of modems, you should realize that not all modems use frequency modulation. The higher-speed modems use more exotic forms of modulation such as combined phase and amplitude or quadrature phase modulation. Some modems use light amplification modulation and pulse two different intensities of light, whereas other modems use direct electrical voltages changing between a plus voltage and a minus voltage (baseband).

Now that we have seen the types of modems and discussed the basic function of a modem (to modulate a signal), let's discuss some of the features that are in various modems. The features listed below may not be in all modems; this is a comprehensive list of the types of features that have been built into modems.

- Loopback functions for diagnostic purposes probably are the single most important feature built into modems. Automatic loopback allows the user to set a remote modem on loopback and send a message to that modem. The mes-

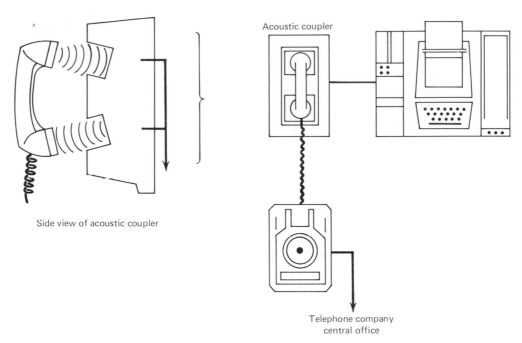

FIGURE 2–10 Acoustic coupler.

sage will be looped back to the original sender, where it can be checked for accuracy to help diagnose where a fault might be in the network. Loopback switches allow you to diagnose whether the problem is in the connector cable between the terminal and modem, whether it is on the digital or analog side (see Figure 2–9) of the modem, or whether the problem lies in the local loop communication circuit itself. The digital side of the modem is the side that plugs into the terminal; the analog side is the side that plugs into the telephone circuits, unless you are on a complete digital network. In that case you would be using a data service unit instead of a standard modem. All modems have loopback facilities available.

• Some modems have the ability to be turned on or off from a remote station. These contain automatic answer and automatic calling capabilities so that a remote terminal can be started from thousands of miles away.

• Some modems allow the simultaneous transmission of both voice and data. One model allows a voice conversation to go over the circuit while simultaneously transmitting a 2400 bit-per-second data stream.

• Some modems allow a reverse channel capability for some of the older protocols. The purpose of a reverse channel is to avoid interrupt of the ongoing message stream but still provide for a path over which message acknowledgments can be sent. For example, the modem could be receiving messages at 2400 bits per second but simultaneously it would be sending back, in the reverse direction, one character acknowledgment for each received message. This tells the original transmitter whether the message was received correctly. Technically this might look like full duplex transmission, but it is not because of the vast difference of speed in each direction. The reverse channel may be transmitting at only 10 to 75 bits per second.

- Multiplexing is built into some modems.

- Microprocessor circuits are built into some modems for automatic equalization to compensate for electronic instabilities on the transmission line. This equalization compensates for attenuation/delay distortion which causes errors and therefore requires retransmission of messages.

- Many modems today have built-in diagnostic routines for self-checking of their own circuits in order to determine where a fault might lie.

- Some modems have alternate speed switches so you can switch up or down in speeds, let's say from 9600 to 4800 bits per second.

- Some modems have split streaming by which the modem transmits three message streams at different speeds. This is achieved by having one message stream transmitting at 4800 bits per second and the other two at 2400 bits per second.

- Some modems are more efficient in that they have a lower clocking or *retrain time*. Retrain time is the turnaround time when the message direction changes. For example, when you are transmitting in half duplex mode (one direction at a time), the modem is in the transmit mode, but when you receive a message, it must switch its electronic circuits to be in a receive mode.

- For efficiency, some modems have a longer flywheel effect for brief drops of the carrier wave (the 2025/2225 hertz signal). When the carrier wave drops for a couple of microseconds, you technically have totally lost the circuit. With a built-in flywheel effect, however, the modem can remain connected to the circuit and pick up where it left off after the carrier wave returns. This assumes the drop was not for too long a period of time.

- Modems can be plugged into other equipment by means of various standard interface cables. For example, the plug and cable that come with the modem should be specified through one of the standards such as RS232 or RS449. (These standards are discussed in Chapter 4.) It would be quite embarrassing to order a modem and learn later that it will not plug into your terminal.

- Integrated modems are those modems that are built into a device. For example, a modem can be built into a terminal or front end communication processor.

- Certification of modems involves the Bell Telephone System. In order for modems to operate over the dial-up telephone network without Data Access Arrangements (DAA), they must be registered and certified with the Federal Communications Commission. Certification is not required for modems operating over lease circuits, only for dial-up. Data access arrangements add a cost to your circuit, so it is wise to look for modems that are certified when using the dial-up network. The purpose of a data access arrangement is to limit the amplitude (power) of the signal presented to the telephone circuit. See the section, Data Protectors, later in this chapter.

- Some modems can perform full network analysis, monitoring such features as the RS232C interface and circuit characteristics (analog), although usually this is performed by other test equipment. Tests are carried out during normal data transmission using out-of-band signaling (unused portions of the bandwidth).

- Some modems provide status indication on the modem's front panel, such as "clear to send" or "request to send." These indicators are helpful but do not constitute comprehensive diagnostics and may seldom be looked at in large data communication network control centers.

- Some vendors emphasize failsafe modems: a spare modem is mounted internally in the same package and tied to the central site control facilities. If the primary modem fails, the spare is automatically switched in.

MULTIPLEXERS

To multiplex is simultaneously to place two or more separate transmissions on a single communication circuit. Multiplexing usually is done in multiples of 4, 8, 16, and 32 simultaneous transmissions over a single communication circuit. Figure 2–11 shows a typical four-level multiplexed circuit.

Multiplexers can be separated into two major categories: *frequency division multiplexers* (FDM) and *time division multiplexers* (TDM). *Frequency division multiplexing* can be looked upon as having a stack of four or more modems which operate at different frequencies so their signals can travel down a single communication circuit. Another way of looking at frequency division multiplexing is to imagine a group of people singing. When people sing together you might have a base, a baritone, an alto, and a soprano. What you hear is the combination of the four people singing, but sometimes you can clearly identify one or more of the individual singers.

In frequency division multiplexing the frequency division multiplexer and the modem are usually a single piece of hardware. Compare Figure 2–12 with Figure 2–13. You will see that the FDM has a single piece of hardware (multiplexer/mo-

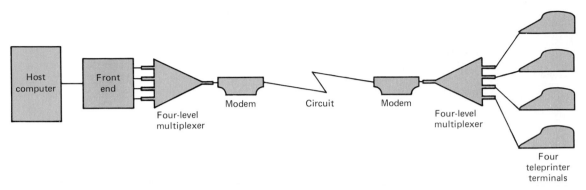

FIGURE 2–11 Multiplexed circuit (TDM).

dem) whereas TDM has two separate pieces of hardware (time division multiplexer and modem).

In frequency division multiplexing, the FDM utilizes the available bandwidth of a voice grade circuit and divides it into multiple subchannels. Bandwidth capacity of a voice grade circuit is discussed in Chapter 3. Looking at Figure 2–12, you can see that FDM takes the available bandwidth of a voice grade circuit and divides it into four separate subchannels. Each subchannel has a pair of frequencies that are utilized to transmit binary 1s and 0s. If you review Figure 2–9, opera-

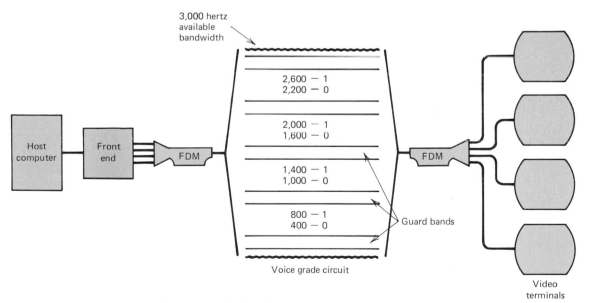

FIGURE 2–12 Frequency division multiplexed circuit (FDM).

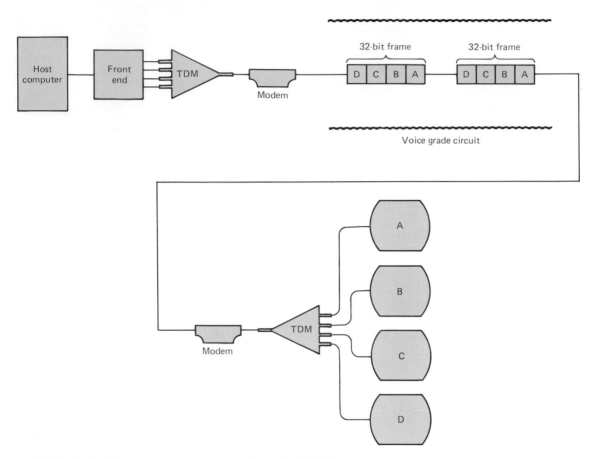

FIGURE 2–13 Time division multiplexed circuit (TDM).

tion of a modem, you will see that there is only one pair of frequencies (2225 and 2025 hertz) to transmit binary 1s and 0s.

The guardbands in Figure 2–12 are the unused portions of bandwidth that separate each pair of frequencies from the others. They thus keep the signals in each of the four subchannels from interfering with the adjacent subchannels and allow space for frequency drift. The guardband serves the same purpose as does a plastic insulator surrounding a copper wire; it keeps adjacent copper wires or subchannels from interfering with the others' transmission.

Another characteristic of frequency division multiplexing is that the subchannels need not all terminate at the same location; therefore, frequency division multiplexing can be used in a miltidrop network where each dropoff operates at a different frequency. This changes the multidrop philosophy of sharing the circuit to one in which all terminals on the multidrop circuit can use the circuit simultaneously.

Frequency division multiplexers are somewhat inflexible in that, once you determine how may subchannels are required, it may be difficult to add more subchannels without purchasing an entirely new frequency division multiplexer that is divided into a great number of subchannels. Also, the maintenance cost on frequency division multiplexing equipment usually is greater than that required for time division multiplexing equipment.

Time division multiplexing (TDM) is really a type of time slicing or sharing the use of a communication circuit among two or more terminals. Each terminal takes its turn. In time division multiplexing, the multiplexer takes a character from each transmitting terminal and puts them together into a frame. The frames are put onto a high speed data stream for transmission to the other end of the circuit. In Figure 2–13 we show a four-character frame for a four-level multiplexer. This is *pure* multiplexing because it is totally transparent to everyone on the network, including the system programmers. In pure multiplexing your messages are never held back or slowed up by the multiplexer, as in statistical time division multiplexing which is discussed later in this section.

In Figure 2–13 a character is taken from each terminal, placed in its frame, sent down the circuit, and delivered to the appropriate device at the far end of the circuit. If each of the four terminals transmits at 1200 bits per second, then the time division multiplex bit stream would have to transmit 4800 bits per second. Notice that there is no terminal addressing here. Each position of the four-position frame gives its character to the appropriate terminal at the other end, even if that character is a blank. When you start addressing each character position of a frame, then you are moving into the world of statistical time division multiplexing.

Time division multiplexing is generally more efficient than frequency division multiplexing, but it does require a separate modem (frequency division multiplexing is in reality a special modem). It is not uncommon to have time division multiplexers that share a line among 32 different low-speed terminals, although these might be replaced with statistical time division multiplexers that can hold 32 higher-speed terminals. It is easy to expand a time division multiplexer from, let's say, 8 to 12 channels. Time division multiplexing channels usually all originate at one location and all terminate at another location, but this does not necessarily have to be the case. Time division multiplexers usually are less costly to maintain than frequency division multiplexers.

Statistical time division multiplexers (STDM) allow the connection of more terminals to the circuit than the capacity of the circuit. In its simplest context, if you have 12 terminals connected to a statistical time division multiplexer and each terminal can transmit at 1200 bits per second, then your total is 14,400 bits per second transmitted in a given instant of time. However, if the STDM/modem/

circuit combination has a maximum speed of only 9600 bits per second, then you would find that there might be a period of time when the system is loaded above its capacity.

The technique of statistical time division multiplexing takes into account the fact that there is some downtime and that all terminals do not transmit at their maximum rated capacity for every possible microsecond that is available. With this in mind, you can start addressing each character in a frame or message and time division multiplex on a statistical basis.

For example, let's assume we have a statistical time division multiplexer that multiplexes individual characters from 12 terminals. In this case, a terminal address is picked up in addition to the character and is inserted into the frame. Using the same four character frame that was used in the previous example, look at Figure 2–14. Notice that, in addition to the 8 bits for each individual character, we have added 5 bits of address space (ADD). These five bits of address allow you to address 32 different terminals using binary counting ($32 = 2^5$). Now the multiplexer takes a character from each terminal only when the terminal has a character to send. The technique is that you scan through the 12 terminals and take characters from, let's say, terminals 1, 4, 5, and 12. These would be sent immediately. You then scan through all 12 again to determine which terminals need servicing. This process is repeated indefinitely. At the other end of the communication circuit, the character is given to the proper device because the 5-bit address that is included with each 8-bit character identifies the terminal device.

Another type of statistical time division multiplexer involves multiplexing entire messages from terminals. With this type of multiplexer you use one of the newer bit-oriented protocols and interleave entire messages rather than characters. For example, when terminal 1 has a message to send, the multiplexer picks up the entire message (all of its characters) and passes it to the modem for transmission, after which the multiplexer immediately scans for the next terminal that has an entire message to send. The primary difference is that the first statistical scheme is for asynchronous character-by-character transmission, while the second scheme is for synchronous or block transmission.

While statistical time division multiplexing may be very efficient, you should be aware that it can cause delays. When traffic is particularly heavy, you can have anywhere from zero to a maximum 30-second delay of your data. Some data will be held back by buffers when too many terminals transmit at maximum capacity for too long a period of time. In fact, some of these statistical time division multiplexers have 32,000 bytes of buffer space memory. One technique that is implemented to improve the throughput of statistical time division multiplexers is to compress the data, thus reducing the number of bits transmitted per character or per message.

FIGURE 2–14 Statistical time division multiplexing (STDM) (opposite).

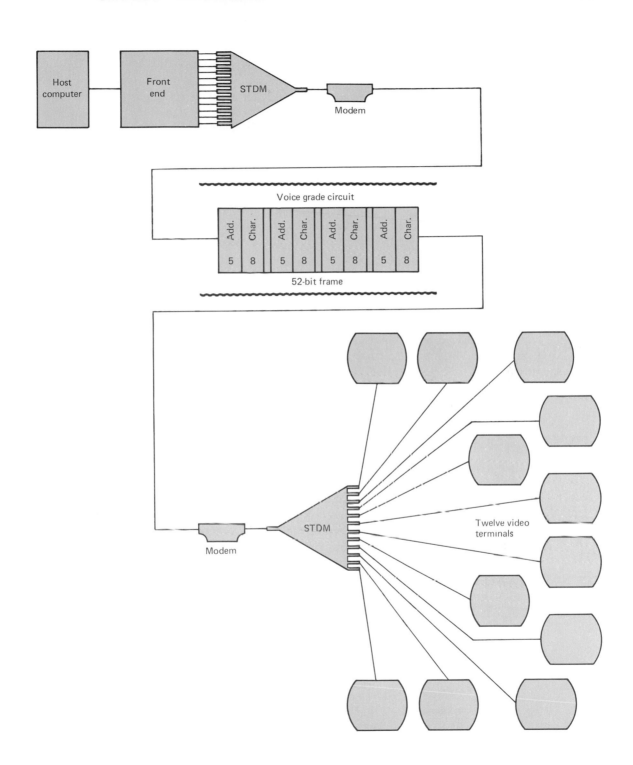

CONCENTRATORS

Concentrators, in today's terminology, are really special forms of multiplexers, or even biplexers. Concentrators are used for the same purposes as multiplexers. In fact, they were originally intelligent multiplexers.

The primary use for a concentrator is to combine circuits. In this respect you can have 16 low- or medium-speed circuits that are concentrated into one or two high-speed lines. For example, you might concentrate approximately twelve 4800 bit per second communication circuits into one 56,000 bit per second digital communication circuit. Even though this does not work out in a perfectly equal way, the statistical intelligence will take care of the small difference. Concentrators, like statistical multiplexers, can buffer or hold back data. Some concentrators can even perform switching functions to switch messages to different communication circuits.

Biplexers are sometimes called *inverse multiplexers* (see Figure 2–15) because a biplexer can take a 19,200 bit-per-second transmission circuit and divide it into two incoming or outgoing 9600 bit-per-second circuits.

In order to avoid confusion, you can assume that when you have *pure* multiplexing, there is no basic programmed intelligence in the device, but when you have intelligence and programming capability, it might be a statistical multiplexer or a concentrator. Also, the newer STDM/concentrators perform other functions such as circuit contention and switching.

PROTOCOL CONVERTERS

Protocol converters may be a type of intelligent multiplexer/concentrator that is used for a specific purpose: to allow a terminal or host computer that uses one protocol to talk to another host computer that uses a different protocol. For example, there are protocol converters that allow an asynchronous terminal to communicate with IBM host computers using Synchronous Data Link Control (SDLC) or Binary Synchronous Control (BSC) protocols. Another protocol converter allows asynchronous terminals to interconnect with public or private X.25 packet switching protocols.

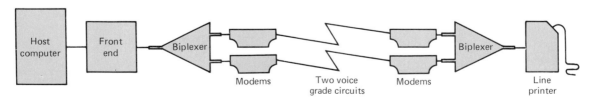

FIGURE 2–15 Biplexer.

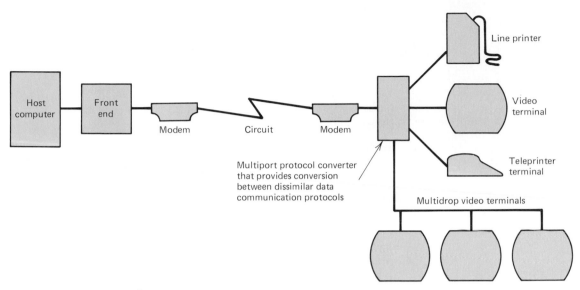

FIGURE 2-16 Protocol converter.

Some of these protocol converters (see Figure 2-16) offer from 1 to 16 ports for the connection of the individual terminals. Incoming information from the terminals have their protocols converted to the host computer's protocol. Some models of protocol converters also include multiplexing, concentrating, and packet assembly/disassembly features for packet switched networks. You can see that protocol converters, concentrators, and multiplexers overlap each other. In fact, the terminology often is used interchangeably by many vendors.

In the next few years there probably will be a collision among the various hardware vendors, caused by the overlap in features and functions between front end communication processors, modems, statistical time division multiplexers, concentrators, protocol converters, and the new digital PBX/CBX switchboards. By using microprocessor chips instead of minicomputers, all of the above pieces of hardware could be combined into a single piece of hardware that performs all the functions now performed separately.

HARDWARE ENCRYPTION

The description of what encryption is and how it works is in Chapter 8, Security and Control. In this section we only want to discuss hardware encryption versus software encryption.

Software encryption involves the use of stored programs to do the encryption. This technique is vulnerable to anyone who can copy the contents of the com-

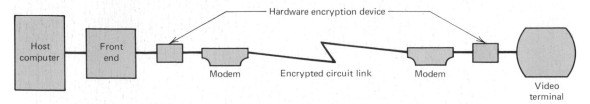

FIGURE 2–17 Hardware encryption.

puter's memory. In data communications, software encryption usually is not used. Software encryption is used for the storage of data on disks/tapes or for other types of programming security problems.

In data communications, hardware encryption devices are employed. They are most often located as shown in Figure 2–17. Data from the host computer's front end or the terminal enters the encryption device, where it is secretly encoded. Most hardware encryption devices have standard RS232 interfaces. These devices expect digital information as input; their output is digital also. This is one reason why the encryption device usually is placed between the information generating device (CRT and front end in this case) and the modem. After encryption, the data are then appropriately transformed by the modem. This is a straightforward "link" encryption setup.

Hardware encryption involves a lockable box that is about the size of a modem. When this box is opened, the secret encryption key is destroyed automatically. Most encryption today is done on a link-by-link basis, except for military type encryption, which is done on an end-to-end basis. For practical purposes, you can assume that *link-to-link* encryption is encryption of the data from modem to modem, whereas *end-to-end* encryption is encryption from user to user.

Most hardware encryption devices today utilize the data encryption standard (DES) that was verified and authorized by the United States National Bureau of Standards. Also, most encryption devices allow the insertion of several secret keys at one time to reduce the cost of key management. Key management cost is the cost incurred when you have to go to the remote end of a communication circuit in order to change the key. By having several keys, you can have a master key that is changed, for example, every two years; by transmitting new keys under the encryption of the master key, you can change the basic key daily if that is desirable.

TERMINALS

There are five major categories of terminals: teleprinter, video, remote job entry, transaction, and intelligent (which is a combination of the above).

Shopping for a terminal can be a real problem if it is not done correctly. For example, a bargain terminal might have the wrong code format or an incompatible protocol, which would require the additional purchase of an expensive protocol converter or other code conversion features. Another consideration is to determine which interface cable can be used with the selected terminal. Some systems might use an RS232 cable and others a RS449 cable. The terminal you select should be compatible with both of these standard cable interfaces. Each vendor's terminal must be examined individually; therefore, the rest of this section on terminals will discuss the general characteristics of each of the five categories of terminals.

Teleprinter Terminals   These terminals produce a paper printout and have a typewriterlike keyboard. In this discussion, they have no programmable capacity (intelligent terminals are a separate category). This type of terminal can perform its printing through an impact, a dot matrix, or even an electronic ink jet. Its characters can print at a speed range from 10 characters per second to 300 or more characters per second. This has always been the most common form of terminal, although the video terminal will replace it in the future because of the cost.

The keyboard on a teleprinter terminal is a typical typewriter keyboard, although it may contain several special function keys for data communication networks. Also, there might be a third function assigned to a single key such as start, stop, delete, or end of transmission. The first two functions were lower case and upper case for an alphabetic character.

Some of the features to look for in a teleprinter terminal are size of the dot matrix, maximum print speed, bidirectional printing, size of the printable character set, number of character positions across the paper, form feed mechanism or individual sheet insertion, and graphics printing capability. Some teleprinter terminals offer a self-test answer-back mode whereby the terminal circuitry can be tested for problems. Some also have a built-in acoustic coupler/modem.

Video Terminals   These terminals have a television screen and a typical typewriter keyboard. Sometimes they are called cathode ray tubes (CRT), video display unit (VDU), or video display terminals (VDT). Alphanumeric video terminals are used in the business office whereas graphic video terminals are used by graphic designers. The intelligent version of a video terminal might be used for computer-assisted design.

In addition to a standard keyboard, the video terminal has a marker on the screen called a *cursor*. It moves about the screen in order to show the terminal operator the next position in which a character will be printed.

When selecting a video terminal, you should consider the transmitting line speed, whether the user can rotate or tilt the screen for easy viewing, whether a split-screen mode is available so two different screens can display simultaneously, whether the character matrix size is large enough for easy viewing, if a detachable keyboard would be advantageous, if there is a self-test mode available, how many characters can be displayed horizontally and how many lines vertically, whether it can accommodate a separate printer that can be turned on and off as needed, and which editing functions are available such as character insertion/deletion, line insertion/deletion, erasing, and paging.

With regard to the individual terminal operator, there are some specific items that must be taken into account when you are ordering a video terminal. These functions will increase operator productivity since eyestrain and fatigue are the most common complaints of video terminal operators.

- Always get the largest dot matrix possible (highest resolution). For example, a 10×14 dot matrix is easier to read than a 5×7.
- Get an anti-glare screen to eliminate glare from overhead lights.
- Ensure that characters do not jitter, because this causes eyestrain. Use a magnifying glass to look at the characters; you should be able to determine if they jitter or shake when magnified.
- The cursor should be visible from eight feet away so it can be easily seen at three feet. Sometimes it is advantageous to have the option of either a blinking cursor or one that is on as a steady light, as desired.
- Detachable keyboards and/or a sloped keyboard may be desirable.
- Reverse video is a nice option to ease eyestrain. With this feature, the operator can change from dark background with light characters to dark characters with a light background.
- Yellow/green tube color is the easiest to see and tends to reduce eyestrain.
- Several adjustments are desirable to accommodate the variety of operators who might use a video terminal. These adjustments are for tube brightness, focus, and contrast.

Remote Job Entry Terminals   Usually these are nodes of a network or terminal stations where there are several types of connected devices. Data often are transmitted from a host computer to a remote job entry terminal that might have a variety of terminal devices such as a video terminal, a high-speed printing terminal, several data entry type devices (such as disk or tape), and perhaps a microprocessor. Terminals in such an area operate at higher speeds, such as 9600 bits per second or greater, because large quantities of data are transmitted from this terminal station to the central host computer.

Transaction Terminals   These are usually low-cost terminals that are used by individuals in their homes, corporations at their business locations, and government agencies. The most common transaction terminal is the Automated Teller

Machine (ATM) used by banking institutions for the purpose of cash dispensing and other related functions. Your telephone is a type of transaction terminal that accepts voice transactions for transmission.

A transaction terminal that should be quite popular in the future is a display telephone or videophone. With it you will be able to transmit both voice and picture from your home or office.

Other transaction terminals are point-of-sale terminals in a supermarket; these enter charges directly from the supermarket to your bank account, or they can be used for checking credit or verifying checks. These terminals can be built into electronic cash registers.

Another transaction terminal will be your personal television set through cable television use of two-way teletex/videotext data communications.

Intelligent Terminals   An *intelligent terminal* is one that has a built-in microprocessor with capability of being programmed and executing stored programs. It might be a video terminal, a teleprinter terminal, or a transaction terminal that has some intelligence in it. Because intelligent terminals usually have memory space and stored program capabilities, they can be viewed as small digital computers. Many intelligent terminals also have external disk capability for the storage and retrieval of data. The world's most popular form of intelligent terminal is the microprocessor microcomputer.

SWITCHES

Even though switching can be built into the front end communication processor, statistical multiplexer/concentrator, PBX/CBX, or host computer, there are many standalone message switching systems. A typical message switching system might be minicomputer or microprocessor based. It functions as a message switch plus a store-and-forward system. Sophisticated message switches utilize stored program techniques and highly reliable third-generation solid state circuitry.

A typical standalone switch might have 12 or more asynchronous communication circuits that can operate at different speeds. The system should be able to support various code structures and should be expandable to handle more communication circuits. The switch should be able to handle half duplex or full duplex and should provide an interface to the Direct Distance Dial (DDD) network, Telex, TWX, and other international networks.

A standalone switch should be able to automatically dial and receive calls, as well as switch messages between any of its circuits. Another feature should be a delivery/verification/confirmation type of response with the message sender knowing when the message was delivered and knowing that there is a positive delivery acknowledgment. Sophisticated switches allow various addressing schemes such as 3, 5, or 7 addressing characters in order to be able to address a multitude of circuits, individual terminals, and individual devices that might be attached to an intelligent terminal/microprocessor. Finally, a good switch should provide for in-transit storage of an adequate number of messages (at least 500). It should have the ability to retrieve messages that were sent during the day, possibly during previous days, and message logging for transaction trails/historical purposes. Some switches also provide alarms in case of circuitry failure and self-diagnostics to locate the failure. The telephone company central office is a switch.

PBX/CBX

Private Branch Exchange (PBX)/Computer Branch Exchange (CBX) switchboards are a piece of hardware. The PBX/CBX is a stored program digital computer. These digital automated switchboards can perform many or all of the functions of the front end communication processor, modem, multiplexer, concentrator, protocol converter, and switches. To understand completely what this piece of hardware and its associated internal computer programming can do, you would have to review the functions of all the above-mentioned pieces of hardware.

Digital automated switchboards will probably replace many individual pieces of communication hardware because of two major changes that are taking place. First, management and control functions of both voice and data communications are being combined into a single department by many corporations and government agencies. Second, it is very cost effective to physically combine voice and data communications over the same communication circuits.

Once both the management functions and the physical movement of voice/data are combined, the digital automated switchboard (PBX/CBX) will become the central piece of hardware for both voice and data networks. This integration also is mandatory before we can truly experience the fully automated office. This integration will also guarantee the most efficient use of both voice mail and electronic mail within the organization. A typical configuration that shows the use of an automated digital switchboard is shown in Figure 2–18.

DATA PROTECTORS

When non-AT&T devices are connected to the standard Direct Distance Dial (DDD) telephone circuits, the devices can be registered in one of three ways. The three types of registration are permissive, programmable, and fixed loss loop. The type of registration determines how the terminal, modem, or telephone connects to the telephone line.

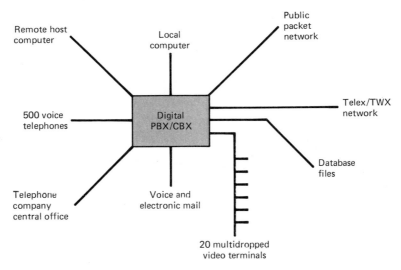

FIGURE 2–18 Digital PBX/CBX.

A *permissive* connection is used to connect a standard telephone to the telephone line or a data terminal to a switchboard. Permissive devices limit the amplitude (power) of the signal presented to the telephone line.

A *programmable* device has two main advantages. First, the modem always transmits at its maximum allowable level, and second, the telephone company installs a "data-quality" circuit.

The *universal* connection or *fixed loss loop* connection has a switch on the jack to prevent the installation of the wrong connector cable. When the fixed loss loop mode is used, the telephone company installs an attenuation pad to ensure that the signal that arrives at the telephone company central office does not damage their equipment.

The purpose of these three devices is to protect the telephone company's circuits from extraneous signals that might damage the switching facilities at the central office.

FACSIMILE DEVICES

Even though they are a type of terminal, facsimile devices are discussed as a separate piece of equipment. This is because transmission of hard copy documents is one of the key items in the automated office. Transmission of an exact picture of a hard copy document, including legal signatures, is one of the most important features in the environment of the automated office. This is especially so in areas such as legal contracts, medical records, and authorizations, and for control of business records.

Although analog facsimile has been used widely in business, government, and professional areas, modern high-speed digital facsimile devices provide many added features such as encryption, security identification codes, and delivery verification. Figure 2–19 shows a picture of a facsimile device. Whereas analog facsimile devices can transmit a page in one or two minutes, digital facsimile devices now transmit a page (8½ × 11 inches) in less than a minute. Also, if you use a standard copy machine that can reduce two 8½ × 11 pages down to one 8½ × 11 page, you have technically doubled this transmission speed again.

Facsimile transmission can be very threatening to post office systems because you can transmit a letter from one area to another for anywhere from $.30 to $.50, depending upon the volume of traffic per month and the cost of your communication circuits. For this reason, facsimile transmission is in direct competition with both electronic mail and voice mail.

Once you have a facsimile device connected to a communication circuit, the basic steps for transmission are:

• Establishing your call by either dialing manually or having an automatic call placed (*physical circuit connection*).

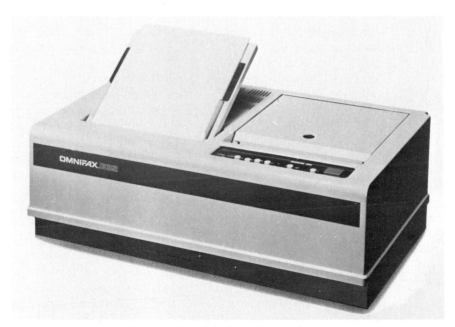

FIGURE 2–19 Omnifax G32 digital desktop facsimile.

- The premessage procedure, or handshaking, which consists of identification of the called station (facsimile machine) and any other procedures that might be required to set up the session.
- Transmission of the message (the *session*), which involves the synchronization between the two devices, any error detection and correction methodologies, and the movement of the message from one facsimile device to the other.
- Postmessage procedures, which include end-of-message signaling at the conclusion of the page, any signaling that signifies more than one page, end of transmission, and anything else required to end the session.
- Manual or automatic disconnection of the call (*physical circuit disconnection*).

Facsimile devices are connected using standard interface cables such as the RS232 interface. Because this device is a special type of terminal that can send a facsimile copy of whatever is on a piece of paper, it can be connected directly to a leased circuit, dial-up circuit, statistical multiplexer, local area network, or one of the new automated switchboards (PBX/CBX).

MICROPROCESSORS

Since Intel's introduction of the microprocessor circuit chip, the microprocessor probably has been the single most important invention with regard to data communications since the advent of signaling over a pair of copper wires. The microprocessor chip is the heart or control point of most of the data communication hardware available today. Hardware such as front end communication processors, modems, multiplexers, protocol converters, switches, and automated switchboards depend upon microprocessor circuits. Besides controlling the data flow of information, these microprocessor circuits form the basis for the data entry, retrieval, and processing terminals on a network. There are also microprocessor-based digital computer terminals. A microprocessor is the central processing unit (CPU) in a microcomputer. A microcomputer includes a microprocessor and the input/output (I/O) as well as the storage devices.

The microprocessor-based computer (microcomputer) has two major categories: the less powerful home computer and the more powerful business computer. These microcomputers operate using 8-bit, 16-bit, and 32-bit circuit chips. The 8-bit machines are the most popular for both home and business computers, although 16-bit chips are here and 32-bit microprocessor chips will dominate the market when the appropriate software is available.

For the business microcomputers there are many different configurations. Mi-

COMPUTER	Radio Shack TRS-80 Model III	Heath H-89	Commodore PET	Commodore VIC 20 Video Computer	North Star Advantage
HARDWARE					
RAM memory	4K to 48K	48K to 64K	16K to 32K	5K to 32K	20K to 64K
ROM memory	4K or 14K	48K	18K	16K	8K
Black/white display	yes	yes	yes	no	yes
Color display	no	no	no	yes	no
Graphics	64 characters and dot graphics	33 graphic characters	128 full-screen character graphics	176 x 184 dot	2- and 3-dimensional
Interface	integral	integral, 3 serial ports	integral connection	integral	integral cable plug
Keyboard	basic plus keypad	special function and editing keys, numeric pad	basic	basic	same as IBM Personal Computer
Storage	5¼″ diskette	5¼″ diskette, Winchester	5¼″ diskette	5¼″ diskette	5¼″ diskette
Capacity/drive	131KB user, 175KB user	100KB to 11MB	170KB, 1050KB	170KB	360KB
Printers	M, I	M, I	M, I	M	M, I
Data communications	asynchronous	asynchronous	asynchronous	asynchronous	synchronous, asynch.
Optional I/O	none	none	cassette recorder	cassette, joysticks, paddles	serial and parallel
SOFTWARE					
Languages	BASIC, COBOL, Pascal, AUTHOR, PILOT, FORTRAN, Quick, Quiz	MBASIC, CBASIC, COBOL, Pascal, FORTRAN	BASIC	BASIC	BASIC, MBASIC, CBASIC, COBOL, Pascal, FORTRAN, CP/M
Operating systems	DOS	DOS, UCSDp, CP/M	1K	own	DOS
Documentation	20 books, manuals	about 40 manuals	about 12 manuals	5 to 10 manuals	about 10 manuals
CONSUMER INFO					
Support	dealer training	none	none	none	dealer training
Maintenance	dealer, on-site	on-site, carry-in	dealer	dealer	dealer, SORBUS service
Warranty	90 days parts and labor on hardware	assembled: 1 yr parts and labor; kits: 90 days parts	90 days parts and labor	90 days	90 days parts and labor
APPLICATIONS					
Personal inventory	yes	yes	yes	yes	yes
Investment management	yes	no	yes	yes	no
Personal finance	yes	yes	yes	yes	no
Office at home	yes	yes	yes	yes	yes
Doing research	yes	yes	yes	yes	yes
Playing games	yes	yes	yes	yes	no
Learning new skills	yes	yes	yes	yes	yes
Writing/editing text	yes	yes	yes	yes	yes

FIGURE 2–20 Microprocessors. (*Source: Forbes Magazine*, August 2, 1982, p. 61.)

Apple II Plus	Apple III	IBM Personal Computer	Atari 400	Atari 800	OSBORNE 1
48K to 64K	128K to 256K	16K to 256K	16K	16K to 48K	60K
12K	4K	40K	10K	10K	4K
yes	yes	yes	yes	yes	yes
yes	yes	yes	yes	yes	no
15 colors	15 colors	line graphic characters	128 colors, graphics	128 colors, graphics	32 graphic characters
integral cable plug	50-pin connector	IBM monochrome, printer adapter	integral cable plug	integral cable plug	RS-232, Centronics 700 compatible
basic	numeric keypad	83 key, numeric pad, function keys	flat	basic	basic plus numeric pad
5¼" diskette	5¼" hard disk w/5MB	5¼" diskette	5¼" diskette, cassette	5¼" diskette, cassette	5¼" diskette
140KB	140KB	160KB, 320KB	88KB	88KB	100KB per drive
M, T, I	M, T, I	M	MI, MT	MI, MT	any RS-232 interface
asynchronous	asynchronous	asynchronous	asynchronous	asynchronous	asynchronous
wide variety	wide variety	cassette jack	tape cassette, joysticks, loud-speakers, paddles	tape recorder, joysticks, loud-speakers, paddles	none
BASIC, COBOL, Pascal, others	BASIC, Pascal, others	BASIC, COBOL, Pascal, FORTRAN	BASIC, PILOT, FORTH	Microsoft BASIC, BASIC A +, Pascal, PILOT, FORTH	CBASIC, MBASIC
DOS 3.3, UCSDp	SOS	DOS, UCSDp, C/PM-86	own	own	CP/M
about 100 manuals	about 100 manuals	2 manuals	about 3 manuals	about 3 manuals	system, language manuals
none	none	dealer training	users group	users group	dealer training
on-site, sales centers	on-site, sales centers	dealer	dealer	dealer, regional service center	dealer
90 days parts and labor	90 days parts and labor	90 days	90 days parts and labor	90 days parts and labor	90 days limited hardware
yes	yes	yes	yes	yes	yes
yes	yes	yes	no	yes	yes
yes	yes	yes	yes	yes	yes
yes	yes	yes	no	yes	yes
yes	yes	yes	yes	yes	yes
yes	yes	yes	yes	yes	no
yes	yes	yes	yes	yes	no
yes	yes	yes	no	yes	yes

DOS: Disk Operating System.   UCSDp: University of California, San Diego operating system.   CP/M, CP/M-86: operating systems designed by Digital Research Inc.   I: Impact.   M: Matrix.   T: Thermal.

crocomputers can have in the area of 40 to 50 million characters of data stored on external disks (a single four-platter 5¼-inch disk can hold 25 million characters). They can be configured to meet any needs of the business office. The new generation of microcomputers that use 32-bit chips can be coupled together to offer even greater processing capabilities.

We would like to discuss the home microcomputer (usually an 8-bit microprocessor with less disk capacity) at this point because of cable TV, rooftop antennas to satellites, videotext systems, the telephone company local loops, and/or cellular radio capabilities that can make every home a node on a worldwide data communication network. This network can be used for electronic mail, processing of data, information retrieval from databases, and hundreds of other uses.

A microcomputer at home can be used for many purposes. Some of your office work can be performed with it to make your day at the office easier. Word processing programs can help you type letters or write articles. You can perform some research, or just retrieve information from various public databases for your own use. Perhaps you might want to learn a new skill such as programming, or use public databases to learn other skills. Of course games such as Pac-Man contribute to your leisure time. Microcomputers are an ideal way to inventory personal possessions, maintain personal financial control, keep a variety of records, and manage your investments portfolio. In the future a basic use of home microcomputers will be to interconnect with various networks that will be available worldwide. In these networks you will be able to retrieve a wide variety of data, send electronic mail messages to anyone in the world who is on the network, carry on two-way dialogues, pay bills, or do banking. And, of course, by using your microcomputer with its disks, you will be able to store the retrieved data.

This section on hardware is not intended to be a book on microcomputers, but Figure 2–20 compares a variety of "home" microcomputers that are now available. These are the terminal devices that will be popular for use in private home networks.

QUESTIONS

1. Review the list of functions of a front end communication processor. Identify what you consider to be the six most important functions of a front end.

2. What are the two most important functions of a modem?

3. If you were buying a multiplexer, why would you choose either TDM or FDM?

4. For data communication transmissions which would you use, hardware encryption or software encryption?

5. Of the five classifications of terminals listed in this chapter, which are available at your organization?

3

THE TECHNICAL DETAILS

This chapter discusses the technical details of data communications in easy-to-understand terminology. It explores the fundamental concepts of how and why a data communication network functions. These concepts include codes, efficiency, capacity, modulation, bits/baud, polling and selecting, throughput, response time, signaling, echo suppressors, TASI, and voice call multiplexing. Each section stands by itself as an independent description of a fundamental data communication concept.

COMMUNICATION CODES/EFFICIENCY

The *bit* (0 or 1) is the smallest unit of data stored. Groupings of these bits, known as *codes*, make up characters, or bytes. Finally, groupings of characters make up words and sentences which form our business ideas and these are the data/information elements with which we work. The codes used for data communications/teleprocessing range from a 5-level code to an 8-level code. By *level* we mean the number of bits that are required to represent a character, also called a *byte*. A character might be an alphabetic letter, a number, or any of the special symbols such as commas and the like.

United States of America Standard Code for Information Interchange (USAS-CII) or more commonly ASCII is the most popular code for data communications. It is the basic standard code and is available on most terminals. This is an 8-bit code that has 128 valid character combinations. The number of combinations can be determined by taking the number 2 and raising it to the power equal to the number of information bits in the code. In this case $2^7 = 128$ characters. The eighth bit is the *parity bit* for error checking on individual characters.

USASCII code is used widely on both asynchronous and synchronous data communication equipment. (The concepts of synchronous and asynchronous transmission are discussed in Chapter 4.) Figure 3–1 shows the code structure for this ASCII code and an explanation of each of the control characters.

Extended Binary Coded Decimal Interchange Code (EBCDIC) is IBM's standard information code. This code has 256 valid character combinations because there are 8 information bits and parity is carried as a ninth bit. You probably have heard people talk about a 9-channel tape drive. The 9-channel tape drive was developed originally to hold the IBM code with its 8 data bits and 1 parity bit. If used in asynchronous transmission, this code has 11 bits per character because there is 1 start, 8 data, 1 parity, and 1 stop bit. In synchronous transmission it would have only 9 bits per character unless the parity bit was stripped off prior to transmission. Figure 3–2 shows the standard EBCDIC code configuration chart.

It should be noted that the bit positional numbering system is different between EBCDIC and ASCII. ASCII numbers its bit positions from 8 to 1 (left to right), while EBCDIC numbers its bit positions from 0 to 7 (also left to right). In fact, IBM addresses everything left to right, such as memory, records, and bits in a byte. Figure 3–3 depicts the positional numbering system differences between EBCDIC and ASCII.

Bit positional difference is mentioned because if you were converting a master file from an EBCDIC system to an ASCII system and the file documentation said that bit number 3 in a status byte reflected an overdraft status of this account, what would you do? To someone who learned on an ASCII system, bit number 3 would be in a different location than to the person who wrote the documentation for the EBCDIC file description.

Binary Coded Decimal (BCD) code is a 6-bit code that has 64 valid character combinations ($2^6 = 64$). The BCD code was the logical extension from the earlier tab card oriented Hollerith code. Depending on the specific hardware, this code can have 1 or 2 parity bits; therefore it can be either 6 bits (if the parity bit is not transmitted), 7 bits (1 parity bit), or 8 bits (2 parity bits) during synchronous transmission. A start bit and a stop bit would have to be added for asynchronous transmission. Figure 3–4 shows the BCD code.

One of the oldest codes of data communication is called the Baudot code. It is a 5-bit code that has only 32 possible character combinations; however, there are also two functions, called *letters* and *figures*. When one of these two functions is used, it sets the equipment so all characters typed after that point are in a different configuration than they were previously. In effect, this raises the number of valid character combinations to 58, just barely enough for simplified data communications. There is a version called International Baudot that has a sixth bit added for parity purposes. Baudot code is used on the earlier teletype equipment and on very slow communication circuits (150 bits per second or less). Figure 3–5 shows the Baudot code configuration.

Self-checking or "M" out of "N" codes also are available. These codes are able to indicate if one of the bits was changed during transmission. For example, with a 2 out of 5 code, there are always two 1s and three 0s representing a character. If this precise ratio (two 1s and three 0s) is not present in each received character, then an error has occurred during transmission of the data. One disadvantage of this system is that it makes many of the possible combinations unusable. For example,

$$C = \frac{N!}{M! \, (N - M)!}$$

shows how to calculate the number of usable code configurations when using one of these self-checking codes. C (the number of combinations) is determined by taking the various factorials of N and M, where N equals the total number of bits in the code and M equals the number of 1 bits in the code. With the example of a 2 out of 5 code we have: $C = 5!/2! \, (5 - 2)!$ or only 10 legal combinations. Obviously this is far too few combinations; therefore, a 2 out of 5 code is totally unusable except for transmitting numbers only.

Several years ago IBM developed a 4 out of 8 code, where 4 of the 8 bits were 1s and 4 of the 8 bits were 0s. Any other combination was an error. This code was of limited value because instead of 256 valid combinations, there are only 70. Also, it does not have the option of stripping off the parity bit before a synchronous transmission and putting it back on upon receipt. On the positive side, this 4 of 8 code detects errors better than a single parity bit. It has the same error checking capabilities as the double parity bit in a BCD code structure.

b7 →					0	0	0	0	1	1	1	1
b6 →					0	0	1	1	0	0	1	1
b5 →					0	1	0	1	0	1	0	1
Bits	b4	b3	b2	b1	column row ▼ ►	0	1	2	3	4	5	6	7
	0	0	0	0	0	NUL	DLE	SP	0	@	P		p
	0	0	0	1	1	SOH	DC1	!	1	A	Q	a	q
	0	0	1	0	2	STX	DC2	"	2	B	R	b	r
	0	0	1	1	3	ETX	DC3	#	3	C	S	c	s
	0	1	0	0	4	EOT	DC4	$	4	D	T	d	t
	0	1	0	1	5	ENQ	NAK	%	5	E	U	e	u
	0	1	1	0	6	ACK	SYN	&	6	F	V	f	v
	0	1	1	1	7	BEL	ETB	/	7	G	W	g	w
	1	0	0	0	8	BS	CAN	(	8	H	X	h	x
	1	0	0	1	9	HT	EM	)	9	I	Y	i	y
	1	0	1	0	10	LF	SUB	*	:	J	Z	j	z
	1	0	1	1	11	VT	ESC	+	;	K	[	k	{
	1	1	0	0	12	FF	FX	,	<	L		l		
	1	1	0	1	13	CR	GS	–	=	M	]	m	}
	1	1	1	0	14	SO	RS	.	>	N	>	n	~
	1	1	1	1	15	SI	US	/	?	O	–	o	DEL

Character Codes
1. Standardized groupings of bits (1s and 0s) to represent alphanumeric and control information.
2. American Standard Code for Information Interchange (ASCII)—ANSI X3.4.
    a. A 7-bit code which yields 128 possible combinations or character assignments.
    b. Ninety-six graphic, i.e., printable or displayable, characters.
    c. Thirty-two control characters, including
        (1) Device-control characters such as Line Feed, Carriage Return, Bell, etc.
        (2) Information-transfer control characters such as ACK, NAK, etc.

FIGURE 3–1 ASCII code structure.

Mnemonic and Meaning		Mnemonic and Meaning	
NUL	Null	DLE	Data Link Escape (CC)
SOH	Start of Heading (CC)	DC1	Device Control 1
STX	Start of Text (CC)	DC2	Device Control 2
ETX	End of Text (CC)	DC3	Device Control 3
EOT	End of Transmission (CC)	DC4	Device Control 4
ENQ	Enquiry (CC)	NAK	Negative Acknowledge (CC)
ACK	Acknowledge (CC)	SYN	Synchronous Idle (CC)
BEL	Bell	ETB	End of Transmission Block (CC)
BS	Backspace (FE)	CAN	Cancel
HT	Horizontal Tabulation (FE)	EM	End of Medium
LF	Line Feed (FE)	SUB	Substitute
VT	Vertical Tabulation (FE)	ESC	Escape
FF	Form Feed (FE)	FS	File Separator (IS)
CR	Carriage Return (FE)	GS	Group Separator (IS)
SO	Shift Out	RS	Record Separator (IS)
SI	Shift In	US	Unit Separator (IS)
		DEL	Delete

FIGURE 3–1 *(Continued)*

Now that we have discussed codes, we must review efficiency with regard to a specific code set. One objective of a data communication network is to achieve the highest possible volume of accurate information through the system. The higher the volume, the greater the resulting systems efficiency and the lower the cost. System efficiency is affected by such characteristics of the circuits as distortion and transmission speed, as well as by turnaround time, the coding scheme utilized, the speed of the transmitting and receiving equipment, the error detection and control methodologies, and the mode of transmission. In this section, we will focus on the coding scheme.

Transmission efficiency is defined as the total number of information bits divided by the total bits.

Each communication code structure has both information bits and redundant bits. *Information bits* are those used to convey the meaning of the specific character being transmitted, such as the character "A." *Redundant bits* are used for other purposes such as error checking. Therefore, a parity bit used for error checking is a redundant bit because it is not used to identify the specific character, even though it may be necessary. As you can see, if you did not care about errors, the redundant error checking bit could be omitted. Also, any message control characters, such as an END OF BLOCK character would be considered redundant bits. These bits are redundant by definition only, because they are needed for accurate data communications, just as the periods and commas are required when you write a letter.

Figure 3–6 shows that the efficiency of the code (E_c) equals the bits of information (B_I) divided by the bits in total (B_T). This means that if you have an 8-level code with 7 of the bits used to represent the character and one of the bits representing parity, then you would have a code efficiency of 87.5%. This would be calculated by taking the bits of information (B_I) and dividing by the bits in total (B_T). For example, $7/8 = 0.875$.

Bits 4567	Hex1	00				01				10				11				Bits 0,1
		00	01	10	11	00	01	10	11	00	01	10	11	00	01	10	11	2,3
	↓	0	1	2	3	4	5	6	7	8	9	A	B	C	D	E	F	Hex 0
0000	0	NUL	DLE			SP	&	–										0
0001	1	SOH	SBA				/			a	j			A	J			1
0010	2	STX	EUA		SYN					b	k	s		B	K	S		2
0011	3	ETX	IC							c	l	t		C	L	T		3
0100	4									d	m	u		D	M	U		4
0101	5	PT	NL							e	n	v		E	N	V		5
0110	6			ETB						f	o	w		F	O	W		6
0111	7			ESC	EOT					g	p	x		G	P	X		7
1000	8									h	q	y		H	Q	Y		8
1001	9		EM							i	r	z		I	R	Z		9
1010	A					¢	!	|	:									A
1011	B					.	$	,	#									B
1100	C		DUP		RA	⟨	.	%	@									C
1101	D		SF	ENQ	NAK	(	)	_	'									D
1110	E		FM	ACK		+	;	⟩	=									E
1111	F		ITB		SUB	|	_	?	"									F

EBCDIC Code as Implemented for the IBM 3270 Information Display System

Extended Binary Coded Decimal Interchange Code (IBM's EBCDIC).

a. An 8-bit code yielding 256 possible combinations or character assignments.

b. A representative subset, that of the IBM 3270 product family.

   Absence of certain functions not usable by 3270 products (e.g., paper feed, vertical tab, back space) which would show up in EBCDIC code charts for other products that might make use of them.

FIGURE 3–2 EBCDIC code.

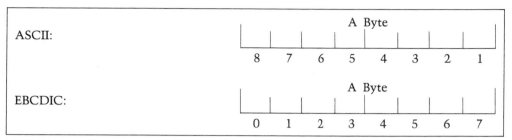

FIGURE 3–3 Bit positional differences between ASCII and EBCDIC.

Zone $\left\{\begin{array}{l} b_6 \longrightarrow \\ b_5 \longrightarrow \end{array}\right.$

					Digit		
0	0	I	I				
0	I	0	I	b_4	b_3	b_2	b_1
SP	␢	—	& +	0	0	0	0
1	/	J	A	0	0	0	I
2	S	K	B	0	0	I	0
3	T	L	C	0	0	I	I
4	U	M	D	0	I	0	0
5	V	N	E	0	I	0	I
6	W	O	F	0	I	I	0
7	X	P	G	0	I	I	I
8	Y	Q	H	I	0	0	0
9	Z	R	I	I	0	0	I
φ	‡	!	?	I	0	I	0
# =	,	$	.	I	0	I	I
@ '	% (	*	□)	I	I	0	0
:	γ	]	[	I	I	0	I
⟩	\	;	⟨	I	I	I	0
√	⧺	△	⧧	I	I	I	I

CHARACTER

6	5	4	3	2	1
ZONE		DIGIT			

CODE	DESCRIPTION
SP	Space
␢	Space (Even Parity)
‡	Record Mark
!	Minus Zero
?	Plus Zero
γ	Word Separator
√	Tape Mark
⧺	Tape Segment Mark
△	Delta (Mode Change)
⧧	Group Mark

FIGURE 3–4 BCD code.

If you have a 6-bit code with 2 parity bits, then the efficiency of the code (E_C) would be 6/8 or 75%. As is evident by this comparison, some codes are more efficient than others. The addition of parity bits lessens efficiency because parity bits are redundant, not being needed to convey meaningful information but only to check for errors.

The same formula (as is shown in Figure 3–6) can be used to get an estimate of

Lower Case	Upper Case	1	2	3	4	5
A	—	•	•			
B	?	•			•	•
C	:		•	•	•	
D	$	•			•	
E	3	•				
F	!	•		•	•	
G	&		•		•	•
H	£		•			•
I	8		•	•		
J	'	•	•		•	
K	(	•	•	•	•	
L	)		•			•
M	.			•	•	•
N	,			•	•	
O	9				•	•
P	0		•	•		•
Q	1	•	•	•		•
R	4		•		•	
S	bell	•		•		
T	5					•
U	7	•	•	•		
V	;		•	•	•	•
W	2	•	•			•
X	/	•		•	•	•
Y	6	•		•		•
Z	"	•				•
Letters (shift to lower case)		•	•	•	•	•
Figures (shift to upper case)		•	•		•	•
Space				•		
Carriage return					•	
Line feed			•			
Blank						
• represents a 1 Blank represents a 0						

FIGURE 3–5 Baudot code.

$$E_C = \frac{B_I}{B_T}$$

E_C: efficiency of the code

B_I: information bits

B_T: total bits

FIGURE 3–6 Efficiency of codes.

the efficiency of an asynchronous transmission system. As an example, assume that there is an 8-bit code structure where 7 bits represent the data and 1 bit is for parity. In asynchronous transmission there is usually one start bit and one stop bit. Therefore, the bits of information (B_I) would be 7, but the bits in total would be 10 (B_T). The efficiency of the asynchronous transmission system, at its maximum, would be 7 bits of information divided by 10 total bits for an efficiency of 70%. It should be noted that if there were any other control characters involved, such as message character counts or other control characters sent at the end of your transmission, the efficiency would drop below 70%.

The same basic formula can be used if an estimate is needed for the efficiency of a synchronous transmission system. In this case, the bits of information (B_I) are calculated by determining how many "information" characters are in the message block. If the message portion of the frame or packet contains 100 information characters, using our 8-bit code, there would be 7 bits times 100 characters, or 700 bits of information. Next, the bits in total (B_T) would be the 700 bits of information, plus all the redundant bits that are inserted for control and security purposes. These other bits include the parity bit (eighth bit) that is appended to each character, the control bits in the flag at each end of the frame, the bits in the control field, the address field, the frame check sequence, and any internal control characters from the packet frame such as format identifiers, logical channel numbers, sequence numbers, and a couple of synchronization (SYN) characters.

For this example assume that there are a total of 11 control characters for our message. Therefore, the number of redundant bits is 11 control characters times 8 bits per character, plus 100 bits (the parity bit for each of the characters in your message) for a total of 188 redundant bits. This shows that the efficiency of this synchronous system is 700 bits of information (B_I) divided by 888 bits in total (B_T) for an efficiency of 79%.

This example shows that synchronous systems are more efficient than asynchronous systems and some codes are more efficient than others. To extend our example further, assume we have compression/compaction software or a hardware scheme that strips off the individual parity bits for each character prior to transmission and puts them back on at the destination. The number of redundant bits (no parity bits) would be reduced by 100; therefore, efficiency would rise to 89% (700 bits of information divided by 788 bits in total). If the frame check sequences and other error checking techniques for synchronous transmission are good enough, you might consider stripping off the individual parity bits for each character prior to transmission in order to gain a 10% increase in throughput efficiency of data bits. For a more accurate method of measuring throughput efficiency, see the section Throughput (TRIB) later in this chapter.

CIRCUITS/CHANNELS

A circuit is nothing more than the path over which data moves. Many people use the word *line* interchangeably with the word *circuit*, although line implies a physical wire or glass fiber connection. Circuit is more correct when speaking of satellite and/or microwave transmission. Sometimes the words *circuit*, *channel*, and *line* are used interchangeably.

Often an individual communication circuit is subdivided into separate transmission subchannels as is done with multiplexing. Some users refer to a channel when they are speaking of a single transmission facility and also when they are speaking of a circuit that has been subdivided into numerous channels (more correctly, *subchannels*).

FIGURE 3–7 Open–wire pairs.

Also, there is the word *link*. There can be many links in a cross-country communication circuit because a link is any two-point segment of a communication circuit. Therefore, a multidrop circuit has many links as it traverses the country. Voice grade telephone lines are sometimes referred to as *trunks* or *trunk lines*.

There are many types of transmission media in use today. There are also very exciting aspects in store for the future as the various transmission media enter our homes for direct communication through such techniques as videotext, electronic mail, voice mail, and two-way cable TV. The discussion below defines each of the types of transmission media or circuits used to transmit data.

Open-wire Pairs   These are copper wires suspended by glass insulators on telephone poles (see Figure 3–7). They are spaced approximately one foot apart. While they are still familiar in many areas, they are being replaced by cables and other more modern transmission media. Open wire pairs are fast becoming a part of the past.

Wire Cables   These are insulated pairs of wires and therefore can be packed quite close together (see Figure 3–8). Bundles of several thousand wire pairs are placed under city streets and in large buildings throughout the country. In fact, your own house probably has a four-wire, wire cable connecting your telephone to the telephone company central switching office. Wire cables usually are twisted (twisted wire pairs) in order to minimize the electromagnetic interference between one pair and any other pair in the bundle of cables. Wire cables also

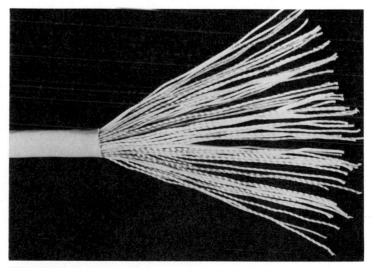

FIGURE 3–8 Wire cables.

are being replaced by more efficient transmission media such as coaxial cable, microwave, satellite, and optical fibers.

Coaxial Cable   Figure 3–9 shows a single coaxial cable and a bundle of coaxial cables. Each individual coaxial cable consists of copper in the middle with an

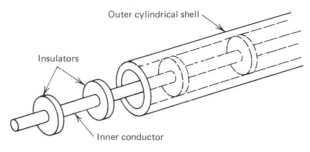

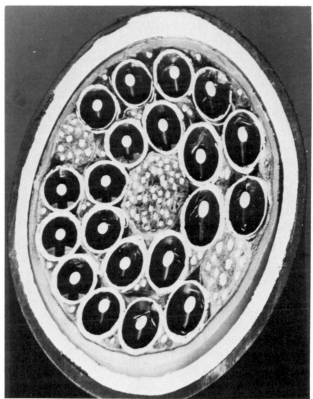

FIGURE 3–9 Coaxial cables showing exterior view and cross section.

outer cylindrical shell for insulation. This type of circuit can transmit using substantially higher frequencies than a wire pair. Therefore, it is far more efficient to use a coaxial cable because it can contain many telephone conversations. In fact, a 2-inch diameter bundle of coaxial cables can handle approximately 20,000 telephone calls simultaneously, either voice or data. Coaxial cables have very little distortion, cross talk, or signal loss; therefore, they are a better transmission media than either open wire pairs or bundles of wire cables.

Microwave Transmission   This transmission medium is the one most used for long-distance data or voice transmission. It does not require the laying of any cable because long-distance dish or horn antennas with microwave repeater stations are placed approximately 25 to 30 miles apart (see Figure 3–10). It is a line-of-sight transmission medium. A typical long-distance antenna might be 10 feet across, although over shorter distances in the inner cities these antennas might get down to 3 feet in diameter. In larger cities we now have microwave conges-

FIGURE 3–10 Microwave tower.

tion; so many microwave dish antennas have been installed that they interfere with each other and the air waves are saturated. This is a problem that will force future users to seek alternate transmission media such as satellite or optical fiber links.

Satellite This transmission medium is similar to microwave transmission except, instead of transmitting to another nearby microwave dish antenna, it transmits to a satellite 22,300 miles out in space. Figure 3–11 depicts a geosynchronous satellite in space. Figure 3–12 shows the satellite in operation, and Figure 3–13 shows a satellite antenna. Whereas the satellite antennas of the 1960s were very large, satellite antennas of the future may be down to 3 feet in diameter.

One of the disadvantages of satellite transmission is that a delay occurs because the signal has to travel far out into space and back to earth (propagation delay). A typical signal propagation time is approximately 0.5 to 0.6 seconds for the delay in both directions. In addition, there may be a delay for going through ground-based switching equipment. This delay is controlled by the common carriers to avoid disruptive voice telephone conversations. In data communications, however, this type of delay would be disastrous if you were using a Binary Synchronous Communications (BSC) protocol with the stop and wait ARQ. This is because when an individual message block is sent, you have to wait for a positive

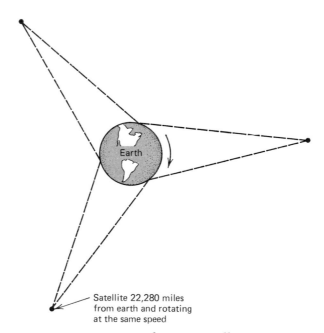

Earth

Satellite 22,280 miles
from earth and rotating
at the same speed

FIGURE 3–11 Geosynchronous satellites in space.

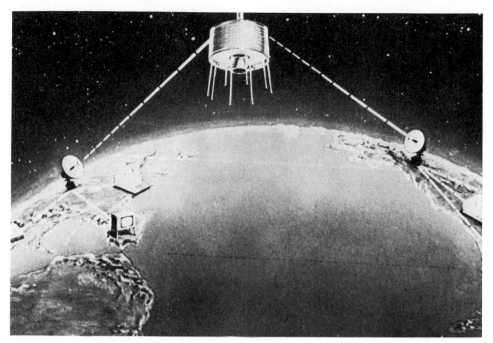

FIGURE 3-12 Satellite in operation.

or negative acknowledgment before sending the next message block. As you will see in Chapter 6, the newer protocols (SDLC, X.25, HDLC) allow you to send a burst or group of messages and then the called party responds to this group or burst (continuous ARQ). Satellites use different frequencies for receiving and transmitting. These frequencies are in the ranges of 4 to 6 GHz, 12 to 14 GHz, and 20 to 30 GHz.[1]

Optical Fibers   This is the newest of our technologies for data transmission over a continuous line (see Figure 3–14). Fiber optics equipment uses pulses of light transmitted through hair-thin strands of plastic or glass. Plastic probably is falling by the wayside, because glass can be made much purer. With greater purity, the signal can be transmitted over a longer distance before an amplifier/repeater station has to be added to increase signal strength. Currently, optical cables need repeater amplifiers every 4 to 6 miles, whereas copper wire pairs require amplifiers every mile.

[1] 1 Hz (hertz) equals 1 cycle per second.
  1 KHz (kilohertz) equals 1000 cycles per second.
  1 MHz (megahertz) equals 1 million cycles per second.
  1 GHz (gigahertz) equals 1 thousand million cycles per second.

FIGURE 3–13 Satellite antenna.

Scientists are now developing superglass. Some tests have been successful with continuous strands of this pure glass in 60-mile lengths with no intermediate amplifier repeaters. Britain today has systems that utilize one optical fiber to carry 2000 simultaneous telephone conversations on a single light beam at a rate of 140 million bits per second. Even higher rates of 576 million bits per second (which can simultaneously carry 7680 telephone calls) have been demonstrated successfully. The typical light sources are either light emitting diodes (LED) or laser light.

AT&T is developing fiber optic undersea cables (the TAT-8 system) and they predict that fiber optics will supplant satellites in the long-term future.

The biggest advantage of fiber optical cables is the bandwidth, because they have the potential for transmitting data at speeds up to 10^{14} bits per second. These frequencies are 10,000 times greater than the upper ranges of radio frequency bands (microwave/satellite).

FIGURE 3–14 Optical fiber.

Optical fibers are immune to electrically generated noise and therefore have a lower error rate. For example, a fiber optic cable might have a bit error rate of 10^{-9} as compared with a 10^{-6} bit error rate found in metallic connectors.

Optical cables have a very high resistance to taps (virtually impossible), they will operate in very high temperature environments that would melt copper cables, and they offer much higher levels of security, as you will see in Chapter 8, Security and Control.

Cellular Radio   This is a form of high-frequency radio where antennas are spaced strategically throughout a metropolitan area (see Figure 3–15). The user (voice or data transmission) dials or logs into the system and their voice or data are transmitted directly from their home, automobile, or place of business to one of these antennas. This is a direct replacement for the copper cable local loops.

This system has intelligence. As you move away from one transmitter while driving an automobile, the signal weakens. With cellular radio, transmission is switched automatically to a closer antenna without communication being lost. Today, cellular radio is used widely by the radio paging companies. By 1990, however, it may become the standard transmission media for local loops in large metropolitan areas. Other competitors for the local loop market (sometimes called DTS, Digital Termination Systems) would be cable TV (coaxial cable) and rooftop antennas (satellite).

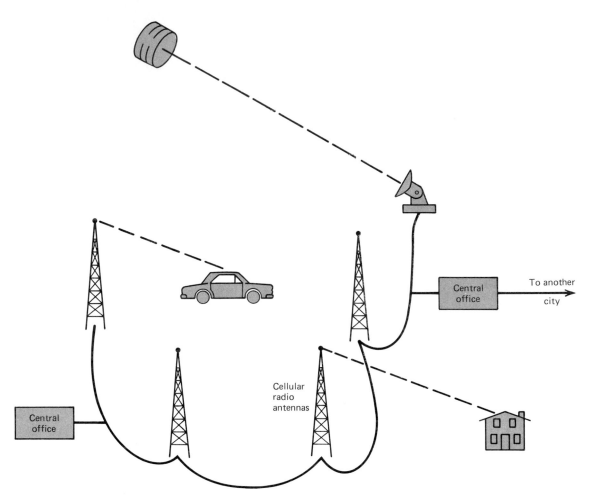

FIGURE 3–15 Cellular radio system.

Other Miscellaneous Circuit Types Other types of circuits are waveguides, tropospheric scatter circuits, short-distance radio, and submarine cables.

A *waveguide* is a conductive tube down which radio waves of very high frequency travel. These tubes may be from 2 to 15 inches across. Waveguides usually are used in the back of microwave towers to transmit the signal from the repeater/amplifier of electronic equipment to the dish antenna. They are used over very short distances, 10 feet or less. The American Telephone & Telegraph WT4 waveguide system can carry 230,000 two-way telephone calls simultaneously.

The *troposphere* (it extends upward from the earth to about 6 miles) scatters radio waves and it can be used for communication links of up to about 600 miles.

Tropospheric scatter is especially useful in the South Pacific, where islands are widely scattered and separated by long distances. One problem with this method of transmission is that it requires very large antennas, on the order of 60×120 feet. Tropospheric scatter circuits can be used to transmit television over shorter distances, and up to several hundred voice communications can be transmitted over a link of approximately 100 miles. Tropospheric scatter circuits are not advised for data transmission because of the high error rate encountered. Obviously microwave links, or even satellite circuits, would be preferable.

Short-distance radio is the type of transmission circuit that is used by walkie-talkies, police radio, taxis, and other community services. Depending on the distance involved, short-distance radio might be used on the cellular radio circuits or it might be a type of independent high-frequency radio.

Submarine cables are transmission circuits that go between the various continents. The most popular form in use today is a bundle of coaxial cables. Currently, Bell Laboratories is developing a high-speed underwater optical fiber cable. This cable probably will come into use in the mid-1980s, and it may be able to handle transmission speeds of 274 million bits per second, which would accommodate 4000 voice channels.

CAPACITY OF A VOICE GRADE CIRCUIT

Analog transmission takes place when the signal that is sent over the transmission media continuously varies from one state to another. This would be similar to having a dimmer switch on an electric light so the intensity of the light can be varied from very bright to very dim, but it would be a continuous varying as contrasted with a switch that just turns the light on and off (the on/off switch is digital).

Most telephone circuits utilize analog transmission because they were developed for voice transmission, not data transmission. New systems today are built in a digital fashion.

In order to understand bandwidth in an analog channel, it may be desirable to review a sine wave and frequency or hertz. Figure 2–7 (Chapter 2) illustrated sine waves that had different frequencies (different number of cycles per second). Hertz means cycles per second; therefore, a 1700 cycle-per-second tone also can be called a 1700 hertz tone.

Figure 3–16 shows that most transmissions over a voice grade circuit are done within the human hearing range. This is because most of the telephone systems were built for human speech and not data. Human hearing is approximately in the range of 20 to 20,000 hertz, although most people cannot hear above 14,000 hertz. Figure 3–16 also shows the frequency range for coaxial cable, microwave, satellite, and laser (optical fibers). Bandwidth refers to a range of frequencies. For

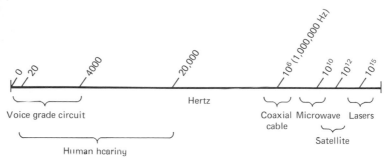

FIGURE 3–16 Frequency spectrum.

example, Figure 3–16 shows that the bandwidth of a voice grade communication circuit is from 0 to 4000 hertz.

A voice grade communication channel or circuit (also known as series 2000/3000 circuit) is the most common communication circuit today. This is the circuit that is used with dial telephones, and it typically is the circuit that is installed when organizations lease private or dedicated communication circuits. The bandwidth of this communication circuit is 4000 hertz.

Figure 3–17 shows how this 4000 hertz bandwidth is divided for data transmission use. To start with, there is a 300 hertz guardband at the bottom and a 700 hertz guardband at the top. These prevent data transmissions from interfering with other transmissions when these circuits are stacked on a microwave or satellite link and thus they prevent the frequencies from overlapping between your communications and the communications of others. They can be viewed as similar to the plastic insulation that is put around a copper cable to keep it from short-circuiting with another copper cable.

Figure 3–18 demonstrates how the guardbands provide 1000 hertz of empty

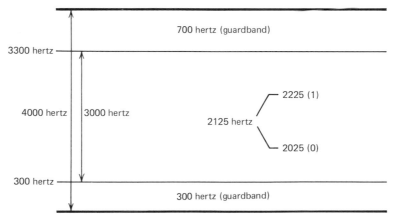

FIGURE 3–17 Voice grade circuit bandwidth.

space between adjacent communication channels. Each voice grade circuit is stacked up using the available bandwidth. This leaves the bandwidth from 300 hertz to 3300 hertz for your data transmission (see Figure 3–17). It is within this 3000 hertz bandwidth that the modem must modulate the carrier wave and transmit the different frequencies. Recall the modem that had a carrier wave of 2125 hertz (see Figure 2–9, Chapter 2). It utilized 2225 hertz for 1s and 2025 hertz for 0s. Returning to Figure 3–17, note that the signal travels right down the middle of the available bandwidth. Modems always try to transmit and receive in the middle of the available frequency bandwidth, where amplitude and phase distortion are the lowest (fewer retransmissions due to errors). Figure 3–19 shows the frequency assignments for a Bell 202 modem.

It is this 3000 hertz bandwidth that limits the maximum transmission speed. For example, the modem controls the speed at which data bits can be transmitted (300 bps to 16,000 bps on voice grade circuits). The 3000 hertz bandwidth limits the maximum speed.

Let's examine the relationship between bandwidth and bits per second. Assume a situation that uses a dial-up circuit, with 2600 hertz out of the 3000 hertz available to the modem. The remainder might be unusable because of noise and distortion or it may be used for ringing operators, signaling the return of coins in a pay telephone, or other telephone company signaling functions.

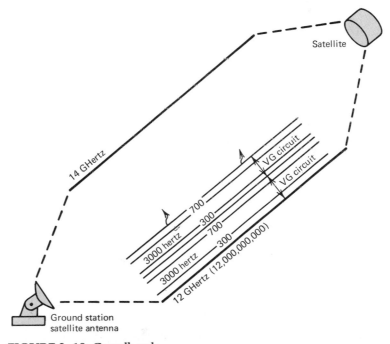

FIGURE 3–18 Guardbands.

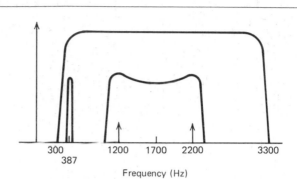

A significant factor to consider when using a 202 modem is line turnaround time required for switching from transmit mode to the receive mode. Echo suppressors in the telephone equipment that are required for voice transmission on long-distance calls must be turned off by the modem to transmit digital data. The modem must provide a 200 ms signal to the line to turn off the echo suppressors every time it goes from transmit to receive mode; hence, if short records are being transmitted, the turnaround time can slow the throughput considerably.

SPECIFICATIONS

Data:

Serial, binary, asynchronous, half duplex on 2-wire lines

Data Transfer Rate:

0 to 1200 bps—switched network

0 to 1800 bps—leased lines with C2 conditioning

Optional 5 bps AM reverse channel transmitter and receiver available for switched-network units (387 Hz).

FIGURE 3–19 Bell 202 (half duplex/1200 BPS).

First Nyquist and then Shannon proved that there is a theoretical maximum capacity, and it is based upon bandwidth. A random stream of bits going across the 2600 hertz bandwidth has a maximum capacity of 25,900 bits per second. This is demonstrated by using Shannon's law:

$$\text{Maximum Bits per second} = \text{Available bandwidth} \quad \text{LOG}_2 \quad \left(1 + \frac{\text{Signal-to-noise ratio}}{}\right)$$

$$\text{Maximum} = 2600 \, \text{LOG}_2 \left(1 + \frac{1000}{1}\right)$$

$$\text{Maximum} = 25{,}900 \text{ bits per second}$$

This calculation uses a 30-decibel signal-to-noise ratio. The signal-to-noise ratio is the strength of the transmit signal in decibels (dB) compared with the level of white noise (Gaussian noise) on the channel/circuit. A signal-to-noise ratio of 30 dB is 1000/1 and a signal-to-noise ratio of 20 dB is 100/1. This means that the signal is 1000 or 100 times more powerful than the background noise on the circuit.

Incidentally, if the signal-to-noise ratio is lowered to 20 decibels in the above example, then the maximum transmission capacity would be 17,300 bits per second. Therefore, the higher the signal-to-noise ratio, the greater the maximum capacity in the channel. Also, the greater the bandwidth, the greater is the maximum capacity.

ANALOG MODULATION

Modulation is the technique that modifies the form of an electrical signal (a carrier wave) so that the signal can carry intelligent information on some form of communication medium. The modulated signal often is referred to as a *broadband signal*. The signal that does the carrying is called a *carrier wave*, and modulation changes the shape or form of the carrier wave in order to transmit 0s and 1s. As you already know, the modem is the device that does the modulation.

Although there are many techniques, the three most common forms of analog modulation are amplitude, frequency, and phase modulation.

As is shown in Figure 3–20, *amplitude modulation* uses a nonvarying frequency. The presence of a 1 bit would make this frequency go to a higher amplitude (more power), and a 0 bit would make it go down to a lower amplitude. This change between a higher and a lower amplitude is how 1 bits and 0 bits are transmitted to the receiving modem.

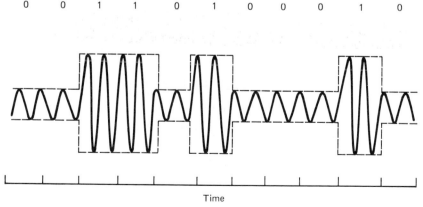

FIGURE 3–20 Amplitude modulation.

If the modem operates at a speed of 1200 bits per second, then the timing between changes from the higher to the lower amplitude is 833 microseconds (1/1200 = 833 microseconds). On the other hand, if there are two 1 bits in a sequence, then the higher amplitude is held on the circuit for a time period equal to 1666 microseconds (2 × 833 microseconds). Of the three types of analog modulation, amplitude modulation may be the poorest because it is the most susceptible to noise during transmission and causes more errors.

Frequency modulation, through the use of *frequency shift keying* (FSK), is a modulation technique whereby each 0 or 1 is represented by a different frequency. In this case, the amplitude does not vary. This type of modulation is represented by different tones. A high-pitched tone (higher frequency) equals a binary 1 and a low-pitched tone (lower frequency) equals a binary 0. Figure 3–21 shows frequency modulation of a bit stream of 0s and 1s. Again, if the modem operates at 1200 bits per second, then there is a frequency change each 833 microseconds, unless there are adjacent similar bits. Modems that use FSK may use three basic frequencies, such as 2125 hertz, 2225 hertz, and 2025 hertz. In this case when no data are being transmitted, the carrier wave signal is 2125 hertz. When a stream of 0s and 1s is transmitted, the carrier wave switches between 2025 hertz and 2225 hertz, depending upon whether it is a 0 or a 1 that is transmitted. You might want to review Figure 3–17.

The third technique of modulation is *phase modulation*. It is the most difficult to understand because there is two-phase (0° and 180°), four-phase (0°, 90°, 180°, and 270°), and eight-phase (0°, 45°, 90°, 135°, 180°, 225°, 270°, and 315°) modulation. Figure 3–22 shows *phase shift keying* (PSK) for a stream of 0 and 1 bits. Notice in Figure 3–22 that every time there is a change in state (0 or 1), there is a 180° change in the phase. A 180° phase change can be seen easily, but it is more difficult to see a 45° change in phase.

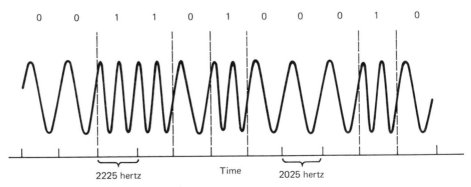

FIGURE 3–21 Frequency modulation.

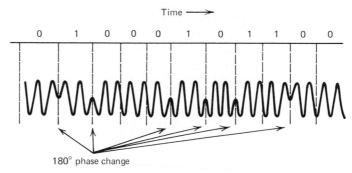

FIGURE 3–22 Phase shift keying (PSK).

The other common type of phase modulation is *differential phase shift keying* (DPSK). In DPSK there is a phase change every time a 1 bit is transmitted; otherwise the phase remains the same (see Figure 3–23).

Different modulation techniques are used to obtain data rates of 2400 bps and above. At 2400 bps, a common technique (used in the Bell 201) is *Quadrature phase shift keying* (QPSK). QPSK involves splitting the signal into four phases, so that a single frequency tone could take one of four values: 90, 180, 270, or 360 degrees of phase shift. Since a single tone can take any of four values, that tone can represent two bits of information. Note that this is more efficient than simple FSK, where two tones are required to represent one bit of information. For speeds above 2400 bps, other techniques are used. For example, most 9600 bps modems use a technique called *quadrature amplitude modulation* (QAM). QAM involves splitting the signal into four different phases and two different amplitudes, for a total of eight different possible values. Thus a signal tone in QAM could represent three bits (two to the third power). The problem with all high-speed modulation techniques, however, it that they are more sensitive to imperfections in the communication channel.

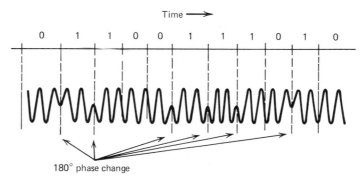

FIGURE 3–23 Differential phase shift keying (DPSK).

Leased communication circuits, as obtained from common carriers, are unfortunately not perfect communication channels. They tend to exhibit several types of imperfections, which limit the amount of information that can be transmitted across them. The two most common types of imperfections are called *envelope delay distortion* and *amplitude deviation distortion*. *Envelope delay distortion* is caused by a phenomenon known as *dispersion*, in which different frequency components of a waveform propagate at different velocities. A complex waveform, such as that produced by a modem, typically is composed of many different frequency components. When the various components of such a waveform begin to progagate at different speeds, they cause the waveform to "spread out" in time. If the waveform spreads out too much, the receiving modem may not be able to interpret it correctly, thus causing line errors. *Amplitude deviation distortion* is an undue amount of variation in the attenuation vs. frequency characteristic on the channel. Again, the result of excessive amplitude distortion is increased error rate due to misinterpretation of the signal by the receiving modem.

In order to transmit data reliably at data rates of 2400 bps and above, modem manufacturers developed a technique called *equalization*. Basically, equalization means that the modem tries to compensate for imperfections in the communication channel. Two types of equalization are commonly used today: compromise and adaptive equalization. *Compromise equalization*, which is used in medium-speed modems such as the Bell 201, involves an equalizer that is set for average, or compromise values based on observed communication line behavior over a large number of lines. Compromise equalization is used at lower speeds (2400 bps) because at those speeds the equalization does not have to be as precise, and the circuitry for compromise equalizers is relatively simple and inexpensive. At higher data rates (4800 and 9600 bps) the results produced by compromise equalizers are not sufficient, so a more sophisticated technique, *adaptive equalization*, is used. An adaptive equalizer is a device that continuously monitors the signal and adjusts the equalization to obtain the best transmission quality at any given time. In most modern high-speed modems, such as the Codex CS 4800 and 9600 or the IBM 3865 (or equivalents), this adaptive equalization is done by a microprocessor with a sophisticated equalization program.

DIGITAL/BIPOLAR

Digital signals are discrete on and off signals as contrasted with the continuous form of analog signal. Figure 3–24 shows various digital signals (baseband) starting with the older unipolar and moving through two types of bipolar signals. Digital signals are drawn as square signals instead of the smooth curve sine wave analog signal. Notice that when there is no signal in unipolar signaling, the voltage

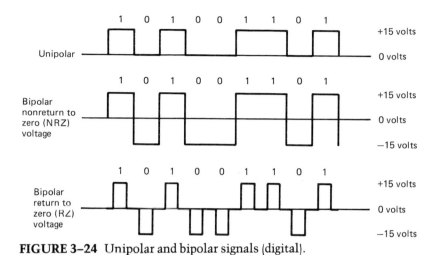

FIGURE 3–24 Unipolar and bipolar signals (digital).

level is 0. In bipolar, the 1s and 0s vary from a plus voltage to a minus voltage. The voltage ranges from 3 to 24 volts, depending upon the equipment utilized. The AT&T/Bell DDS (Digital Data System) system uses bipolar signals. So does the RS232C connector cable. Figure 3–25 shows the RS232C voltage levels. In Europe bipolar signaling sometimes is called *double current* signaling because you are going between a positive and a negative voltage potential.

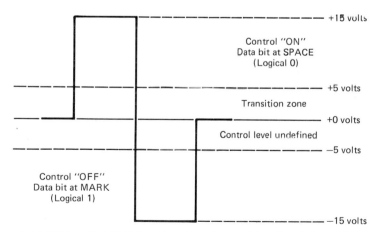

FIGURE 3–25 RS232C voltage levels. RS232C also defines the level and polarity of the signals going to and from the modem. A logical "1" is referred to as a "MARK" and a logical "0" is called "SPACE."

DIGITAL MODULATION

One digital technique of modulation is called *pulse modulation*. It is used for transmitting digital information directly. In other words, an electrical digital signal is not converted to an analog signal. One form of a digital signal might be converted to another form, such as NRZ to RZ, as shown in Figure 3–24, but it is still digital pulses that are being sent down the circuit. Pulse modulation requires a different set of modems.

Another type of digital modulation is *pulse amplitude modulation* (PAM), which gives a different height digital pulse for each different plus or minus voltage (see Figure 3–26). A second type is *pulse duration modulation* (PDM), which gives a longer timed pulse relative to the signal it is measuring. A third type, called *pulse position modulation* (PPM), has more pulses per unit time, depending upon the power of the signal that it is measuring. These types of digital modulation normally are used to convert analog voice signals to a digital form for transmission on local loops and long-distance interexchange channels/circuits (IXC).

Let's explore digital modulation to see how a voice signal can be converted to a binary pulse train. Figure 3–27 illustrates how the original sine wave is converted into digital pulses of 0s and 1s.

As demonstrated in Figure 3–27, three binary numbers (such as 011) will code each pulse amplitude height and will be sent down the communication circuit to the distant station/node. This serial transmission of binary pulses is referred to as *pulse code modulation* (PCM). Figure 3–27 is self-explanatory, so study it care-

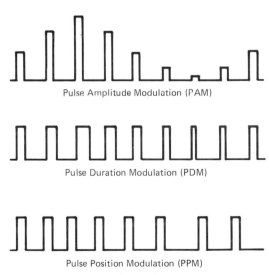

Pulse Amplitude Modulation (PAM)

Pulse Duration Modulation (PDM)

Pulse Position Modulation (PPM)

FIGURE 3–26 PAM/PDM/PPM.

The signal (original sine wave) is quantized into 128 pulse amplitudes. We have used only eight amplitudes for simplicity.

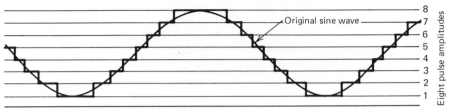

After quantitizing, samples are taken at specific points to produce amplitude modulated pulses. These pulses are then coded. Because we used eight pulse levels, we only need three binary positions to code each pulse.[1] If we had used 128 pulse amplitudes then a 7-bit code plus 1 parity bit would be required.

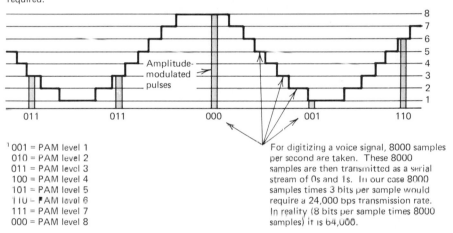

[1] 001 = PAM level 1
010 = PAM level 2
011 = PAM level 3
100 = PAM level 4
101 = PAM level 5
110 = PAM level 6
111 = PAM level 7
000 = PAM level 8

For digitizing a voice signal, 8000 samples per second are taken. These 8000 samples are then transmitted as a serial stream of 0s and 1s. In our case 8000 samples times 3 bits per sample would require a 24,000 bps transmission rate. In reality (8 bits per sample times 8000 samples) it is 64,000.

FIGURE 3–27 Pulse code modulation (PCM).

fully. You should be able to figure out how the transformation takes place from the analog voice signal to the digital (0 and 1) bit stream.

A pulse code modulation technique such as this can be used to convert an analog voice signal to a digital stream of pulses for digital voice transmission. As might be expected, it is much easier to transmit "data bits" digitally because they do not have to be converted from an analog sine wave to a digital pulse. Some of the most recent equipment that is being marketed allows you to digitize an analog voice signal and then send the digital transmission over a single voice grade communication circuit with a bandwidth of 4000 hertz. The AT&T Communications T-1 carrier uses pulse code modulation to carry 1.544 million bits per second of either voice or data information.

Incidentally, as you may have guessed by now, other 0s and 1s can be interleaved between the 0 and 1 digital signals from the transmission that we digitized in Figure 3–27. This means that a common carrier can convert from analog transmission to digital transmission and send two or more separate transmissions over a circuit that previously may have handled only a single voice call. In other words, bit multiplexing is used.

AT&T makes extensive use of PCM internally and actually transmits quite a bit of its information in PCM digital format. The basic unit of information in the AT&T digital hierarchy is the voice channel. The AT&T voice channel is derived from an analog channel that is sampled at a rate of 8000 samples per second. Each sample is composed of 8 bits of information, meaning that a voice channel requires a data rate of $8000 \times 8 = 64,000$ bps. In the analog AT&T system hierarchy, 24 analog voice channels are combined to form what is called a *group*. Similarly, in the digital hierarchy, 24 digital voice channels are combined into one highspeed channel, usually called a "T" channel. The T-1 channel data rate is 1.544 Mbps, which includes 1.536 Mbps of data and 8000 bps of synchronization. The actual device that performs the conversion to a T-1 channel is called a *D Channel Bank*. As you might imagine, T-1 channels are then combined into T-2 channels, with a data rate of 6.312 Mpbs, and so on. The hierarchy goes up to a T-4 channel at approximately 96 Mpbs, but it is very rare to see such a channel outside the AT&T Communications backbone network.

The most common type of digital channel available to the user is the T-1 channel, and that is available only in the major cities. However, if a user has a great deal of digital data to send, a T-1 channel can be very attractive because it is economical and very "clean" since it plugs right into the AT&T Communications backbone network without going through any digital-to-analog and analog-to-digital conversion. Many of the newer digital PBXs (e.g., Northern Telcom SL-1 and Rolm CBX) provide T-1 interfaces that can be used by customers with large communication networks.

BITS/BAUD

Bits and *baud* are terms that are used incorrectly much of the time. The two terms quite often are used interchangeably, but there is a technical difference. In reality, the network designer, or network user, is interested in bits per second because it is the bits that are assembled into characters, characters into words, and thus business information.

Baud actually measures the signaling rate (pulse rate). The signaling rate is the number of times per second that the signal on the communication circuit changes. In other words, if a modem is able to change between any two frequencies 1200 times per second, then the signaling rate is 1200 baud.

$$\frac{1\ \text{second}}{1200\ \text{baud}} = 833\ \text{microseconds}$$

Figure 3–28 shows a modem that has a carrier wave of 1700 hertz. It goes up to 2200 hertz when transmitting 1s and down to 1200 hertz when transmitting 0s (here our example uses different frequencies). If this modem operates at 1200 baud, it means the modem is able to switch between 2200 hertz and 1200 hertz at a speed of 1200 times each second. This in turn means that the modem in Figure 3–28 has a switching rate of 1200 baud and also a bit-per-second rate of 1200.

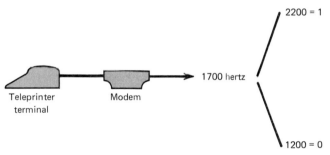

FIGURE 3–28 Single–bit transmission.

Assume we want to transmit 2 bits at a time instead of 1 bit at a time. Figure 3–29 shows that if transmission is 2 bits at a time (dibits), you must have 4 different combinations. Transmission of 3 bits (tribits) at a time has 8 combinations, and 4 bits (quabits) at a time has 16 combinations.

Now we can determine where bits per second and baud are different from each other. Looking at Figure 3–30, notice that the modem now has four different frequencies that it can switch between (2200, 2000, 1400, and 1200 hertz). Also notice that each frequency now corresponds to a pair of bits rather than a single bit. This is still a 1200 baud modem, which means that it can switch between any of the four frequencies at a speed rate of 1200 times per second. The difference now is that transmission is two bits at a time.

When the modem operates in the fashion of transmitting dibits, a 1200 baud modem is capable of transmitting 2400 bits per second. Other dibit modulation techniques are shown in Figure 3–31.

Dibits	Tribits	Quabits
00	000	0000
01	001	0001
10	010	0010
11	011	0011
4 Combinations	100	0100
	101	0101
	110	0110
	111	0111
	8 Combinations	1000
		1001
		1010
		1011
		1100
		1101
		1110
		1111
		16 Combinations

FIGURE 3–29 Dibits/tribits/quabits.

You might even imagine a modem that has 16 different levels of frequency. If it is able to switch among any of the 16 different levels at a 1200 baud rate then it is transmitting at 4800 bits per second. This is because you need 16 different discrete identifiable signals to transmit 4 bits at a time. It should be noted that modem makers would never use 16 different frequencies for this type of transmission; they probably would use a combination of two or more types of modulation such as combining phase with amplitude modulation.

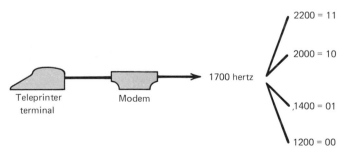

FIGURE 3–30 Dibit transmission using frequency shift keying (FSK).

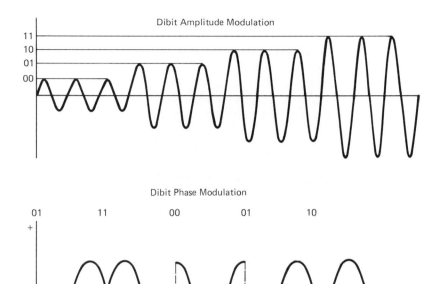

A message with bit sequence 0111000110 was divided into the dibit sequence 01 11 00 01 10, which modulated a carrier wave once per cycle and produced the phase modulated signal.

FIGURE 3–31 Dibit modulation schemes.

In summary, a baud is the signaling rate that tells you how many times per second the signal changes. By transmitting 2, 3, or 4 bits with each change of signal, the modem designer is able to transmit more bits per second than there are baud on the circuit.

The goal of a modem designer is to keep the baud rate as low as possible while making the bit rate as high as possible. The reasoning behind this is that, if the baud rate is 1200, then the time available to identify the signal at the receiving modem is 833 microseconds (1/1200 = 833). If the baud rate is increased to 2400, then the time available to identify the incoming signal at the receiving modem is only 416.5 microseconds (1/2400 = 416.5). As a result, the receiving modem only has one-half the time to identify the signal when the baud rate is 2400, as compared to 1200. Finally, most vendor literature misuses the word baud, so it has become common practice to think of baud as equal to bits per second, even though this is not technically correct.

POLLING/SELECTING

Polling and selecting take place in a centrally controlled system. *Polling* is the process of sending a signal to a terminal in order to give it permission to send messages that it might have ready. *Selecting* is the process of sending a signal to a terminal in order to determine if it is in a current status that will allow it to accept a message from the central computer site. Most people tend to refer to both of these conditions as polling.

Polling is performed by the front end communication processor, although it can be performed easily by the host computer or a remote intelligent terminal controller. There are several types of polling such as roll call, fast select, and hub go-ahead.

With *roll call polling*, the front end communication processor consecutively works through a list of terminals, first polling terminal 1, then terminal 2, then terminal 3, and so on, until all are polled. Roll call polling can be modified to prioritize terminals in the following sequence: 1, 2, 3, 1, 4, 5, 1, 6, 7, 1, 8, 9, and so on. Terminal 1 may have priority because of extremely heavy usage.

Typically, roll call polling involves excessive "wait time." The front end has to poll a terminal and then wait for a response. The response that the front end is waiting for might be the incoming receipt of a message that was waiting to be sent, a negative response indicating there is nothing to be sent, or the full "time-out period" may expire because the terminal is temporarily out of service. Usually there is a timer that "times-out" the terminal after waiting, for example, one-tenth of a second without getting a response. If some sort of fail-safe time-out is not used, the system poll might lock up on an out-of-service terminal indefinitely. Incidentally, more sophisticated systems totally remove an individual terminal from the polling list after getting three consecutive time-outs.

Fast select polling schemes were developed in order to eliminate the time of waiting for a response when a terminal does not have a message to send, and to eliminate the time-out wait when a terminal is not operating correctly. In a fast select type of poll the front end polls the terminals until the first incoming message is received. Assume we have 20 terminals and use fast select polling. The front end sends a poll character to terminals 1, 2, 3, 4, 5, and 6, and then the first incoming message is received. If the first incoming message is from terminal 2, the front end stops polling and waits long enough to receive any incoming messages from terminal 6 (the last terminal polled). Next, the front end resumes polling at terminal 7 and polls terminal 7, 8, 9, etc., until the next message is received. After waiting an appropriate time for receipt of any message from the last terminal polled, the front end once again resumes polling where it left off and continues to terminal 20. After polling all 20 terminals, it may automatically start the polling list again at terminal 1. With this type of polling scheme, the front end does not wait for a negative response from terminals with nothing to send and, if a terminal is not operating correctly, it does not waste time waiting for a time-out before proceeding to the next terminal.

Hub go-ahead polling is used in multidrop circuit configurations. The front end passes the poll character to the farthest terminal on the multidrop circuit. That terminal then sends its message and/or passes the polling character to the next inbound terminal. That terminal also passes the poll back to the next inbound terminal, and so on until it reaches the terminal closest to the front end. The closest terminal passes the poll back to the front end and it restarts by again passing the poll to the farthest terminal. This technique relieves the front end of many polling tasks because the terminals themselves undertake the process of sending messages in a sequential fashion. It should be noted that hub go-ahead polling assumes more intelligence in each of the terminals in order to properly handle the poll. Intelligent terminals also are necessary because there must be a means of bypassing a terminal that is out of service, since it would be a polling break in the multidrop link.

THROUGHPUT (TRIB)

There are many factors that affect the throughput of a data communication system. Probably the most important is the transmission rate of information (bps) and the communication circuit bandwidth because bandwidth denotes the absolute upper limit of speed. If terminals are multidropped, that is another factor because the circuit must be shared. If terminals are multiplexed, that becomes a factor because each terminal uses a reduced bandwidth (a subset of the total bandwidth on the circuit). Another factor is the capability of the front end communication processor to handle multiple incoming and outgoing communication circuits. If the front end cannot handle circuits or messages simultaneously, the capacity of the system is degraded. Software design is also a factor, because it determines which protocol is used and whether transmission is in full duplex or half duplex. Propagation time, especially on satellite circuits, affects throughput. Also, the time required for the host computer to process a request and/or perform a lookup or update in a database is a factor in throughput. Error rates in hardware, in software, and on the communication circuit affect throughput because of possible retransmissions of the same message. The polling scheme (central control) or whether the system is an interrupt system will affect throughput. Obviously, many items affect throughput. It is appropriate at this point to examine one of the major parameters of importance: how many usable characters of information can be transmitted per second.

Transmission Rate of Information Bits (TRIB) is the term normally used to describe the effective rate of data transfer. It is a measure of the effective quantity of information that is transmitted over a communication circuit per unit of time.

The American National Standards Institute (ANSI) provides definitions for calculating the transfer rate of information bits. TRIB calculations may vary with the type of protocol used because of different numbers of control characters required and different time between blocks. The basic TRIB equation is shown in Figure 3–32. Compare Figure 3–32 with Figure 3–6. Figure 3–33 shows the calcu-

$$\text{TRIB} = \frac{\text{Number of information bits accepted}}{\text{Total time required to get the bits accepted}}$$

$$\text{TRIB} = \frac{K(M-C)\,(1-P)}{M/R+T}$$

K = Information bits per character

M = Block length in characters

R = Modem transmission rate in characters per second

C = Average number of noninformation characters per block (control characters)

P = Probability that a block will require retransmission because of error

T = Time between blocks in seconds such as modem delay/turnaround time on half duplex, echo suppressor delay on dial-up, and propagation delay on satellite transmission.

FIGURE 3–32 TRIB equation.

lation of throughput assuming a 4800 bit-per-second half-duplex circuit. If all factors in the calculation remain constant, except for the circuit, which is changed to full duplex (no turnaround time delays, T = 0), then the TRIB increases to 4054 bits per second.

$$\text{TRIB} = \frac{7\,(400-10)\,(1-0.01)}{(400/600)+0.025} = 3908\ \text{BPS}$$

K = 7 bits per character (information)

M = 400 characters per block

R = 600 characters per second (derived from 4800 bps divided by 8 bits/character)

C = 10 control characters per block

P = 0.01 (10^{-2}) or one retransmission out of one hundred blocks transmitted—1%

T = 25 milliseconds (0.025) turnaround time

FIGURE 3–33 TRIB calculation.

Look at Figure 3–33, where the turnaround value (T) is 0.025. If there is a further propagation delay time of 475 milliseconds (0.475), that figure changes to 0.500. Now, to demonstrate how a satellite channel affects TRIB, the total delay time is now 500 milliseconds. Still using all the original figures as they were in Figure, 3–33 (except for the new 0.500 delay time), the TRIB for our half duplex, satellite link is reduced to 2317 bits per second, which is almost one-half of the full duplex (no turnaround time) 4054 bps.

RESPONSE TIME

Response time in its simplest form is the elapsed time between the generation of an inquiry at a remote terminal and the receipt of the last character of the response at the same terminal. Therefore, response time includes transmission time to the computer, processing time at the computer, access time to obtain any needed database records, and transmission time back to the terminal.

The best indicator of response time for a system not yet developed is a set of statistics drawn from another operating system that supports the same application and uses the same protocol. In other words, examining a similar network with similar operating characteristics and applications is the best indicator of how a planned network will perform. The problem is that finding such a duplicate system is almost impossible. If no like network exists from which to draw performance data, then some predictive techniques must be used.

When using these predictive techniques, do not rely on average response times. Instead, always state the question as: "X" % of all response times must be less than or equal to "Y" seconds. In other words, a typical statement might be that 95% of all response times must be less than or equal to 3 seconds. Mean and standard deviation of these figures might be used in order to identify the reliability of the final response time figure.

A typical cost versus response time curve is shown in Figure 3–34. When the response time is shortened, the cost increases; when it is lengthened, the cost decreases. Factors which affect system cost include host computer speed and capacity, speed and size of front end communication processor, capacity of the communication circuits, remote intelligent control devices, software programs/protocols, and the like.

Queuing theory allows for the definition of such elements as service time, facility utilization, and wait time at the host. A queue is a line of items to be handled or serviced. Single server and multiserver queuing relationships must be understood within the environment of system priorities. We cannot begin to explain, in this text, the techniques of statistical and queuing formulation. That should be reserved for a more advanced course, although modeling is introduced in the next section. Estimations can be of the best or worst case. These techniques often

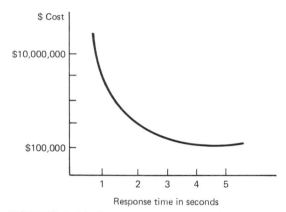

FIGURE 3–34 Cost versus response time.

yield average results that describe the average operational performance of a network. Statistical views of network performance based on queuing theory can vary from real performance by as much as 20%, but they can provide estimates of their own accuracy.

Simulation is a technique to model the behavior of the communication system or network. Response time is viewed as an elapsed time incurred, which is part of the accumulation of the elapsed times of a series of individual events. Sophisticated programs can be written to simulate the action of a series of events, and these programs add up the elapsed times of each event. Simulation programs run on large machines and can execute several thousand polls within a few seconds in order to generate a statistical view of the projected network. Simulators typically ignore error conditions because error conditions are the exception and not the rule. They can be built into sophisticated simulators, but this vastly increases the complexity of the programs. Queuing analysis can be used to verify the predicted results of simulation. Most vendors offer simulators to assist in this area. These simulators examine the effects of many parameters of a communication system on projected performance such as:

- Number of intelligent control units per circuit
- Number of terminals per intelligent control unit
- Printers/printer buffer size/printer speed
- Modem delay
- Propagation delay
- Statistical time division multiplexer delay (if any)
- Line protocol overhead

- Message lengths/occurrences/rates
- Host computer processing delays
- Database access delays
- Multiple queues or single queues
- Polling/selecting

The specific components that contribute to response time are message input time, application processing time, and message output time.

The *message input time* is the sum of the polling time, transmission time (including modem turnaround time and time for acknowledgment), and queuing time in the front end communication processor or host computer. The transmission time component usually is stable, but the other factors are determined statistically according to traffic volume. A typical input time might be 0.85 seconds.

The *application processing time* includes all program processing time and all input/output accesses to the database. As might be expected, these timings are variable, depending upon message traffic and the number of transactions being handled by the host computer. An example of a typical application processing time might be 0.75 seconds.

The *message output time* is the sum of the internal queuing in the host computer and/or front end communication processor and the transmission time (including all modem turnaround, selection, and acknowledgment time). Again, the transmission time component usually is stable, whereas internal queuing is a variable figure depending on the current volume of transactions at the host/front end communication processor. An example of a typical message output time is 0.90 seconds.

If the sum of these typical average times is approximately 2½ seconds, imagine what would happen if another half second is added for propagation delay time or other delays. In a typical communication application, the component that becomes the most sensitive to increased volume is the application processing time/database handling in the host computer. Response time on a current network is easy to measure by use of a network analyzer (or even a stopwatch). Predicting it during the design stage, however, requires detailed network analysis involving queuing theory or simulation and a lot of common sense.

Queuing theory/simulation programs break the process into more segments than just message input time, application processing time, and message output time. Simulation takes into account such factors as terminal buffering, effect of an intelligent terminal control device, statistical time division multiplexers, mode of transmission used by the modem, communication circuit speed, error rates on communication circuits, queuing at transmission nodes/front end/host computer, line configurations such as point-to-point/multidrop/multiplex, message lengths, expected arrival times of messages, propagation delays, any priorities built into the system, average versus peak loads, central control verses interrupt, type of applications, speeds of output devices, and the intrinsic factors within the host computer itself, such as its hardware architecture and the software/protocols utilized.

MODELING NETWORKS

The complexity of designing data communication networks presents the designer with many problems. The overall goal of the designer is to construct a system that will provide adequate response time to the end user while ensuring that the cost to deliver that response time is reasonable. In order to accomplish this goal, the designer must understand the response time and cost trade-off issues associated with a data network.

In this section we will deal with the issues related to designing multipoint polled networks. This type of network was chosen because it is the most common class of data network and it presents difficult issues related to how user transactions are handled in a system with multiple queues.

As an example, when a user is ready to send a message to the host computer for processing, that message passes through many queues (wait times) or delay periods, as illustrated in Figure 3–35.

If there are other terminals sharing the circuit (multipoint), the queues and delays can compound or interfere with each other because usually only one terminal on a given communications circuit can communicate with the computer at a given time. Therefore, if there are five terminals on a circuit and each has a message ready to send, it is easy to see that the last terminal may have to wait until the others are finished. Also, the computer may accept a message from Terminal 1 and then Terminal 2 and, before going to Terminal 3, send the response to Terminal 1's message, causing an additional delay for the others waiting for service. From this example, some of the complexities can be demonstrated, but understand that these are only a few of the delays that are encountered within data communication systems.

Fortunately, computer modeling techniques have been developed to help the designer evaluate the problem and determine a solution. These modeling software programs allow the designer to examine the issues without having to get involved with all the complexities of the problem. They are, if you will, "tools of the trade" and must be looked upon as *tools*, not push-button programs. A skilled user can derive good solutions by applying the tools successfully, while an unskilled user may apply the tools incorrectly and get totally different results. For example, the MIND system by Contel Information Systems is such a tool.

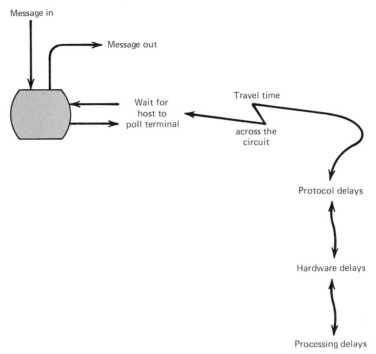

FIGURE 3–35 Message delay.

A *model* is a body of information about a system gathered for the purpose of studying the system. A *mathematical model* describes the entities of a system with the attributes being represented by mathematical variables; the activities are described by mathematical functions that interrelate the variables. Given a mathematical model of a system, it is sometimes possible to derive information about the system by *analytic means*. Where this is not possible, it is necessary to use numerical computation for solving the equations of the mathematical model. *System simulation* describes the technique of solving problems by following the changes over time using a dynamic model of a system. A *dynamic mathematical model* allows the changes of system attributes to be derived as a function of time.

The analytical models generally make several simplifying assumptions to make the mathematics easy and provide results that can be evaluated easily and cheaply. Simulation models, on the other hand, can reflect the working of a system to any finer desired level of detail. The time and cost of development and program runs are directly proportional to the level of detail incorporated.

Let us illustrate some of the analytical and simulation modeling techniques used in a data network design. Consider a set of terminals connected to a host computer (CPU) via a multidrop line (Figure 3–36). The CPU and the terminals bear a master/slave relationship in the sense that the transmission from the terminals are controlled by the CPU or front end. Evaluating the performance of the system under the polling discipline is the task of the systems designer. The systems designer uses modeling as a means to predict the performance of the system and answers such questions as: How does the response time of the system vary as a function of the terminal load and the number of terminals? How is the response time affected by a specific polling discipline such as giving priority to outbound traffic (CPU to terminal) over inbound traffic?

With regard to *analytical* models, in order to analytically predict the response time and throughput of a polled system, the designer resorts to queueing models. A general queueing model of a polled system is shown in Figure 3–37 and may be described as queues served in cyclic order with walk times (*walk time* is the time

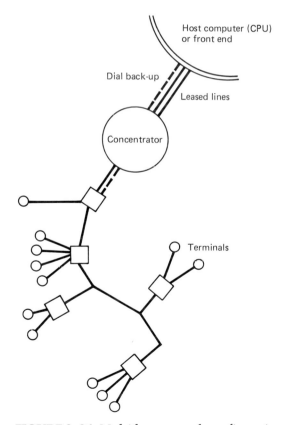

FIGURE 3–36 Multidrop network configuration.

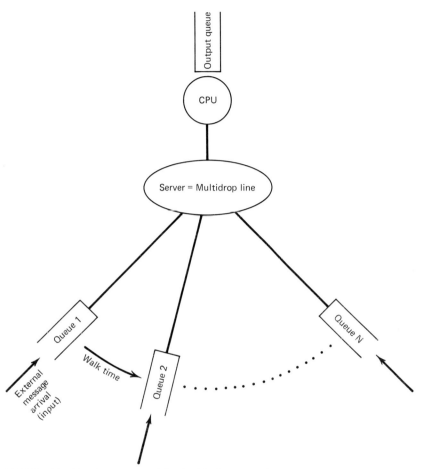

FIGURE 3–37 Queues model of polled terminals.

to switch service from one queue to another and includes the overhead time attributable to polling messages, propagation delay, modem synchronization time, etc.).

Messages arrive at a terminal according to a random process, which may be terminal dependent, and are queued up for transmission. The server in this case is the transmission medium and is made available to each queue periodically, as defined by the polling protocol. The polling discipline also defines the amount of service received when the server arrives at a queue. Other characteristics of the system, such as poll message length and modem turnaround time, are modeled in the switch-over time. Outbound response messages may be included as arrivals to an output queue at the CPU. The major difficulty in solving the above queueing model is the interrelationship between the queues at the various terminals. An exact model, therefore, has to solve an N-dimensional queueing process, which is a formidable task.

An analytical model requires a sophisticated user who is competent in mathematics for its development and use. When the user is discriminating, it can provide preliminary insights, but seldom can it yield numerical values of sufficient accuracy for the operational design of a system. To obtain a more realistic model capable of providing accurate numerical answers, we must resort to *simulation*. The attempt to produce a better mathematical model leads to complications. A point is reached at which a simulation model begins to appear the easiest way of finding circuit utilization and response time.

Because of the complexity of the analytical models, it is often easier to use a simulation model to answer many of the questions posed by the systems designer. The Multipoint Line Simulation System (MLSS) by Contel Information Systems is one example of a simulation tool for the design of multipoint lines.

It allows the user to enter a model of network traffic load and obtain an analysis of waiting times and response times in various parts of the network. The model may be tailored to the user's network by entering parameter values specific to the system at hand. Alternatively, the user may prefer to rely primarily in some applications upon the set of average values provided by the system. This set of values allows the simulation model to operate under a specific circuit discipline and is representative of many popular protocols currently in use. It is not limited to bisynch or SDLC-like protocols. With specification of appropriate parameters, very different circuit disciplines can be modeled. Additionally, operation of most terminal-oriented systems under full duplex protocols such as SDLC can be modeled.

Its focus is on modeling the multipoint line. Polling (the determination of whether or not a terminal is ready to send input), selection (the determination of whether or not a terminal is ready to receive output), positive and negative acknowledgment, error occurrence and recovery, and hardware delays in all devices (e.g., terminal, modem, terminal controller, concentrator, front end communications processor or FEP) are all modeled directly. The output provides response times on an overall basis, by terminal type and by transaction type, and the delay component at the various devices.

Experience with simulation of complex data communication systems has led to the conclusion that generally more harm than good is done when the designer attempts to model and simulate the entire system in detail; therefore, it is best to model a single line at a time. The most popular models use a hybrid approach in which a simple analytical model of the operation of the communication processor and central processor is embedded within the simulation. This enables the designer to examine the issues of response time in the context of a complete network.

Analytical queueing models of real systems using polling are extremely complex and can be made workable only if simplifying assumptions are made. This frequently (and often unpredictably) results in inaccuracies in the response times obtained and/or restricts the applicability of the model to certain traffic levels. Consequently, these models should not be relied upon unless the designer is interested only in gross answers. Pure simulation models, while more accurate, are cumbersome and costly to run.

A hybrid technique in which analysis is embedded within the simulation can yield a flexible model which provides accurate results. The input to the model can be detailed empirical data, if it is available, or it can be defined approximately by a well-known statistical process (e.g., Compound Poisson Process). A major limitation of simulation models, that of development cost, can be avoided by using the design packages developed and marketed by reputable organizations.

SIGNALING ON A VOICE NETWORK

There are two basic aspects to consider in the use of a telephone: the information that is conveyed during the conversation (the session) and the coded control signals that set up and terminate the call (control signaling). Data transmission also has a "session" and "control signaling." Some of these control signals convey information, such as the telephone number that might be passed from office to office, or the status of certain equipment can be signaled, such as whether it is busy or whether a person has answered a call.

The control signals that are passed throughout the telephone network are signals that do some of the following:

- The dial tone indicates that dialing can begin.
- A busy signal indicates to the caller that the call cannot be completed (60 pulses per minute), implies that the called person's telephone is in use, or that the central office to which they are connected is overloaded).
- A high-speed busy signal (120 pulses per minute) indicates that the long-distance trunk lines (IXCs) are busy or overloaded.
- A ringing tone indicates that the called number is ringing.
- A pulsing loud noise indicates that something has gone wrong and the caller should restart the process.
- The tones heard when dialing a number indicate that the numbers are being transmitted from your telephone to the telephone company central office.

- Many other control signals for the telephone company, such as signals that pass a number between central offices, signals for recording billing information, signals that give the status of certain equipment, signals to ring long-distance operators, signals for diagnosing and isolating system failures, signals to control special equipment such as echo suppressors, signals for coin return in pay phones.

A description of signaling with regard to data transmission is in Chapter 4 in the section on Data Signaling/Synchronization using an RS232C cable.

ECHO SUPPRESSORS

Two-wire circuits (dial-up circuits) have a problem of echoes. When people talk on a two-wire circuit, echoes may occur under some conditions. Echoes arise in telephone circuits for the same reason that acoustic echoes occur: there is a reflection of the electrical wave from the far end of the circuit.

Telephone companies install echo suppressors on the two-wire circuits in order to prevent the echoing back of your own voice when you are talking to someone else. The echo suppressor permits transmission in one direction only. When you talk, your voice closes the echo suppressor. When the other person starts talking, his or her voice closes the echo suppressor in his or her direction; because you have stopped talking, your echo suppressor opens. Obviously if you are both talking, the power of your voices so overwhelms the echos that you do not hear them. This is because the echo suppressors close in both directions when you both talk. Figure 3–38 shows a picture of echo suppressors.

Lease data circuits (private dedicated lines) do not have echo suppressors; therefore, echo suppressors are not present on a four-wire lease circuit used for data transmission. Echo suppressors are present in dial-up circuits.

When a computer is dialed using an acoustic coupler, there is a loud ringing tone when the computer answers. The reason for this tone is that it disables the echo suppressors. The tone is held on the line for approximately 200 to 400 milliseconds in order to disable or close the echo suppressors in both directions. Immediately after the tone ceases, the carrier wave signal comes up, perhaps a 1700 hertz tone, and it is this carrier wave signal that keeps the echo suppressors closed in both directions. If the carrier wave is lost (maybe an electrical failure) for approximately 150–300 milliseconds, the echo suppressors reopen. This reopening causes the dial-up data transmission to have many garbled or destroyed characters. The only choice at this point may be to redial the call and start over in order to close the echo suppressors in both directions.

The reason there are so many garbled characters after the echo suppressors reopen is because echo suppressors take approximately 150–200 milliseconds to

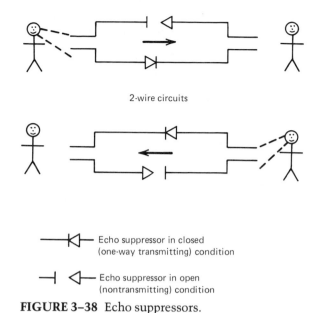

2-wire circuits

——K⊢— Echo suppressor in closed
        (one-way transmitting) condition

—⊣ ◁⊢— Echo suppressor in open
        (nontransmitting) condition

FIGURE 3–38 Echo suppressors.

open and close. To illustrate this effect, assume we were transmitting when such a lapse occurred. The first 150 milliseconds of the next transmission would be lost because the echo suppressor would not be fully closed and the data bits could not get through. If we changed directions, the first 150 milliseconds of your data transmission also would be lost while the echo suppressor closed. It is for this reason that dial-up modems disable the echo suppressors before transmission.

TASI (VOICE CALLS)

Time Assignment Speech Interpolation (TASI) is a technique used on some of today's long-distance, frequency division multiplexed voice lines. It allows for the packing of extra voice conversation into a fixed number of circuits.

Usually when two people are conducting a telephone conversation, both parties do not speak at the very same moment, and for a small portion of the time neither speaks. Most long-distance voice circuits are four-wire circuits. When each person speaks, only one pair of the four wires will be used. This means that two of the four wires are always empty, unless both people speak simultaneously, which cannot last for a very long period of time.

TASI electronic switching equipment is designed to detect a user's first word, and within a few milliseconds the equipment assigns a communication circuit to that speaker. Actually, an almost undetectable portion of the first syllable may be

FIGURE 3–39 Before and after TASI.

lost; but it is seldom noticed in voice communications. When a person ceases talking, the circuit is switched away and given to someone else. When the person speaks again, the TASI equipment assigns a new circuit path. Occasionally, if the circuits are overloaded, the TASI equipment may be unable to find a free path. Even though it is for a very brief period of time, several words might be lost when this occurs (see Figure 3–39).

The benefit of TASI equipment is that, if there are 100 circuits, more than 100 voice calls can be handled simultaneously.

VOICE CALL MULTIPLEXING

Multiplexing of data communication transmission was discussed in Chapter 2 in the section on Multiplexers. The telephone company further multiplexes received signals, whether they are voice or data. We will refer to the process as *voice call multiplexing*, even though technically it may be data that is being further multiplexed.

AT&T Communications has devised the Bell System Group multiplexing function. The telephone company takes groups of calls that are destined for the same area and multiplexes them together. This permits groups of calls to travel on a single coaxial cable, microwave system, satellite transmission, or optical fiber transmission media. Figure 3–40 shows the names given to these groupings and the number of voice grade circuits that are involved in each group.

This illustrates how your company's network may multiplex data transmission so that 16 simultaneous data conversations can take place on one circuit.

Name	Number of Voice Grade Circuits
Group	12
Super Group	60
Master Group	600
Jumbo Group	3600
Jumbo Group Multiplex	10,800

FIGURE 3–40 Bell System Group multiplexing.

The telephone company may further multiplex this one circuit into a group (12 voice grade circuits) for transmission to its destination. In addition, the telephone company may again multiplex it into a jumbo group of circuits. Multiplexing done by the telephone company is *pure* multiplexing (not statistical), and it is totally transparent to users. There is no holdup or delay of messages.

QUESTIONS

1. How efficient would a 6-bit code be if it had two parity bits, one start bit, and two stop bits (some very old equipment utilizes two stop bits) in asynchronous transmission?

2. Which circuit type has the greatest capacity for carrying voice or data transmissions, satellite or optical fiber?

3. If the signal-to-noise ratio is 10 dB, how much more powerful is your signal than the background noise?

4. Which modulation technique is more error-free during transmission: amplitude, frequency, or phase modulation?

5. Draw a digital signal that has a characteristic of returning to zero between consecutive data bits.

6. Draw a situation where frequency division multiplexing is utilized and the modem transmits 3 bits per baud.

7. Utilize the TRIB calculation in Figure 3–32 to identify throughput for the current transmission circuit that you utilize at your university or business office.

8. Draw a picture showing the difference between PSK and DPSK (review the figures).

9. Go to the library and read a copy of Nyquist's and Shannon's papers.

4

THE BASIC COMPONENTS OF COMMUNICATIONS

This chapter defines and discusses the basic concepts of data communications/teleprocessing. The technique used is to examine technically the flow of data/information from a remote terminal through the connector cable, the modem, the local loop, the telephone company central office, the circuits, and into the front end/host computer. As the message moves, each technical aspect is defined and described.

THE BASIC CONCEPTS

In order for us to discuss the basic components of data communications, Figure 4–1 depicts a standard point-to-point data communication network between a remote terminal and a host computer. For simplicity, this configuration omits the more complex design configurations such as multidrop, multiplex, and others. The reason for this is that the basic explanation of how data are transmitted from terminals to host computers would be obscured by complex design methodologies; we prefer a simple straightforward flow approach of the data transmission concepts from a terminal to a host computer.

As you can see in Figure 4–1, each component is identified across the top of the figure starting with the terminal and going on through connector cables, modems and, ultimately to the host computer. Below each component the related concepts that are involved in moving data from a terminal to a host computer are identified.

As we move through the components of Figure 4–1, you can follow the movement of your data signal (message) and the basic concepts that allow the signal to move from the terminal to the host computer.

THE TERMINAL

First we have to discuss the basic input/output component that will be used on the network (follow along with Figure 4–1 as you read this chapter). Obviously, the terminal could be a microprocessor, a teleprinter, a video terminal, or any other type of terminal. In order to define the basic concepts here, we will be discussing baseband signaling, ASCII code, and asynchronous versus synchronous transmission (see Figure 4–1, below the terminal).

BASEBAND SIGNALING

The transmission of signals from their original source (the terminal) usually is done in direct electrical voltages called *baseband signaling*. When you type a character or message, each character is represented by 8 bits.

Basically, baseband signaling is the digital transmission of electrical pulses. This digital information is binary in nature in that it only has two possible states, a 1 or a 0 (sometimes called *mark* and *space*). The most commonly encountered plus and minus voltage levels range from a low of 3 to a high of 25 volts. These baseband digital signals must be modulated onto the telephone company's interexchange channels by the modem.

The binary 0s and 1s that are transmitted (groupings of these 0s and 1s make up

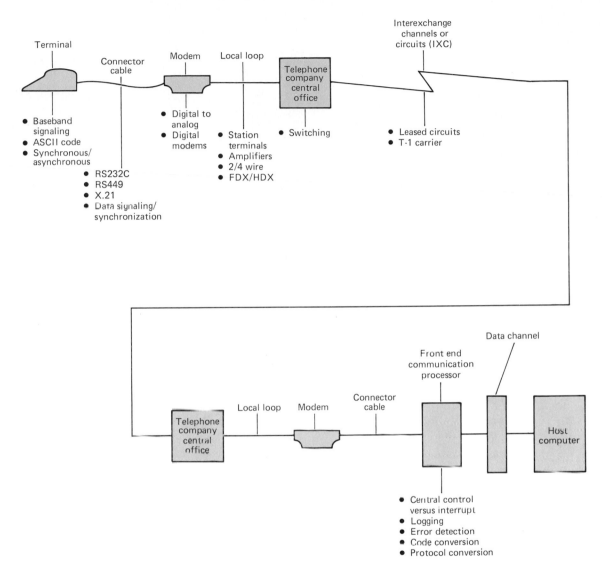

FIGURE 4–1   The basic technical concepts.

a character) are represented by different levels of voltage such as + 5 volts for binary 1 and − 5 volts for a binary 0. These signals may be unipolar or bipolar (Figure 3–24 showed this). The terminal outputs these plus and minus voltages (baseband signal) and they pass over the connector cable and into the digital side of the modem. This is called a digital signal (review Figure 2–9); the modem then converts it to a broadband signal for transmission.

Leakage of the electrical current either at the terminal connector plug or as it passes over the connector cable, causes distortion of your signal and this in turn causes errors. Capacitance and inductance also cause distortion that can result in errors.

With *capacitance*, as we raise the voltage at one end of a communication circuit there is some delay before the voltage at the other end rises by an equal amount. The copper wire acts rather like a water hose in that it needs to be filled to capacity before the electricity applied at one end is received at the other end. The problem here is that if the pulses (+ and − voltages) are too short in duration, or if there are too many pulses sent per second, then they become indistinguishable when they are received. The faster the pulse rate (baud), the more difficult it becomes to interpret the received signal.

Inductance in a circuit resists the sudden buildup of electric current. This resistance can be a cause of distortion, and therefore errors during transmission.

ASCII CODE

The terminal must have some sort of code with which to represent the individual alpha and numeric characters that we use in our everyday business communications. The most common code of data communications is the American Standard Code for Information Interchange (ASCII). This is just one of many different code structures. With ASCII, when you type the character "A", the terminal will output the baseband digital pulses for the following series of bits: 10000010. These will be represented by plus and minus voltages of electricity and will be passed over the connector cable.

ASYNCHRONOUS/SYNCHRONOUS

Even though microprocessors and host computers transmit the data within their memory in a parallel fashion (8-bit to 32-bit parallel transfer), on a data communication circuit the bits are transmitted in a serial fashion, one after the other. In other words, the terminal and the front end communication processor must have the proper electronics to serialize and deserialize these bit streams. Inside the terminal, it may be transmitting the 8 bits of a character in parallel, but it must serialize them and put them out on the connector cable as a serial bit stream (one bit after the other). Now that it is recognized that the bits go down a circuit in serial fashion, you must realize that there is an asynchronous transmission mode and a synchronous transmission mode. *Asynchronous* transmission is character-by-character transmission (start and stop); whereas *synchronous* transmission is block transmission (entire message).

With asynchronous transmission, each character is transmitted independently of all other characters. In order to separate the characters and synchronize transmission, a start bit and a stop bit are put on each end of the individual 8-bit character. This means that our example of asynchronous transmission has 10 bits per character in transmission as shown in Figure 4–2. There is no fixed distance between characters because, if the terminal is one that transmits the character as soon as it is typed, then the distance between characters varies with the speed of the typist. If the asynchronous terminal holds the entire message (buffers) and transmits the message when the operator hits the "send" key, then there will be a fixed distance between characters, because the terminal unloads its buffer on a character-by-character basis in a fixed timing sequence. All asynchronous transmission has a start bit and a stop bit in order to separate characters from each other and to allow for orderly reception of the message. Some terminals may have multiple stop bits.

In summary, the synchronization (timing) takes place for an individual character because the start bit is a signal that tells the receiving terminal to start sampling the incoming bits of a character at a fixed rate so the 8 data bits can be interpreted into their proper character structure. A stop bit informs the receiving terminal that the character has been received and resets the terminal for recognition of the next start bit. Synchronization of the character is reestablished upon reception of each character.

Synchronous transmission (see Figure 4–3) is used for the high-speed transmission of a block of characters, sometimes called a *frame* or *packet*. In this method of transmission, both the sending and receiving devices are operated simultaneously, and they are resynchronized for each block of data. Start and stop bits for each character are *not* required. In other words, if you had 100 characters using an 8-bit ASCII code structure, the message part of the block of data would be 800 bits long.

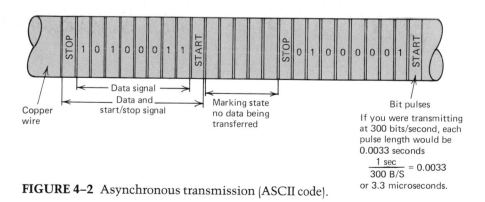

FIGURE 4–2 Asynchronous transmission (ASCII code).

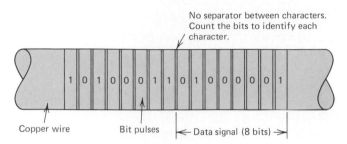

FIGURE 4–3 Synchronous transmission.

For synchronous transmission the synchronization is established by passing two SYN characters between the sending and receiving devices. In fact, there are really two levels of synchronization here. The first level is the data signaling between the terminal and the modem; this passes over the connector cable. The second level involves the two SYN characters that go between the terminal and the modem at the other end of the IXC circuit. This will be described in the next section, Connector Cables.

In summary, there are two methods here, asynchronous and synchronous. Synchronous transmission is by far the most popular on high-speed data circuits (4800 bps or greater), whereas asynchronous is quite adequate (300–2400 bps) and widely used for interactive timesharing and interactive conversations between a terminal and a host computer.

CONNECTOR CABLES

Go back to Figure 4–1 on the basic technical concepts of transmitting data. The next component to review is the connector cable, the standards (such as RS232/RS449/X.21), and the data signaling or synchronization for the movement of synchronous data. The synchronization is achieved by having a start bit and a stop bit on each character or sending SYN characters (01101001) to the remote modem. Our discussion will center more on the synchronization/data signaling between the terminal and its local modem and the remote modem.

RS232C/RS449/X.21

The RS232C is the connector cable that is the standard interface for connecting data terminal equipment (DTE) to data circuit-terminating equipment (DCE). The newer standard is RS449. RS232C/RS449 are the Electronic Industries Association (EIA) standards. Basically, the RS449 standard was built by combining the

RS442 and RS423 standards. The X.21 interface cable is the CCITT mini-interface standard. (CCITT stands for Consultative Committee on International Telephone & Telegraph).

Figure 4–4 shows a picture of the RS232C interface plug and provides a description for each of its 25 protuding pins. This standard connector cable is a 25-wire cable that passes control signals and data between the terminal (DTE) and the modem (DCE). The RS232 has a maximum 50-foot cable length but, by means of special low impedance cable, it can be stretched to 100 feet or greater. This is not

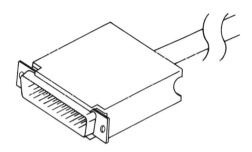

Pin	Function	Pin	Function	Pin	Function
1	Frame Ground	10	Negative dc Test Voltage	19	Sec. Request To Send
2	Transmitted Data	11	Unassigned	20	Data Terminal Ready
3	Received Data	12	Sec. Data Carrier Detect	21	Signal Quality Detect
4	Request to Send	13	Sec. Clear to Send	22	Ring Indicator
5	Clear to Send	14	Sec. Transmitted Data	23	Data Rate Select
6	Data Set Ready	15	Transmitter Clock	24	Ext. Transmitter Clock
7	Signal Ground	16	Sec. Received Data	25	Busy
8	Data Carrier Detect	17	Receiver Clock		
9	Positive dc Test Voltage	18	Receiver Dibit Clock		

The terminal connection to the modem is defined by the Electronic Industries Association (EIA) specification RS232C. RS232C specifies the use of a 25-pin connector and the pin on which each signal is placed.

FIGURE 4–4 RS232C interface.

advised, however, because some vendors will not honor maintenance agreements if the cable length is stretched beyond the 50-foot standard. When data signaling/synchronization is discussed later, you will want to recall pins 4 and 5 from this figure.

The RS449, shown in Figure 4–5, has many advantages over its predecessor, the RS232. A 4000-foot cable length can be used, there are 37 pins instead of 25, and various other circuit functions have been added, such as diagnostic circuits and digital circuits. Also, secondary channel circuits (reverse channel) have been put into a separate 9-pin connector. As an example of some of the new features, look at pin number 32, "select standby." With this pin, the terminal can instruct the modem to use an alternate standby network such as changing from a private lease line to a public packet switched network, either for backup or simply to access another database not normally used. In other words, a terminal can be connected to two different networks and the operator can enter a keyboard command to switch the connection from one network to another. With regard to loopback (pins 10 and 14), the terminal can allow basic tests without special test equipment or the manual swapping of equipment or cables.

The X.21 interface is based on only 15 pins (wires) connecting the DTE and the DCE. This requires an increased intelligence in both the DTE and the DCE. At present it is not suitable for the analog private lines, switched lines, or half duplex lines that constitute a considerable majority of the existing communication circuits. X.21 is a international standard, however.

DATA SIGNALING/SYNCHRONIZATION

Recall the discussion on synchronous (block) transmission. Let's look at data signaling/synchronization through the eyes of an RS232C connector cable. When the terminal operator presses the "send" key to transmit a block of data to the host computer, the "request to send" pin (pin 4 of the 25 wires in the cable) transmits the signal "request to send" from the terminal to the modem. This informs the modem that a block of data is ready to be sent. The modem then sends a "clear to send" signal back to the terminal by using pin 5. This tells the terminal that it can send the synchronous block of data.

The terminal now "out-pulses" a serial stream of bits that contain two 8-bit SYN characters (01101001) in front of the message block. This bit stream passes over the connector cable to the modem (where it is converted from digital baseband to analog broadband), and out on the local loop to the telephone company central office. From there it goes on to the IXC communication circuits and to the receiving end, where it goes through the same type of equipment, such as telephone company central office, local loop, modem, connector cable, front end,

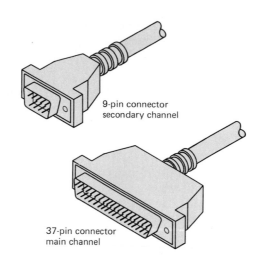

9-pin connector
secondary channel

37-pin connector
main channel

37-PIN CONNECTOR				9-PIN CONNECTOR	

First Segment Assignment		Second Segment Assignment			
Pin	Function	Pin	Function	Pin	Function
1	Shield	20	Receive Common	1	Shield
2	Signaling Rate Indicator	21	Unassigned	2	Sec. Receiver Ready
3	Unassigned	22	Send Data	3	Sec. Send Data
4	Send Data	23	Send Timing	4	Sec. Receive Data
5	Send Timing	24	Receive Data	5	Signal Ground
6	Receive Data	25	Request to Send	6	Receive Common
7	Request to Send	26	Receive Timing	7	Sec. Request to Send
8	Receive Timing	27	Clear to Send	8	Sec. Clear to Send
9	Clear to Send	28	Terminal in Service	9	Send Common
10	Local Loopback	29	Data Mode		
11	Data Mode	30	Terminal Ready		
12	Terminal Ready	31	Receiver Ready		
13	Receiver Ready	32	Select Standby		
14	Remote Loopback	33	Signal Quality		
15	Incoming Call	34	New Signal		
16	Select Frequency/ Signaling Rate Selector	35	Terminal Timing		
17	Terminal Timing	36	Standby Indicator		
18	Test Mode	37	Send Common		
19	Signal Ground				

RS449 is a new EIA specification replacing RS232C. This specification calls for the use of a 37-pin connector. For those devices using a side, forward, reverse, or secondary channel, a second 9-pin connector is specified. RS449 provides for additional control and signaling.
FIGURE 4–5 RS449 interface.

data channel, and into the host computer. You can follow this movement across Figure 4–1.

This process is repeated for each synchronous message block. The data signaling that takes place between the terminal and the modem involves the "request to send" and "clear to send" wires (pins) of the 25-wire RS232C interface cable. The synchronization of message blocks takes place because they are transmitted sequentially and each message block contains two SYN (synchronization) characters in front of it for the receiving modem to use. Timing is everything in data signaling and synchronization. In other words, the timing between bits (Figure 2–9 shows timing between bits) and the timing of when signals and blocks of data are being sent are critical.

MODEMS

Next, looking at Figure 4–1, our signal passes into the modem. You will recall that a modem is a device that changes digital signals to analog signals, or digital signals to a different and more precisely timed form of digital signal. This signal could be a frequency tone, digital pulse, or the modification of an optical beam of laser light. Here we will discuss only the modem's basic function, the conversion of a baseband signal (digital) to a broadband signal (analog).

DIGITAL TO ANALOG

On the digital side of the modem (the side connected to the terminal), we have just received a synchronous block of data. It is represented as a sequence of direct electrical pulses (+ and − voltages) which, when counted into their individual 8-bit ASCII code structure, represent individual characters. The modem now proceeds to take each of these plus and minus voltages of electricity and converts them to an appropriate frequency (carrier wave).

For our example, we will use frequency shift keying (FSK), which involves modulating the carrier wave between two different frequency levels. Assume our modem is transmitting at a speed of 1200 bits per second, which means it can change between either of two different frequencies 1200 times each second of time. Therefore, every time the digital side of the modem receives + 5 volts of electricity, it transmits a frequency tone signal of 2200 hertz. Conversely, every time the digital side of the modem receives a − 5 volts of electricity, it transmits a signal of 1200 hertz. When no signals are being received by the terminal, the modem falls to a middle frequency (1700 hertz), called its carrier wave (see Figure 4–6).

In order to transmit at 1200 bits per second, our modem would have to attain a frequency of 2200 hertz and then hold that frequency pulse on the communica-

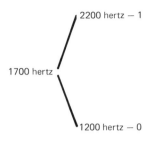

2200 hertz — 1

1700 hertz

1200 hertz — 0

1. When there is no signal to transmit, the modem goes to its carrier wave (1700 hertz). Some modems use either the 2200 or the 1200 hertz as the carrier wave, instead of a third frequency.
2. When a + 5 volt signal is received, the modem out-pulses a 2200 hertz signal, which is a binary 1.
3. When a − 5 volt signal is received, the modem out-pulses a 1200 hertz signal, which is a binary 0.

FIGURE 4–6 Modem (digital to analog).

tion circuit for a time equal to 833 microseconds. The 833 microseconds timing factor is achieved by taking one second and dividing it by 1200 pulses per second; 1/1200 = 833 microseconds. Also note that, as the modem is out-pulsing these different frequencies, it is doing it on the analog side of the modem. In other words, the modem has carried out its basic function of converting data from digital to analog or baseband to broadband.

DIGITAL MODEMS

If the communication circuits use digital transmission for their entire length, instead of converting to analog transmission as we do with the normal telephone circuits, you would have a digital modem. Their purpose is to shape the digital pulses and to perform all of the auxiliary functions such as loopback testing and checking the circuit diagnostics. Their special function is to convert a digital signal to a more precise and more accurate digital signal. For example, a digital modem can take a weak electrical signal, put very precise timing characteristics between the pulses, put it out at a certain strength, and control its electrical characteristics. This is done to reduce noise, distortion, and errors (digital modulation was described in Chapter 3). Digital modems are much simpler, as evidenced by their cost. Whereas a 9600 bit-per-second digital-to-analog modem might lease for $150 per month, its digital counterpart might lease for $15 per month, a factor of 10 times.

LOCAL LOOPS

Local loops are sometimes called *subscriber loops*. These are the circuits that go between your organization and the common carrier facility. The newest digital termination system (DTS) local loops entail rooftop antennas that transmit directly from a home or office, microwave radio, cable TV handling digital signals, or radio frequencies such as those used in cellular radio local loops. In other words, the millions of miles of wire pair local loops that are strung along telephone poles or buried underground will be augmented or replaced over the next 20 years or less. One advantage these newer digital termination systems have over wires is that the cost of installation and maintenance is much lower.

Now our signal is on the local loop (see Figure 4–1) and on its way to the telephone company central office. At this point we have to learn about station terminals, amplifiers, 2/4-wire circuits, and FDX/HDX as they relate to the movement of a signal between a terminal and the central office which might be several blocks or miles away.

STATION TERMINALS

A station terminal is the *terminal block* (leased circuits) or the voice jack (dial-up circuits) that terminates the local loop at your home or business. It is where you connect your modem. The upper half of Figure 4–7 shows the connection to a station terminal for a leased circuit; the lower half shows the connection for a dial-up circuit.

Our signal has now passed over a major boundary point with regard to responsibility for errors and the quality of circuits. It has left our modem and is now using the public local loop circuit of the telephone company or other common carrier.

AMPLIFIERS

Now that our message has left the modem and is in the local loop, the signal suffers attenuation. Attenuation is the weakening in strength of a signal as it passes down a wire. It is caused by resistance. For example, copper wire pairs can experience an attenuation or loss of signal strength because of the weather. This is because electrical resistance of wires rises with the temperature. Wet/humid conditions likewise increase attenuation because signal leakage occurs at insulators when they are wet. It should be noted, however, that there is an inherent resistance in any communication media, whether it is a copper wire pair, coaxial cable, microwave link, or the like.

The telephone company places repeater/amplifiers at 1- to 10-mile intervals in order to increase signal strength lost to attentution. The amplifier and its associ-

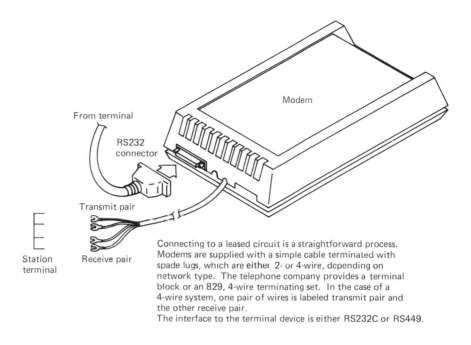

From terminal

RS232
connector

Transmit pair

Station
terminal

Receive pair

Connecting to a leased circuit is a straightforward process.
Modems are supplied with a simple cable terminated with
spade lugs, which are either 2- or 4-wire, depending on
network type. The telephone company provides a terminal
block or an 829, 4-wire terminating set. In the case of a
4-wire system, one pair of wires is labeled transmit pair and
the other receive pair.
The interface to the terminal device is either RS232C or RS449.

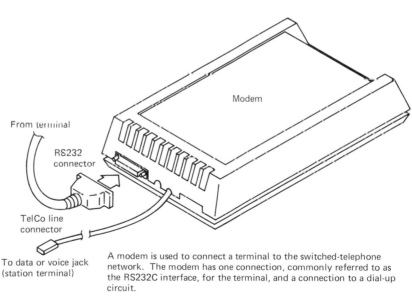

From terminal

RS232
connector

TelCo line
connector

To data or voice jack
(station terminal)

A modem is used to connect a terminal to the switched-telephone
network. The modem has one connection, commonly referred to as
the RS232C interface, for the terminal, and a connection to a dial-up
circuit.

FIGURE 4–7 Connecting the modem to the local loop.

ated circuits are referred to as a *repeater* because it repeats the signal while increasing the signal's strength. The distance between the repeaters/amplifiers depends upon the degree of attenuation because the signal strength cannot be allowed to fall too low. If too much attenuation occurs, it is increasingly difficult to distinguish it from other noise or distortion that is always present on communication circuits.

On analog circuits it is important to recognize that the noise and distortion that are always present on the circuit are also amplified, along with the increase in signal strength. This means that the noise from a previous circuit link is regenerated and amplified each time the signal is amplified. This has a definite effect on errors, which creates the need to retransmit messages. Analog circuits are not as clean or error free as digital circuits. A digital circuit with its associated digital amplifier recreates a new signal at each amplifier station. For this reason, the noise and distortion that are on the previous link of the network are not amplified each time the digital signal goes through an amplifier. This gives a much cleaner signal and results in a lower error rate for digital circuits. The benefit to users is that fewer messages have to be retransmitted because of errors.

TWO/FOUR-WIRE CIRCUITS

On the local loop, data can travel on either a two-wire circuit, where there are only two wires from modem to modem, or on a four-wire circuit, where there are four wires from modem to modem. The normal dial-up circuits are two-wire circuits, whereas private dedicated lease lines are four-wire circuits.

On a four-wire circuit it is easy to perform full duplex transmission (simultaneous transmission in both directions). If you want full duplex transmission on a two-wire circuit, you must use a special modem that creates two different frequency channels on the two wires, thus simulating a four-wire circuit (see Figure 4–8).

If you have a four-wire circuit, you also can keep both the send and receive signals from a modem operating simultaneously. In this way, you avoid the normal turnaround time for the modem circuits. This is because the request-to-send signal is leaving the modem on two of the four wires while the receive side of the same modem is connected to the other two wires. This is *not* full duplex transmission; it is only a technique that is used to reduce the modem circuitry turnaround time (sometimes called *retrain* or *clocking time*) to zero. Your messages would still be in half duplex with this technique.

FULL DUPLEX/HALF DUPLEX (FDX/HDX)

At this point your message can be traveling over the local loop and, ultimately, the IXC circuit in either full duplex or half duplex communications. Actually, there are three modes of transmission: simplex, half duplex, and full duplex.

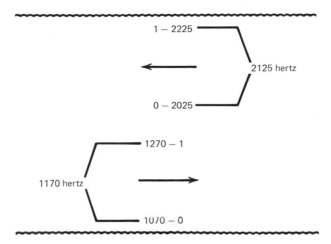

In one direction there is a carrier wave of 2125 hertz, using 2225 for 1s and 2025 for 0s. In the other direction there is a carrier wave of 1170 hertz, using 1270 for 1s and 1070 for 0s.

FIGURE 4–8 FDX on a two–wire circuit.

Simplex is one-way transmission as you have in a local area network (LAN), where all the data travels around the loop in a circle (one direction only).

Half duplex is two-way transmission; however, since you can transmit in only one direction at a time, you encounter turnaround time. For example, assume a message is sent from your modem to a distant modem. At this completion of that message transmission, there is a certain amount of turnaround time while the modem at the receiving end changes from receive to transmit, as well as request-to-send and clear-to-send delay.

With *Full duplex* transmission, you can transmit in both directions simultaneously. Full duplex transmission does not involve turnaround time because the transmit and receive are simultaneous.

Avoid equating full duplex and half duplex with four-wire and two-wire, because they are different. Full duplex and half duplex are communication modes. Full duplex requires a front end processor that has the proper software for simultaneous two-way communication. Also the local loop and interexchange channels must be a four-wire circuit or you must have special modems that can create a four-wire circuit through a type of multiplexing (see Figure 4–8). Finally, the terminal at your node must have the proper devices and functions so it can simultaneously receive and transmit data.

On the other hand, a two-wire or four-wire circuit is nothing more than a configuration that is supplied to you by the common carrier. All voice grade leased circuits are four-wire circuits. All dial-up circuits are two-wire circuits.

TELEPHONE COMPANY CENTRAL OFFICE

The telephone company central office (see Figure 4–1) is the switching center of the telephone company or other special common carrier. At the central office, the common carrier might perform circuit switching, or you can have your circuits wired around the electronic switching system. Dial-up circuits are switched, and leased circuits are wired around the switching equipment.

SWITCHING

Most common carriers today can only offer circuit switching such as we have in the dial-up telephone system. By this method you cannot complete your call if the other telephone or circuit is busy. On the other hand, if you use a public packet switching common carrier, it can perform packet or message switching because it can hold your message in a store and forward area and forward it when the other terminal is available. In our example we are not using dial-up, so switching is not a factor in the movement of our message from the terminal to the host computer; our leased circuits are wired around the switching equipment.

INTEREXCHANGE CHANNELS (IXC)

Our signal (see Figure 4–1) has now moved from the telephone company central office to the interexchange channels or long-haul communication circuits. At this point, we could be using microwave circuits, satellite circuits, wire pairs, coaxial cables, or any other type of circuit media that is available.

LEASED CIRCUITS

In our example, we assume use of a leased circuit. Because of this assumption, our communication circuit is wired around this switching equipment. In other words, we have a direct circuit path from station terminal to station terminal. This offers a much cleaner circuit with less noise and distortion, and with fewer retransmissions of messages because of errors. Our signal will probably travel over wire pairs, coaxial cables, and microwave as it moves across the country or around the world. You will know if it goes on a satellite circuit because of *propagation delay*, which is the time necessary for the signal to travel from the earth to the satellite and back (approximately 0.50 second for both directions).

T-1 CARRIER

One of the most widely used transmission systems is the AT&T Communications T-1 carrier. The T-1 carrier uses pairs of wires with digital repeater/amplifiers spaced approximately 6000 feet apart to carry 1.544 million bits per second. In this 1.544 million bit per second digital stream, 24 channels are encoded, using pulse code modulation and time division multiplexing. Your message occupies one of these 24 voice grade channels/circuits.

The T-1 carrier is used for short-haul transmission over distances of up to 50 miles. It uses a pulse code modulation system in which each individual sine wave of a signal is sampled 8000 times per second. These samples are called *frames*; therefore, 8000 frames per second travel down the line. Each frame contains 8 bits (7 to encode the frame and 1 for control). It is almost like drawing a picture of a waveform by using a series of bars, as you would when drawing a bar graph. Figure 3–27 depicted this form of digital transmission.

FRONT END COMMUNICATION PROCESSOR (FEP)

Our signal has now passed through similar facilities/equipment at the other end of the IXC (telephone company central office, local loop, modem, and RS232C connector cable) on its way to the front end communication processor (see Figure 4–1). The front end communication processor is the basic network control point because network control functions that used to be in the host computer are now in these processors.

We will discuss briefly a few of the front end's functions, but you might wish to review the entire list of front end functions that were discussed in Chapter 2.

CENTRAL CONTROL VERSUS INTERRUPT

In our example, the front end processor probably is a cental control system. This means that, when the "send" key on the terminal is hit, the data signaling/synchronization cannot commence until the terminal is polled: the front end sends a signal asking, "Do you have anything to send?" Because the "send" key has been hit, the request to send (pin 4 of the RS232C connector) can begin the data signaling/synchronization process. If this were an interrupt system, the terminal would interrupt the front end immediately when the send key was hit.

LOGGING

When our message reaches the front end communication processor, the first task the front end performs is to double-log the message. Logging messages onto a disk for short-term storage protects them in case of system failure or interrupt. It gives the recovery software instant access to the last few messages received, thus enabling proper recovery and restart.

Messages also are logged onto a magnetic tape for a long-term transaction trail and/or historical purposes. These magnetic tapes might be saved for a few days and possibly be converted to microfilm. They can provide a long-term history of all transactions entering the system and all returned messages after processing in the host computer.

ERROR DETECTION

The front end also examines messages for parity errors. If the transmission is asynchronous, it can look only for parity errors in individual characters and/or check the total character count per message. Since our message transmission was synchronous, the front end can perform an extensive error check, such as a Cyclical Redundancy Check (CRC), to identify any errors that may have occurred. When an error is detected with synchronous transmission, a signal is sent back to the terminal, instructing it to retransmit the message. If this happens, we are back at the beginning, where a baseband signal is sent out of the terminal, over the connector cable, through the modem, local loop, telephone company central office, the IXC circuits, and onto the front end communication processor. All the previous functions are automatically repeated when this happens. The terminal operator is unaware of the retransmission.

CODE CONVERSION

Our terminal uses ASCII code. At this point the front end communication processor may have to convert that code structure to the host computer's code structure. We will assume that it has to be converted from ASCII to EBCDIC.

PROTOCOL CONVERSION

The front end also may have to convert from an ASCII type of protocol to either SNA/SDLC or BSC protocol. *Protocols* are the set of rules or conventions by which two machines talk to each other.

DATA CHANNEL

The data channel (see Figure 4–1) actually is part of the host computer. It takes a message that has been stripped of its communication control characters by the front end and moves it into the memory of the host computer for processing.

HOST COMPUTER

Our message has now been passed from the front end communication processor, through the data channel, and into the host computer. The host computer processes the message as requested. It may retrieve information from a database, or it may update databases, depending upon the message instructions. Any security restrictions placed on you through individual passwords, terminal to host identification codes, or restrictions placed upon the front end communication processor port through which your message entered will be validated.

Finally, examine Figure 4–1 one more time in order to understand fully and visualize the basic concepts used in moving a message/signal from a terminal to a host computer.

THE ADVANCED CONCEPTS

After examining Figure 4–1 again, we can look at it in terms of the other factors that might affect data transmission.

The Terminal   At the terminal level, there might be any of the five major types of terminals or a facsimile device, or a microprocessor. It might be a multifunction terminal in which the operator tells the system to which function the message should be returned (the disk, the video screen, or whatever is available).

Remote Intelligent Controllers   At the point where the terminal/connector cable is shown there might be a remote intelligent terminal controller that can control a group of terminals. This increases the complexity of the message during its transmission flow because it requires additional addresses and/or control functions. This is especially true if an intelligent controller is used that can hold back messages during peak periods and assign priorities to message transmissions.

Multiplexer   Also at the terminal/connector cable there might be a multiplexer that connects four or more terminals to the single modem. If it is a pure multiplexer without intelligence, not much complexity is added to the system. A statistical time division multiplexer that can hold back messages increases complexity quite a bit because of delayed messages held back during periods of peak message volumes.

Hardware Encryption   Again at the terminal/connector cable there can be a hardware encryption box that secretly encodes data so that no one can use it, even if it is intercepted. This adds the complexity of periodically modifying the encryption keys.

Multidrop Circuits   At the local loop/interexchange channels, there can be multidrop or drop-off points. This means that the terminal shares the circuit with several other terminals. This probably does not add much more complexity to the message except for additional addressing. If a break occurs in this multidrop circuit close to the front end communication processor, this eliminates the ability to transmit to any terminals.

Interexchange Channels If the interexchange channel is a satellite circuit, this complicates transmission procedures because of the 0.5 to 0.6 second delay that is experienced on many satellite circuits. In fact, this could have the greatest impact on the entire network because you might be forced into migrating to one of the newer bit-oriented protocols such as X.25, SDLC, or HDLC, and software changes are expensive.

SUMMARY

We have shown the basic flow of a message from terminal to host computer and explained the technical concepts of the process. You might remember this discussion when you read Chapter 6 on Software. There we outline the various software packages, explain what they do, and describe how the software handles movement of a message into the host computer and back out for its return to the originating terminal. Whereas this chapter was a hardware-oriented and a technical concept discussion, Chapter 6 traces the message flow from the software viewpoint.

QUESTIONS

1. Looking at Figure 4–1, can you define the type of signal (whether it is baseband or broadband) as it passes over or through the following devices: terminal, connector cable, modem, local loop?
2. Is asynchronous or synchronous transmission used at your organization?
3. What is the difference between RS232C and RS449?
4. Diagram the data signaling/synchronization for transmitting a message.
5. What is the difference between a station terminal and an RS232C connector plug?

5

CONFIGURATIONS AND APPLICATIONS

This chapter describes network configurations and selected applications that are dependent upon data communication networks. The configurations described in this chapter are the basic building blocks that would be interconnected when you are developing multiorganization or international networks. Each configuration could stand alone to become a single application network. The selected applications in this chapter are those that utilize the various network configurations in order to become a viable business application system.

NETWORK CONFIGURATIONS

This chapter describes numerous network configurations and selected network applications. Each configuration will be explained in enough detail so that you will understand how it operates.

Have you ever thought of the group of friends with whom you closely associate as a network? These people are an informal information network who exchange various bits of data (voice communications) among themselves. Networks have always been a part of your life even though you did not associate them with the formal networks of data communications and voice transmission.

We intend to describe the different network configurations. These include the basic voice communication network, point-to-point leased networks, the dial-up network, multidrop/multipoint configurations, multiplexed networks, timesharing and remote computing services, packet switching networks, public networks, private home networks (microprocessor-based), local area networks (the fastest growing segment), PBX/CBX digital switchboard systems, and DTS. Applications for data communications (automated office, electronic mail/voice mail, banking, airlines, rental cars) are also described.

VOICE COMMUNICATION NETWORK

The basic voice communication network is not only the largest, but it is one of the oldest, of our twentieth-century electromagnetic data communication networks. This system involves the telephone that you have in your home and which you use to call friends, relatives, or other people. In fact, in today's business communications cost structure, the basic voice telephone system commands three-quarters of the business revenues spent for communications, whereas only one-quarter is spent on true data transfer type of communications.

This network involves the interconnection of your telephone to the telephone company central office (also called cnd office or exchange office) where switching is performed.

Switching includes identifying and connecting independent transmission circuits to form a continuous path from your telephone to the telephone that you are calling. Figure 5–1 shows the interconnection of telephones through the central office in our basic voice network. You will note that the local loop on this figure is nothing more than the copper wire pairs that go from your home to the telephone company central office. The central office is where the electromagnetic or the electronic switching system equipment is located. The central office uses the telephone number that you dialed as an "address" and searches out the other telephone so the two can be connected.

Calls made within the same central office or interoffice trunks are known as *local calls* (see Figure 5–2). Calls that use the tandem trunks and the tandem office

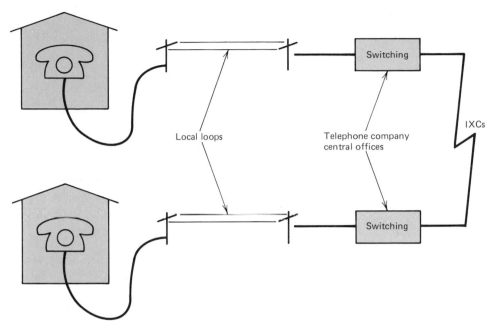

FIGURE 5–1 Telephone interconnection.

are known as *unit calls,* for which there may be an extra charge for each minute of time. The calls that use intertoll trunks (those that require an area code) usually are known as *long-distance calls.*

The telephone system network is so widespread throughout the United States and the entire world that it would be difficult to give a single drawing showing its total configuration. Within the United States, officials in the Department of Defense have indicated that this network is so widespread and so integrated into the entire fabric of the country that it offers a unique backup capability in that it would be virtually impossible to entirely destroy the network and, therefore, totally eliminate all communications throughout the United States. It should be noted that even though Figure 5–2 shows telephones at the end of the local loops, terminals or other data transmission equipment might also be attached to the dial-up telephone network.

POINT TO POINT NETWORK

Figure 5–3 demonstrates a leased circuit that goes from point to point. A point-to-point circuit or network means that an organization builds a private network and, in doing so, has a communication circuit going from its host computer to a re-

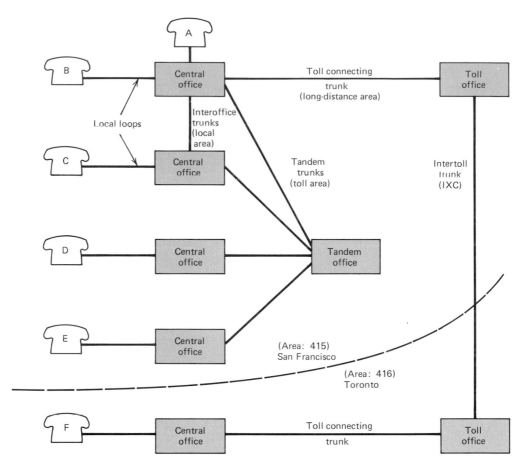

FIGURE 5–2 Central office hierarchy.

mote terminal. Point-to-point circuits are sometimes called *two-point circuits*. This type of configuration is quite advantageous when the remote terminal has enough transmission data to fill the entire capacity of the communication circuit. When an organization builds a network using point-to-point circuits, there

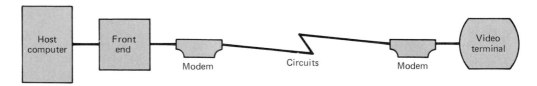

FIGURE 5–3 Point–to–point configuration.

might be many point-to-point circuits emanating from the front end communication processor ports to the various remote terminals wherever they are located. The front end may have anywhere from 1 to over 200 ports.

DIAL-UP NETWORKS

Dial-up is a network in which the organization utilizes the public telephone system, whereas private leased circuits are for your exclusive use only. Dial-up networks are usually point-to-point, but in some areas you can now build a three-point dial-up connection which is known as a *conference call*. In other words, the user might dial up two other terminals and simultaneously interconnect three terminals or two terminals with a host computer site. The basic configuration for dial-up circuits in a data communication network is the same as in Figure 5–1 except there is a terminal instead of a telephone. Besides the standard dial-up network, those who utilize dial-up connections between their terminals and computers might also consider Wide Area Telecommunications Services (WATS), SPRINT, MCI, or other specialized value-added networks that sell dial-up voice capability at a price below that of regular telephone companies.

When using the dial-up telephone system for voice communications only, the basic procedure would be to:

- Lift up the telephone receiver in order to place the telephone in an off-hook position. This signals the central office that you want to make a call. The central office responds with a dial tone to indicate that you may dial the number.
- Dial the telephone number.
- Receive back one of three signals; either a busy signal, a ringing signal, or a loud pulsing signal that identifies a system malfunction.
- When someone answers the telephone, begin your verbal conversation. Two data terminals talking to one another is called a *session*.

When the device at the end of the line is not a standard telephone, but a data type device such as a video terminal, the procedures to sign onto the system and begin working are as follows:

- Lift up the telephone receiver in order to place the telephone in an off-hook position. This signals the central office that you want to make a call. The central office responds with a dial tone to indicate that you may dial the number.
- Dial the telephone number.
- Wait for the tone; when you hear the tone, place the telephone receiver in the acoustic coupler cradle.

- When the on-line light flashes on, depress the carriage return on your terminal keyboard (this step may vary in different systems).
- The central computer may ask for any of the following information (this is the user sign-on procedure):

User ID *enter your ID and hit carriage return.*

Password *enter your password and hit carriage return.*

Account Number *enter your account number and hit carriage return.*

  ⋆ *You must now tell the computer which programming language you are going to use, such as BASIC.*

- The system responds with another asterisk and you are now ready to use the timesharing system through the "dial-up network." This procedure varies with different systems; there may be fewer or more steps.

Even though the above example may not match perfectly the sign-on procedures for the dial-up network at your organization, it is typical of the procedures used when entering a computer-based dial-up system.

LOCAL INTELLIGENT DEVICE

Figure 5–4 shows a local intelligent device configuration that allows a point-to-point circuit to be connected to a local intelligent device controlling one or more terminals. Local intelligent controllers frequently control 16 terminals simultaneously. The purpose of a local intelligent terminal controller is to load the point-to-point circuit more efficiently. Local intelligent controllers also can serve as a security restrictor. In addition, they save costs because some of the intelligence that would have to be built into each terminal can be built into the controller,

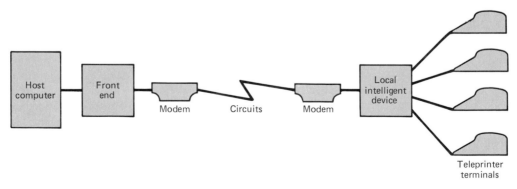

FIGURE 5–4 Local intelligent device.

thereby allowing the use of terminals that are less intelligent and lower in cost. As the next section shows, local intelligent device configurations also might be used in a multidrop network.

MULTIDROP CONFIGURATION

Figure 5–5 shows a multidrop circuit configuration. Notice that the first dropoff (a point where a terminal is attached to the circuit) has only a single teleprinter terminal, but that the second dropoff has a local intelligent terminal controller that is managing a cluster of terminals. Either of these configurations is possible on a multidrop circuit. Organizations that design multidrop configurations do so in order to load the communication circuit more efficiently, reduce circuit mileage, and thus save money. Also, when there are various branch offices or government agencies throughout a city or country, multidrop configurations can be a very efficient method of interconnecting these various branch offices or agencies.

In a multidrop configuration, each "drop-off point" shares the line and is serviced or responded to in some sort of sequential fashion. In other words, only one drop-off point can be using the circuit at the same time. It is only because we can switch between the various drop-off points so fast that it appears to each user that he or she has the entire circuit to himself. It is not uncommon to use 50 or 60 terminals interconnected on a single multidrop circuit. There might be five to seven

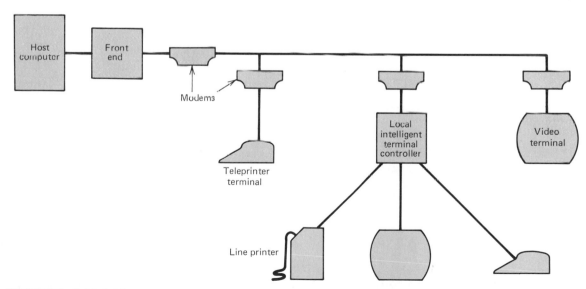

FIGURE 5–5 Multidrop circuit.

local intelligent devices (one per drop-off point) and perhaps six to ten terminals connected to each local intelligent device where that circuit has been dropped off at a branch or agency.

MULTIPLEX CONFIGURATION

Figure 5–6 shows a typical multiplexing configuration. To *multiplex* is to simultaneously place two or more signals on the communication circuit. Multiplexing may be performed by using "frequency division multiplexing" or "time division multiplexing."

The primary benefit of multiplexing is to save communication circuit costs between the host computer or business entity and the many far-flung remote sites. Figure 5–7 shows how several levels of multiplexing might be used in order to save on communication costs. Reading from right to left, the first level of multiplexing is where four terminals are multiplexed onto a single IXC (interexchange channel) for transmission to a distant site. The second level involves multiplexing the resulting 12 signals over a single IXC circuit and, finally, the third level involves multiplexing 24 signals over a single IXC circuit. The modem pairs were left out to simplify this figure.

As you look this over, think about how much more circuit mileage would be involved if you had a point-to-point circuit going from the host computer site to

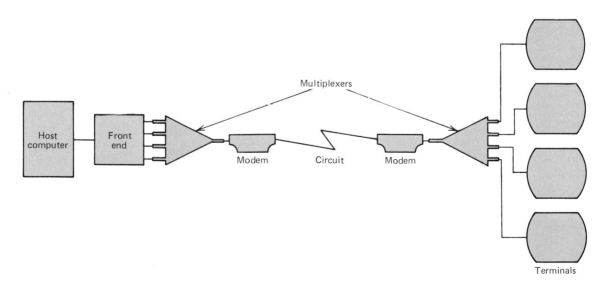

FIGURE 5–6 Multiplexing.

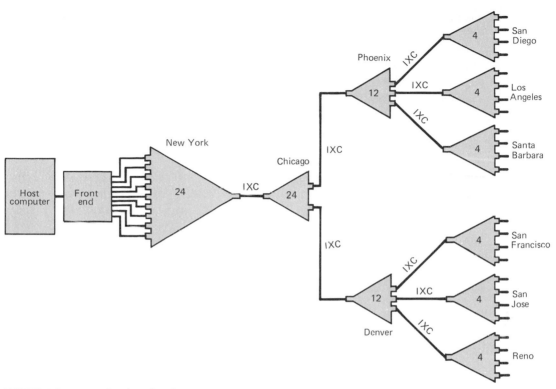

FIGURE 5–7 Levels of multiplexing.

each of the 24 terminal device locations. Also, think about how an alternate configuration, such as multidropping, might be used to interconnect the various terminal device locations, in contrast to one that uses multiplexing.

TIMESHARING SYSTEMS

Timesharing and remote computing services are now available for virtually any data processing application that otherwise might be performed by an in-house data processing center. The network configuration used here might be point-to-point lease, dial-up, or multiplexing. Organizations that offer timesharing and remote computing services offer the users the service, the hand-holding necessary to get their applications up and running, and the specialization in certain areas and specific applications. Some of the larger firms that offer this service are McAUTO (McDonnell-Douglas Corporation), Boeing Computer Services (BCS),

Automatic Data Processing (ADP), Tymshare, Control Data Corporation, General Electric, and Tymnet.

The configuration evolves when the organization decides to use timesharing or remote computer services. Once this decision is made, the user must decide which network configuration is most appropriate for transfer of data to the timesharing service bureau.

PACKET SWITCHING

Figure 5–8 shows a typical packet switching network. Note how there may be built-in redundancy between the different cities because each of the switching nodes (SN) is interconnected to other cities by single or multiple (IXC) circuits. The switching nodes (SN) route the packets through the network.

Also, you will notice that there is a packet switching front end for each user.

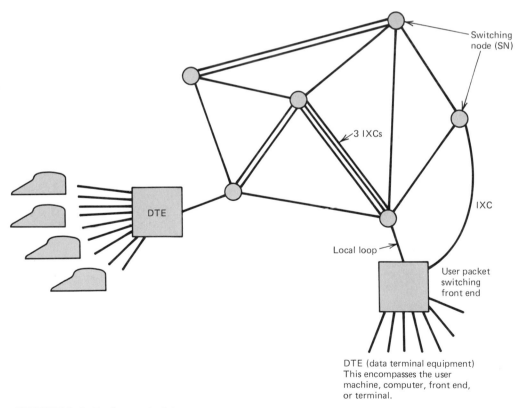

FIGURE 5–8 Packet switching.

Some of the users have connected their packet switching front end to several of the switching nodes in order to gain further redundancy.

The purpose of a packet switching network is to interconnect various cities (or other areas) with several circuit paths that interact to form a complete network. The user may find lower cost data communications with public packet networks, because their charges accrue on the basis of "how much data" a user sends across the network plus a flat monthly connection fee. Charges are not related to the distance between points, as is true of point-to-point, dial-up, multiplex, and multidrop leased networks.

The basic protocol, X.25, defines the standard interface between a user and a public packet switching network. X.25 is discusssed in Chapter 6, Protocols and Software, and the controls related to the X.25 protocol are discussed in Chapter 8, Security and Control. Sometimes packet switching networks are referred to as *value-added networks* because the vendor sets up a complete network, maintains it, and leases it on a volume-of-data basis.

The three most important benefits with regard to the use of packet switching networks are lower cost of data transfer, increased network reliability (uptime), and a well-defined standard interface such as X.25.

To understand how packet switching works, you must first understand that all information passing over the network must be broken up into fixed-length "packets." A typical packet might be a 128-character message block. Once your message leaves your terminal and gets to your packet switching front end, it is broken into separate packets and passed on to a switching node (SN). The switching node wll choose the most efficient path to get it to the other terminal. There it will be joined back together, from the individual packets into a meaningful message. Then it is passed on to the user's terminal. Your data might go by several routes, or they might even go on different communication circuits (IXCs) between the same two switching nodes.

The reason that reliability of packet switched networks is much higher is that there are many different circuit paths the data might take, and there are even different circuits between the same two switching nodes. This means that if a switching node fails, an entirely different path might be used. Similarly, if a specific circuit between any two nodes fails, a different circuit might be used.

Finally, the benefit of having a standard defined protocol (X.25) is that it is very easy to connect any vendor's equipment to the packet network.

At this point, a distinction might be made between circuit switching, message switching, and packet switching. In *circuit switching*, communication circuits are switched between each other. If the entire circuit path is not available, you get a busy signal, which prevents completion of your circuit connection (this is like a dial-up telephone circuit). In *message switching*, the entire message that you are transmitting is switched to the other location. This might not involve connecting the circuits together, because some systems have message switching with store and forward capability. Store and forward, in the context of message switching, allows the temporary storage of your message and subsequent forwarding of it to its destination when the other half of the circuit becomes available (the other party). In *packet switching*, your message is broken into finite-size packets that are always accepted by the network. The message packets are forwarded on to the other party over a multitude of different circuit paths. At the other end of the circuit, the packets are reassembled into the message, which is then passed on to the receiving terminal. Obviously, this means that you must have Packet Assembly and Disassembly (PAD) software at each of the switching nodes or front ends. Notice that packet switching always accepts your message, message switching may not accept it if the other terminal is busy (unless there is a store and forward mechanism), and circuit switching will never accept your message when the receiving terminal device or circuit is busy.

One other point should be made. The organization that does not want to develop its own network and hire the required technical expertise is a prime candidate for public network packet switching because it eliminates many of the in-house technical expertise requirements.

PUBLIC NETWORKS

Various governments and private companies now offer network services to organizations that might wish to subscribe. The network itself is owned by the network operator, but the communication service provided to the customers is called a *public network*. Most public networks in the future will use a packet switching configuration. As an example, the public telephone system today uses a circuit switching methodology, but in the future this same public network will be moving toward packet switching of voice data.

Some of the networks that are available are:

- Tymnet (United States)
- Telenet (United States)
- Datapac (Canada)
- Transpac (France)
- PSS (England)
- Euronet (connects major European cities)

As an example of usage of these nets, let's compare the usage on Tymnet. Between 1973 and 1980 the number of communication circuits increased from 120 to 1300. The hours of usage per month increased from 150,000 to 844,000. The number of characters transmitted per month increased from 2 billion to 27 billion. Finally, the number of host computers attached to Tymnet increased from 8 to over 400.

Public networks and packet switching mix together very well. A corporation or government agency could have its own private packet switching network. Alternately, the same corporation/government agency also might be a user of a public network, whether it be packet switching, message switching, or circuit switching. Increasing private circuit charges may force users to migrate to "usage sensitive priced" packet networks.

PRIVATE HOME NETWORKS

According to popular thought, we may become a society of home-based computer commuters. If this comes about, then we will have various networks that emanate from private homes. These networks could be dial-up, packet switching, public networks, rooftop antennas to satellites, cellular radio loops, or cable television.

Ironically, the United States is lagging behind the rest of the world with regard to home networks. Currently, our most promising vehicle is cable TV for the interconnection of various terminals/microprocessors in an individual's home to other microprocessors, host computers, or a multitude of databases. While readily available, dial-up is too costly for home use.

Videotex is being used experimentally for the dissemination of current news and other items. Outside the United States, government agencies are spearheading videotex networks such as Prestel in England, Teletel in France, Bildschirmtext in West Germany, Captain in Japan, and Telidon in Canada. These networks are being developed by the governments of the respective countries, whereas in the United States videotex development is largely within the domain of private industry.

Figure 5–9 shows three of these public packet networks that are used for videotex into private homes.

LOCAL AREA NETWORKS (LAN)

What is a local area network? The first part of the definition, *local*, is easy; it is within a "small" confined area. It is not completely confined like a computer input/output bus, but it definitely is more restricted than long-haul interexchange channel (IXC) networks. It is the piece of the communication net that interconnects equipment such as word processors, computers, and telefacsimile machines and ties them to the long-haul (IXC) networks.

Prestel network. The British Post Office has implemented a two-way home information network.

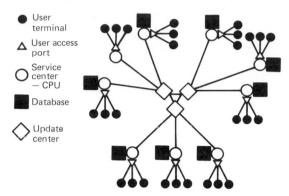

Bildschirmtext network. The West German Bundespost adds to the British implementation by supporting third-party data centers and databases.

Teletel network. France's approach differs in that its public network handles all switching and network management, but not database control.

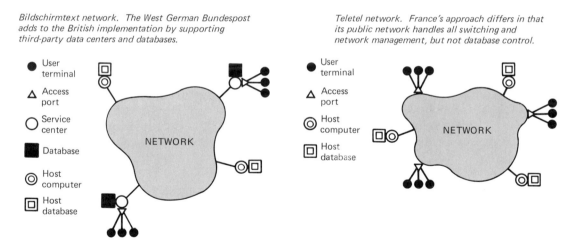

FIGURE 5–9 Public packet networks for private home networks.

LAN is a network covering a clearly defined "local" area: a single building, a group of buildings within a business firm's property or a campus, or if the conditions are correct, a confined area within a city. Ideally it is not limited to communications within that defined area, but provides easy access to the outside world through a gateway.

Continuing with our definition, aside from its limited distance characteristic, the local area network has a capability for high-speed transmission. Since low-speed transmission currently is available on standard telephone networks, our

LAN gives added functionality, providing speeds of 250 kilobits per second, to 10 megabits per second and higher.

The local area network operates outside the "regulated" environment of the federal or state regulatory bodies. Bandwidth now becomes available to support high-speed transmission for full animated motion, full color video, or other high data rate analog or digital signals.

In a utopian environment you would be able to attach an infinite number of users without major modifications to the system. Access would be simple, and the network would be user friendly. It would have protocol conversion allowing any device to talk to any other device, and it would be available 100% of the time, sending messages that were completely free of errors. Since this is not Utopia, we have to compromise this ideal definition by making trade-offs that optimize the network to meet specific user requirements.

Figure 5–10 shows a local area network. This network is built to interconnect a variety of electronic office equipment, miocroprocessors, minicomputers, mainframe host computers, databases, and any other terminal equipment within an organization. Without a local area network, the automated office cannot function. Local area networks are owned and operated by the individual organization. Currently there are about three dozen vendors of local networks. As was noted previously, local networks usually are of a limited distance (a few miles) and they are

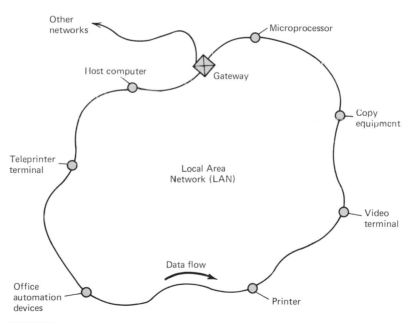

FIGURE 5–10 Local network.

interconnected by intelligent communication technology such as packet switching, bus or tree structures, or I/O (input/output) channels.

There is one bone of contention concerning LANs which should be noted. The local area network designer's view is that the PBX/CBX is simply one component of the network, providing the gateway function between the local area network and the outside world via long-haul networks. The PBX/CBX manufacturer's view is slightly different. They believe that PBX/CBX is the hub of the local area network and that everything else attaches to it.

Which view will prevail? Probably both. At this point, the advanced PBX/CBXs are moving toward a distributed switching architecture that is very similar to the architecture being used by most local area networks. Perhaps this means that the techniques ultimately will combine into a single configuration. Whichever view prevails, we will see a distributed network either within the guise of a PBX/CBX or as a local area network facility which will provide us with our future communications capability.

To convince management that this advanced communication system should be added to the repertoire of equipment and services, we have to prove that it will be cost effective. This is necessary because corporations are in business to make a profit; unprofitable firms are only in business a short time.

New communication networks, then, either have to save money directly or have to provide a means for making more money faster. In the case of the LAN, it can do either or both, depending on current corporate communication capabilities.

One way that the LAN helps in corporate economics is in the centralization of communications management. Right now, we may have a voice network, a word processing network, a computer network, and a telefacsimile network, each one being administered by a separate department. Generally there is little if any contact between the departments (regarding the networks) since the networks tend not to interact. With an integrated corporate network viewpoint, made possible by the LAN, we can centralize the management of these resources, operate them more efficiently, and do it with fewer people.

A second economic benefit is the sharing of resources. Again, many corporate networks are separated functionally with terminals tied to specific jobs. Many organizations also have standalone word processors with built-in memory and microprocessor controls. With an integrated local area network, a host computer can be accessed by any terminal or workstation. Its computing capability can provide text editing, calculations, management information, or data storage and retrieval to everyone in the organization, thereby combining the independent functions into a single integrated activity. This simplifies control and data access and provides a single resource to meet many needs.

A third economic benefit is that our voice network (telephones) can be incorporated into our LAN.

Some LAN functions require specific network access methods (protocols) while others can use any method, from one end of the spectrum (dedicated channels) to the other end (random access). The user can select either of these or any point in between and therefore must consider carefully the advantages and disadvantages of each method.

Dedicated channels may be frequency channels (such as found in television broadcasting) or dedicated time slots (for example, the channel is yours between five after the hour and ten after the hour, every hour). The concept of dedicated channels is very simple, but it can be inefficient. Assume the network is yours for five minutes every hour. It remains yours even during the lunch hour when it is not utilized.

In some situations dedicated channels are necessary. For example, if you were holding a video teleconference, a standard 6 megahertz channel would be used to provide a constant update of your full motion, full color picture. Dedicated channels, then, have their special place in the network of the future but should not be chosen as a general approach to network access, since they simply are not efficient for typical office or distributed data processing functions.

Contention or random access schemes, on the other hand, appear to provide a simple, flexible access method that can dynamically accommodate changing user needs. Pure random access, which was used in the Aloha system in Hawaii, is commonly called the "transmit and pray" approach. What this means is, if you want the channel, take it. If no one else is on it, there is no problem. If someone else is already there, keep trying—you'll get through eventually. In actuality, this seemingly chaotic approach works well for a small number of users dealing in short messages that are sent on an intermittent basis.

Some variations on this system, which might be termed "somewhat more ordered chaos," would encompass one of the access methods that has been supported by the IEEE 802 Local Area Network Committee. This is the *Carrier Sense Multiple Access* (CSMA) approach, which is very simple in concept. Don't get on the network and transmit at random—listen first! If anyone is there, refrain from transmitting. When it gets quiet, take the network and send your message. This eliminates a certain number of collisions, messages' interfering with each other, but a small problem remains. Because of network propagation delay, users who are located some distance from each other can both listen to the channel, find it empty, and begin to transmit at the same time. They follow the carrier sense rule, but their messages still collide. Therefore, let's add another piece to this.

Instead of just listening before we talk, let's listen while we talk. If we detect a collision during transmission, wait for the other message to end, and retransmit. This is *Collision Detection* (CD). Two users still can attempt to retransmit at the same time. To rectify this problem we add an algorithm into the network. Instead of each user retransmitting immediately after the end of the message, it is held back until some random time interval after the colliding message disappears.

This does not eliminate collisions completely, but it reduces them to manageable proportions. This approach allows rather high utilization of the network by many users while still providing the flexibility necessary to accommodate intermittent traffic. This access method is called *CSMA/CD*.

Let's go back and look at a central-control type of protocol where the host computer or front end communication processor performs polling. With polling, the front end polls each individual terminal and gives it permission to transmit. In other words, central control polling is analogous to a classroom situation in which the professor calls on students who raise their hands. The professor polls each student in turn. This type of protocol is used because it maintains both the order and the sequence of communications.

In an interrupt type of protocol (collision detect interrupt) each terminal listens and can speak out or interrupt whenever it wants. This is similar to a real life situation in which three of four people have an informal conversation. Whoever wants to talk just starts talking. If two or more people start talking at the same instant, everyone stops for a short period of time, waits, and tries again. The person who "gets the floor first" is the one who has control of the conversation for a short period of time. This is identical to the way a collision detect system works. A terminal listens, interrupts, and sends its message. As long as no other terminal interrupts during the short time period that the message is traveling on the coaxial cable of the local network, everything is alright.

Follow along with Figure 5–11 while we examine how you can interrupt and still get messages through. Notice that if you had a message that was 100 characters per message and if there were 10 bits per character, that would give you 1,000 bits per message to transmit.

Next, if the local network had a transmission capacity of 10 million bits per second and you divided that by your 1,000 bits per message, 10,000 messages per second could be transmitted.

Assume that the designer specified only a 10% network utilization; therefore,

100 Characters/message $\times$ 10 Bits/character   _____   1,000 Bits/message
$\dfrac{10{,}000{,}000 \text{ Bits/second}}{1{,}000 \text{ Bits/message}} = 10{,}000 \text{ Messages/second}$
10,000 Messages/second $\times$ 0.10 utilization percentage   _____   1,000 Messages/second

FIGURE 5–11 Local network capacity.

90% of the time the network would be empty. Then you would be able to transmit 1,000 messages every second.

The point to be made is that if you had 100 terminals on the local network, then each terminal could send 10 messages each second. This would amount to using only 10% of the network's capacity. Thus, data collisions are reduced severely. When they do occur, the system withdraws the transmission and tries again a short time later.

There are some areas, however, such as process control, where guaranteed access to the network is required. For example, suppose that one chemical must be added to a mixture to produce a special plastic or rubber. If you cannot gain access to the network to add this ingredient at the exact time it is necessary, the network is worthless to you. Because of the need for guaranteed access, *ring networks* have become very popular in the process control area. These networks are not as flexible as the random access networks, but they do accommodate a number of users, provide a certain amount of flexibility, and give a guaranteed maximum access time. The access method used on the ring is the second major standard that is supported by the IEEE 802 committee and it is called the *token access method*.

This method can be likened to a relay race where the track belongs to you as long as you have the baton. When your run is finished, you hand the baton to the next runner, and the track belongs to that runner. In the ring network, the baton is called a *token*, a particular short message which is generated when the network is turned on. If you want the network, you take that message off the network, send your traffic through the network, and add the token message to the back end of your message. The next user down the line who wants to send a message, grabs the token and the network is now his or hers. In the meantime, the first message continues around the ring, until it's taken off by the user and acknowledged. The token gives the terminal permission to send. In this case, either message send, completion, no message to send, or a time-out due to a problem causes token passing to the next terminal.

Currently it appears that the collision detection protocol is the most popular. It is used by Ethernet, Wangnet, Net/1000, Localnet, ConTelNet, HyperChannel, Decnet, and others. Token passing is used by Pronet, and IBM is advocating a "standard" for token passing. This apparently commits IBM to token passing.

The current vendor offerings can be divided into two categories—baseband and broadband. At this point in the discussion, we can define *baseband* simply as a single channel whereas *broadband* provides multiple channels. The full capabilities will be discussed later. We should note that most of the baseband networks utilize the Carrier Sense Multiple Access/Collision Detection approach (CSMA/CD). Most of the newer vendors are following the CSMA/CD specification. As was mentioned, the ring networks continue to use a polling or token approach, which provides guaranteed access to the network.

Since the network is selected on the basis of the overall network performance characteristics, the protocol is only one criterion provided by the vendor. In general, we select a network vendor and then specify that any equipment purchased between now and the foreseeable future match that vendor's protocol. However, we really do not want to discard all of the equipment that has been gathered over the years. Instead, protocol conversion must be provided by the network vendor. This allows the use of most, if not all, of the existing equipment. Compatibility is one of the biggest problems in local area network selection. The development of protocol conversion hardware and software will become a large business area in itself.

The network hardware is available today. The major limiting factor in network availability or capability is software. It is the software that provides the applications, and it is also the software that provides the conversion from protocol to protocol or equipment to equipment. As a result, it is the software vendors who ultimately will drive the local area network market. Manufacturers will provide hardware but, just as happened in the computer industry, small software houses will emerge. They will specialize in application software, protocol conversion software, and operating systems software which, in turn, will enhance network capability and make the integrated network a reality.

In the broadband world, we find slightly more variation, since with multiple channels a vendor can offer a combination of access methods such as frequency division multiplexing (television type channel selection), with a CSMA/CD or token, time division multiplexing (TDM), or any other approach on a given channel. The commercial networks again generally use the CSMA/CD access method, with token access under development.

Baseband involves electrically pulsing the cable directly as with a voltage or current switched on or off. Broadband involves dividing the signaling into allocation slots, such as we see in TV channels. Broadband may involve frequency modulation, amplitude modulation, phase modulation, or pulse modulation techniques. Baseband uses only a small portion of the capacity of a coaxial cable and achieves a highly effective throughput with a straightforward circuitry. However, this technique precludes the use of the cable for other services (e.g., video).

By contrast, broadband uses relatively complex modulation devices to subdivide the coaxial cable into many channels, each of which can have the capacity of a baseband network. As an analogy, think of these modulation devices as being similar to those multiplexers that allow a cable TV system to carry 25 different channels of television over a single coaxial cable. The primary drawback of broadband lines is the expense of the modulation devices/multiplexers.

Baseband versus broadband has become a bone of contention among vendors, although the approach really should be a simple user selection. The basic difference between baseband and broadband is that in a broadband system the information is placed on a multiplexed signal, which allows the selection of multiple channels. It also means that additional equipment, such as a modem, is required in order to place this signal on the transmission medium.

The advantages of baseband are clear. Essentially, it is less complex than a broadband network, and it could be less expensive. Without the need for frequency translation, modems, or amplifiers, the cost for this type of system should be lower than the more complex broadband system.

Baseband, however, is limited to a single channel. This means that we may not be able to integrate all signals, such as voice, data, and video, on the network unless we can dedicate the channel to one function, say a video signal, for as long as necessary. Data and voice can be time division multiplexed, however, so a single channel does not necessarily mean a single user.

There are a few other problems with baseband systems, some of which are caused by the laws of physics, others by vendors. In the laws of physics category, we find a lot of natural and man-made noise in this band as well as cable losses (attenuation) which limit the network length. These restrictions affect network layout and flexibility of use.

The vendors have caused further problems, since they have taken a position that makes most baseband networks incompatible with broadband networks. They have done this quite simply by selecting a different cable impedance and a tapping method (the method of entering the cable) that limits the frequency characteristics of the medium so that high frequencies cannot be impressed efficiently on the cable. These limitations must be offset by cost, but with CATV (cable television) and LAN requirements for modems, amplifiers, splitters, and the like, the cost of broadband will continue to drop, thereby making broadband systems more competitive with baseband, as well as more flexible.

One broadband advantage is "unlimited" distance. Since the baseband network does not use any amplifiers, it is limited to the distance that is allowed by the loss characteristics of the cable. The broadband network, on the other hand, has the capability for much longer distance because the data can be amplified and the signal level can be maintained. Thanks to the amplifiers, the cable layout is very flexible. If another branch is needed, we simply put an amplifier in to bring the branch signal up to a usable level.

The network, then, can be baseband, which may be somewhat cheaper but limited to a single channel, or it can be broadband, which gives multichannel capability but is more expensive today and more complex.

There are essentially three transmission media that dominate today's network installations: *twisted pair* (which is current telephone technology), *fiber optics*, and *coaxial cable*.

Twisted pair, although already in place, can be eliminated as a long-term LAN support medium because it is too limited in its capacity/distance relationship to support future high volume data requirements. At any practical distance, twisted pair will not support the high-speed data transmission that is unique to the local area network.

Fiber optics uses a glass fiber that is drawn into a long cylinder which acts as a "wave guide" for light. Once the light is put in one end of the fiber, it becomes trapped in that fiber, bounces back and forth as though it were hitting a series of mirrors, and eventually comes out at the other end. The light injectors generally are lasers or light emitting diode (LED) transmitters. The laser provides higher power than the LED, but the LED is less expensive.

The advantages of fiber optics are straightforward. It has tremendous bandwidth capability, is impervious to noise, and is more secure against taps. It has more bandwidth available than we could fill with everything currently transmitted and anything we can think about in the future. The fibers are small, they are almost indestructable, and since they use light as a transmission medium, they do not pick up electrical noise. In short, it would seem to be the ideal medium.

The problem with fiber optics, however, is that it is an immature technology. At the moment, it is difficult (and expensive) to cut into the medium for multiple users. It is a fine approach for massive data transfer on a point-to-point basis, but it becomes expensive as a multidrop line. It is noisier than cable, so we need a higher signal power in order to get the same signal-to-noise ratio, but this is not really a practical limitation today. What it does mean is that the higher powered and, therefore, more expensive laser injector may have to be used.

This leads us to *coaxial cable.* Coaxial cable, like people, comes in all sizes, shapes, and characteristics. Each has some unique properties that can either help or hinder design efforts. Cables are available in 50 ohms, 75 ohms, 93 ohms, and a number of impedances in between. Coaxial cable supports frequencies up to 300 or 400 megahertz. In fact, if the correct cable is selected, it will support frequencies into the gigahertz range as well. Coaxial cable, then, fits the requirement for almost unlimited bandwidth support for our LAN, and immediate availability in the marketplace.

The vendors now pose the next question. Will we use a *single* broadband cable, which dictates a frequency division and thus limits the number of channels? Or should we use a *dual* cable approach, which allows full use of the frequency spectrum but may cost more because of the dual cable and installation requirements.

Early networks selected the dual cable approach (Figure 5–12) because bidirectional amplifiers were difficult to find. In the dual cable approach, single-direction amplifiers bring the incoming traffic up to some head-end point which also amplifies the signal, turns it around, and places it on the other cable. Outgoing traffic uses unidirectional amplifiers heading out.

A dual cable configuration with standard CATV components allows some 50 usable channels on the network, which should cover all data or analog signals. The 50-channel calculation is derived very simply by taking the normal 300

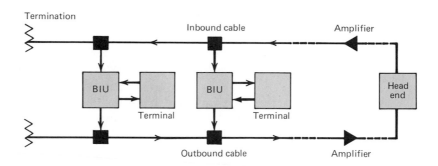

BIU: Bus Interface Unit

FIGURE 5–12 Two–cable configuration.

megahertz band and dividing it by 6 megahertz, assuming the use of standard TV channels. Each 6 megahertz channel should be good for a data rate up to 5 megabits, so that we can have 50 channels—each of which is equivalent to a 5 megabit baseband channel—on the dual cable network.

If a single cable approach (Figure 5–13) is used, we have to replace the signal path of the second cable with an isolated frequency band. In this approach outgoing (transmitted) signals use the bottom half of the frequency band, and incoming (received) signals the top half. For example, outgoing signals are from zero to 150 megahertz, and incoming are from 150 to 300 megahertz. This means one of the cables has been eliminated, but 25 potential data channels have thus been given away. Since we are not channel limited, however, we can afford to lose this num-

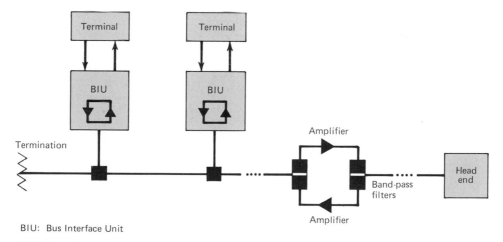

BIU: Bus Interface Unit

FIGURE 5–13 Single cable configuration.

ber of channels and still maintain the needed level of service. This allows the use of smaller conduits and simpler installations, and it takes advantage of the fact that the television industry is gravitating toward a two-way single cable approach. This means we will be buying standard components, which should have the lowest cost.

The single cable configuration really does not look all that different from the dual cable. The exception is that the BIU (Bus Interface Unit) must have bidirectional components and the cable amplifiers themselves must be bidirectional with bandpass filters. This adds some loss and design complexity to avoid degradation in signal-to-noise ratio.

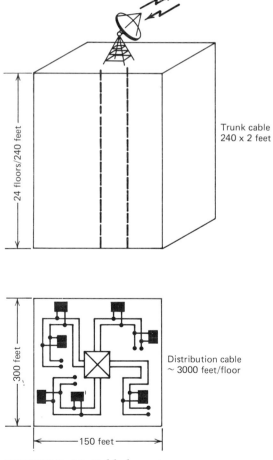

FIGURE 5–14 Cable layout.

What, then, will we use—single cable or dual cable? The answer lies in what the vendor will supply. Since this is not a critical characteristic, generally it will not impact our network selection criteria. The building layout illustrated in Figure 5–14 shows that there is not very much cable in a network design for a high-rise facility. The cost of the cable for either configuration is small compared with the cost of the equipment that will be placed on the cable, so this presently is not an issue. Complexity also is becoming a nonissue because the the CATV manufacturers have solved the single cable problem and equipment is available off the shelf. The key parameter, then, is the functional capability of the vendor offerings.

Let us look at those vendor approaches to baseband cable, shown in Figure 5–15. In the baseband area, the networks that are commercially available are essentially single cable coaxial networks. Hyperchannel, which is a computer-to-computer high-speed network, uses a single cable and accommodates multiple two-way cables as well. In the broadband area (Figure 5–16), we find that MITRE, which is essentially an old technology, still uses dual coaxial, whereas most of the newer offerings have gone to a single bidirectional approach. (Wang, because of extreme bandwidth requirements, also retains the dual cable configuration.)

What can we do in this early period in the growth of intrafacility networks? Users do not know what they want to put on the networks, vendors are not sure what the user wants and have provided very limited vendor offerings, adequate software is not yet available, standards are confused, and we have heard little from IBM, although, they seem to be moving toward the token approach. Despite the fact that everything said here has been general, we must consider the fact that there is a lot of IBM (and IBM-compatible) equipment out in the market. If our office has equipment from many vendors, all of it must interface to the intrafacility

Name	Vendor/ Organization	Access	Speed	
*Ethernet	Xerox (Dec, Intel)	CSMA/CD	10.0	MBPS
*Z-Net	Zilog (Exxon)	CSMA/CD	0.8	MBPS
*Net/One	Ungermann Bass Inc.	CSMA/CD	4.0	MBPS
†Hyperchannel	NSC	CSMA with priorities	50.0	MBPS
*Hyperbus	NSC	CSMA with priorities	1.544 TO	
			6.176	MBPS
*Contelnet (Model 700)	Contel Information Systems	CSMA/CD	2&10	MBPS

  *Single coax bus
  †Single coax—up to 4 in parallel

FIGURE 5–15 Some local networks (baseband).

Name	Vendor/ Organization	Access	Speed
†Mitrix II	Mitre	TDM/Random Access (Hybrid)	7 MBPS
*Localnet System 20	Network Resource Corporation	CSMA/CD	128 KBPS (120 channels)
*Localnet System 40	Network Resource Corporation	CSMA/CD	2 MBPS (2 channels)
*Videodata	3M/ISI	FDM	Multiple low-speed channels
*Amdax	Amdax	⎧FDM ⎨TDM	Multiple low-speed channels 7 or 14 MBPS
*Contelnet (Model 800)	Contel Information Systems	CSMA/CD	2 or 10 MBPS

*Single coax bus
†Dual coax

FIGURE 5–16 Some local networks (broadband).

network. If we have no equipment, the problem is simplified. With IBM's dominant position, however, its approach to this industry could be the basis for de facto standards.

Users have a unique position in the LAN market today. They actually can drive the vendors toward providing the equipment, the capabilities, and the applications that most users want. Vendors are undecided as to how to attack the market. If enough users select the same applications and the same approaches to local area networks, then vendors will begin to supply these approaches. They are in business to service the user! If they do not, their profits will disappear and so will their businesses! Figure 5–17 shows the user's steps toward the future.

The first step for the user is to prepare a plan for communications over the next two years, three years, five years, and even ten years, if possible. The key element is to define the requirements. Is it clerical traffic that will be placed on the network? Will it be executive traffic? Do we want process control capability, or will the network have to accommodate a mix of these plus video teleconferencing and voice?

Once the requirements have been defined, we should develop a functional specification. This allows the vendor to understand the requirements and respond to them. We have to be sure to develop a basic total approach to communications. Do we want a single network covering the entire facility or do we have communities of interest that might be better served by being on their own network with a gateway to the main network? For example, do we want financial information isolated from the main network for security reasons, or because 99% of the traffic

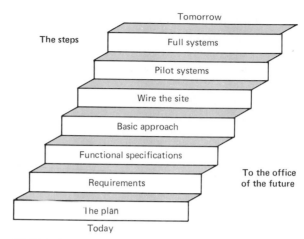

FIGURE 5-17 User's steps to the future.

remains within the financial department and doesn't have to be placed on the main network? If so, the other 1% easily can be handled by a gateway to the main network.

The next step is one that all users are reluctant to take today, and we can understand their reluctance. If we were to wire a facility today, what would we use? Twisted pair? Fiber optics? 50 ohm coax? 75 ohm coax? If you are an IBM user and already have 93 ohm coax all over the building, what do you do with it?

If we are in an existing facility, we can afford to wait before wiring the building. If, however, we are in the midst of a renovation or are building a new facility, we should wire the building now with 75 ohm coax—the standard in the CATV industry. This will cover most needs over the next decade and the wiring can be put in without necessarily committing to the amplifiers, the modems, and the added equipment for broadband operation. It is not that expensive at perhaps $1.50 or $2.00 a foot, and it could save a lot of money over postconstruction installation in the future.

The next item, which is the key to eliminating both vendor and user uncertainty, is to install a pilot system. We have all seen changes in our approach to office operations when new equipment, facilities, or communications capability was added. If requirements are defined today, those requirements will have changed the day a local area network is put into operation. Why wait until then? The best approach is to put a pilot network in today. This allows the staff to use the network to find out what they really need or want. It further provides a chance to evaluate a vendor (or a number of vendors) in a situation where our risk, and theirs, is limited. Finally, it gives us a chance to define our needs at the time when LAN technology can still be customized.

The last step, installation of a full system, need not be implemented until requirements have been firmly defined. Because the technology is still immature, few vendors can provide a network that will meet all customer needs. In fact, the LAN market may gravitate toward a multivendor approach. If the vendors are compatible, if they all adhere to a single specification, integration will be easy. It is unlikely, however, that the vendors will be totally compatible. We probably will use internal staff or software houses to provide conversion software. A system integrator who provides the translation between the vendors and assumes responsibility for system installation, initial support, and ongoing maintenance also will eliminate the difficulties of installing a multivendor system.

In short, we can take advantage of the situation by having the LAN vendors provide low-cost pilot systems for user experimentation, education, and (ultimately) operation. The LAN is the key to the automated office and the integration of office and data processing functions. We should use this key to unlock captive information and free the staff from the shackles of outmoded communications.

PBX/CBX

A Private Branch Exchange (PBX), sometimes called a switchboard, can also be called a Computer Branch Exchange (CBX) in today's technology.

The cord telephone switchboard was still being manufactured as late as 1958. Since that time technological advancements have been substantial. The first major step was the development of automated PBXs which utilized electromechanical switches. The next step was the development of computer-controlled PBX units which were capable of electronic switching. The most recent step was the introduction of a new generation of PBXs which have the ability to convert analog voice signals into digital form for transmission as binary bits. This allows digital switching of both digitized voice or data signals.

This last development is important because digital switches can process more data than analog switches; they also are less expensive and easier to maintain. The digital PBX is capable of switching traffic among a variety of equipment including data terminals, intelligent copiers, facsimile transceivers, and word processing equipment. In addition, digital switches allow the user to integrate data and voice transmission over the same telephone circuit. This can also be accomplished through the use of a telephone handset that digitizes the voice.

This new technology has opened the door for the automated office to be planned around a digital PBX/CBX instead of a local area network. If such capabilities as video conferencing or high-speed document distribution are required, however, then a local area network may be needed.

These new PBXs are built around microchips, microprocessors, and various switching electronics that allow for integration of the voice and data functions through one automated switching facility. In fact, network designers may not have to choose between local area networks and PBXs, because vendors are starting to set up an interface to connect local area networks with digital PBXs. As this occurs, we will be able to interconnect data terminals on a local area network with any telephone in the organization. Instead of just being the central focal point for voice communications, the digital PBX/CBX will become the integrated front end communication processor that interconnects everything, including microprocessors, the host computer, voice communications, local area networks, facsimile machines, public packet switching networks, and so on.

This interface point (PBX/CBX) is probably the point at which the giant communications companies such as AT&T Information Systems and International Business Machines will fight most fiercely.

As you can see from Figure 5–18, the computer-based switchboard (PBX/CBX) may be the central switch connecting point for all communications in the organization. This would be very advantageous both from a cost-saving viewpoint (combining voice and data communications) and from a control viewpoint (all corporate communications controlled from a central site). This cost factor was exemplified at one major bank when they compared communication expenses and learned that data communications equalled $31,000 per month while voice communications equalled $200,000 per month.

Some of the features that these new switchboards can add when combining voice and data are the following:

- Distributed switching capability controlled by programming.
- Reduced wiring because both voice and data can go on the same wires.
- Nonblocking switching whereby, even though all the current ports are busy, an incoming call is not blocked.
- Both synchronous and asynchronous data communications with up to 1.544 million bits per second on the digital switchboards.
- Simultaneous voice and data transmission.
- Modemless switching for digital and local area networks.
- Various features found in front end communication processors such as code conversion, protocol conversion, and the like.
- X.25 protocol compatibility.

Basically these digital switchboards may become the central control switching point of all networks in a corporation or government agency. In other words, there might be extensive competition between the vendors of digital switchboards and vendors of front end communication processors. This might be especially true as more organizations build in-house local area networks.

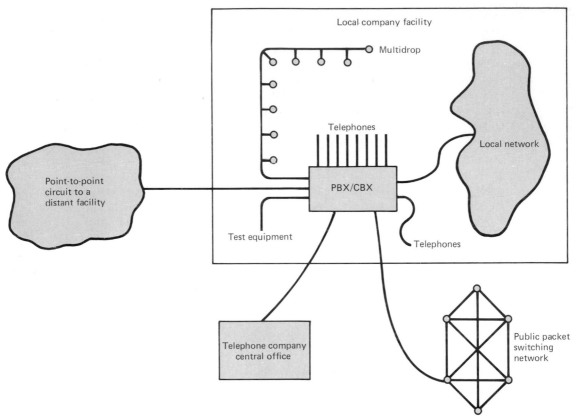

FIGURE 5–18 PBX/CBX network.

DTS

The Digital Termination System (DTS) should be closely related to the newer digital PBXs. This network configuration involves the telephone company and the many other vendors who will soon offer local distribution of data received from satellite carriers. Data will be received from satellites at earth stations and central switching facilities located atop urban skyscrapers. From there it will be broadcast to users, via a radio antenna on their roof, to a transceiver device on the user premises.

This configuration is nothing more than a local loop configuration within a city. As has been stated, this is the "last mile" to the user premises whether it is the home or a large corporation/government agency.

The local loop configurations have long been dominated by the telephone company and their copper wire pairs that go either underground or on overhead telephone poles. Now we will be moving to rooftop antennas and radio frequency transmissions as in cellular radio transmission which will be "in addition to" the current copper wire pair local loops.

APPLICATIONS FOR DATA COMMUNICATIONS

The applications for data communications are so wide and so diverse that we could not possibly cover even 10 percent of them in a book such as this. In fact, an entire textbook could be devoted to applications for data communications. In the first half of this chapter on configurations some applications were mentioned. In the following sections we would like to discuss selected applications such as the automated office, electronic mail/voice mail, banking/finance communication networks, airlines, and rental cars.

Each of the following applications might use any one configuration or combination of the configurations that were discussed previously. The purpose here is to provide a frame of reference for the uses of data communications and the way they tie together our business systems, government agencies, and personal lives.

THE AUTOMATED OFFICE

The automated office is a concept that will have far-reaching effects on the world as we know it today. The term "paperless office" has been used for about ten years. Society has a way to go before it is ready for this envisioned paperless office—probably not until the twenty-first century. However, the technological means to support this capability exists today, and data communication is the key component.

During the last decade the world began to realize that it was becoming buried in paper. Most people do not recognize that more than half of the work force in advanced countries work in offices. Little was done to improve the productivity in this area until recently. In the last ten years productivity of manufacturing operations has been improved by at least 90% on the average, but during the same time frame the productivity of the office function has been improved by less than half that amount. On the average, it is estimated that one-half of the total costs of U.S. corporations are for office-related work. In some specialized industries such as banking, insurance, and government the figure is much higher, ranging up to 70%. The United States alone is spending approximately $800 billion each year on these office functions supported by managers, professionals, and clerical staff.

Office automation is primarily in the hands of the data communication user; the computer technology already exists. All that is needed is data communications to link together all of the diverse business equipment that is required. Currently available hardware devices allow users to perform their tasks more easily and more efficiently. The part that is still missing is resource sharing, which requires fast and reliable data communications. With the new office networks, especially local area networks and digital PBXs, users will soon be able to achieve this resource sharing and to interconnect diverse operations in a manner that could not be achieved in the past.

Figure 5–19 shows how the business office can be tied together, probably with a local network or PBX/CBX, in order to act as a single integrated company business office. The local networks probably will be tied with other local networks or PBX/CBXs across the country or around the world by use of a public switching network interconnected through an X.75 gateway switch.

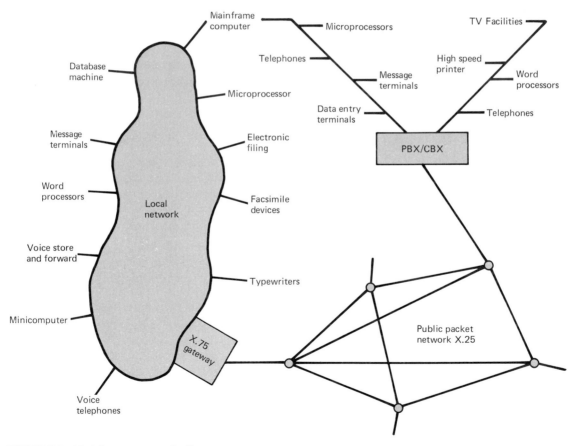

FIGURE 5–19 The automated office.

The general requirements for what might be classed as the ideal type of terminal workstation would be one that had the following features:

- An ability to communicate with all other terminal devices or workstations.
- Easy access to multiple databases.
- Easy attachment of various types of peripheral devices.
- A standalone data processing capability.
- Both voice and data capabilities.

With that in mind, a typical automated office might be interconnected just like the one shown in Figure 5–19 and might contain the types of equipment shown in this figure.

By having such a wide diversity of equipment with the ability to share resources and talk to each other, the office of the future can easily carry out such basic functions as:

- **Preparation:** Create business letters, schedules, interoffice memoranda, and reports.
- **Computation:** Compile data to complete forms, generate graphics, develop management charts, and calculate various figures.
- **Dissemination:** Perform local or remote distribution of all of the above in an electronic fashion.
- **Storage:** Store and retrieve documents, maintain mailing lists, and utilize any databases that are available.
- **Voice/Electronic Mail:** Utilize the communication capability for voice communications and/or electronic mail movement.

As was stated earlier in the section on configurations, the digital PBX/CBX automated switchboards may become the central focal point or central switch for the automated office. It would be easy to integrate both voice and data communications within the automated office and externally to other organizations with whom we have to communicate or transmit data. The communications decision we face is which system will become the central network, the local area network or the PBX/CBX?

Probably the one component that will affect office automation the most is the microprocessor. Microprocessors with their floppy disks will probably become the working tool for one out of every three people who work in the office. Because these devices are individually programmable, they will become the workhorse of the office. The biggest drawback of the automated office will be the fear of automation and the massive changes required in "how you perform your daily work."

Clerical workers, middle management, and upper management currently have a methodology for working *without* a totally automated, paperless type of terminal workstation. It is this inherent fear of automation and change of methods that will be the most difficult to overcome. Today's technology is far ahead of the average person's ability to accept it, so for this reason we probably will move into the twenty-first century before the paperless office is achieved.

The first thing that comes to mind for most individuals when they hear the term *office automation* is "word processing." It is important to recognize, however, that word processing is more important for increasing the productivity of secretaries than of other staff. Additional capabilities are needed in order to satisfy the objectives of the automated executive office. The most important objectives in making managers more productive include:

- Minimizing the amount of time spent on paperwork
- Improving decision-making capabilities
- Improving local and remote voice and data communications

With these objectives in mind, researchers have concluded that automated office technology can be applied to a number of executive functions including:

- Management of the executive work load
- Information access and retrieval
- Word processing
- Sending, receiving, reviewing, and disposing of mail
- Performing computations, and voice and data communications

In managing the work load, the executive will be able to work with an electronic in-basket where paperwork can be viewed and worked with via a desktop video terminal. Items will be placed directly in the work queues of subordinates and automatically be followed up to ensure timely completion. In addition, the executive will have the capability to maintain a diary/action list (to-do list).

In the area of information access and retrieval, the executive will have immediate access to data needed to process paperwork. This includes access to such resources as:

- Database systems
- Information search and retrieval systems
- Mail files
- Document files
- External information sources including public networks

The executive will need some access to word processing capabilities to support such activities as:

- Checking and modifying materials typed by the secretary
- Retrieving text for composing semistandard letters, contracts, proposals, and other documents
- Simultaneous viewing and changing of text between the executive and the typing pool

The executive will work extensively with the electronic mail and message functions. Electronic mail is covered in more depth in the next section of this book. As a minimum, the automated office will provide the capabilities for receiving and viewing mail/messages as soon as they arrive or when time is available, as well as for sending mail messages.

Almost every executive needs some support in the area of performing computations. The automated office will make available personal computational and calculation capabilities, access to computer modeling software tailored to individual decision-making situations, access to timesharing systems, and the ability to utilize graphics.

The automated office will provide the executive with an opportunity to dramatically improve voice and data communications by supporting electronic mail, voice mail, conference calling, and teleconferencing.

The basic functions supported by office automation systems are shown in Figure 5-20. In the future:

- By 1990 as much as three-quarters of the total document transmission will be done by electronic mail.
- By the year 2000 every office worker in the United States will have a computer terminal.

ELECTRONIC MAIL/VOICE MAIL

The public telephone system that handles voice messages has been around for over 100 years. Because this system is primarily a circuit switching system, it has one tremendous disadvantage. When the remote telephone is already in use or no one is present, the telephone call (the message) cannot be completed. Both voice mail and electronic mail overcome this disadvantage. During the next few years two of the major applications for data communications will be voice mail and electronic mail.

With regard to *voice mail*, great technical advances have occurred to make the telephone more accessible, easier to use, more attractive, and a true message

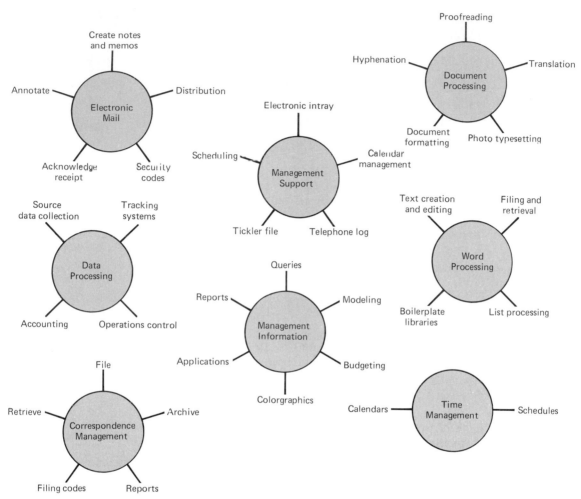

FIGURE 5–20 Functions supported by office automation.

switching system. Voice mail is a flexible means of sending a spoken message to someone even when they are not at the remote terminal (the telephone). The sender speaks into the telephone and the message is stored for later forwarding to its recipient. In effect, this turns the telephone system into a message switching system (instead of circuit switching), and also into a store and forward system.

Actually, voice mail is the transmission of a voice message to a recipient voice mailbox. Using a touchtone telephone with its standard 12-key dialing pad, the caller can record a message, listen to the message before transmitting it, and even change it if necessary. The message then can be sent to one or more recipients or even to a predefined group such as a department within a corporation or government agency. When it is is convenient, the recipients will check their voice mailboxes, scan to see who sent the incoming messages, and choose to listen to some now while saving others for later. Recipients can listen to the message, stop playing it if they are interrupted, skip ahead or back, or replay the message at will. After hearing the message, the recipient can immediately generate a voice reply and send it back to the person who sent the original message. Other options might be to forward the message to a third party and, of course, to discard the original message.

Voice mail has five major advantages over the traditional telephone. With voice mail it is *no* longer necessary to:

- Place several calls to a person in order to find him or her near the telephone.
- Move meeting schedules and match time zone differences around the world.
- Place a number of calls to get a similar message to many different people.
- Know where a person is located geographically in order to complete a call to the person (all you need to know is the voice mailbox number which he or she will check each day).
- Type your messages

With voice mail, the sender can place a call without interrupting the recipient, without having to know if the recipient is in the office or traveling around the world, and without regard to the time of day or night.

If a two-way information exchange is necessary because complex points must be discussed, then voice mail has a nice feature that can be used to set up a precise time, date, and telephone number in order to ensure that the connection can be made the first time. It is at this point that the newer telephones will have the most advantage because of their ability to transmit simultaneously both voice and picture. Video display telephones (display phones) will allow users to see the person with whom they are speaking (possibly for security reasons) and to actually show documents over the system.

Voice mail will be offered by the major telephone companies and the special common carriers such as Sprint and MCI.

Actually, voice mail will be more readily accepted by the general public telephone users than will electronic mail. This is because electronic mail requires, first, the ability to type and second, access to a keyboard in order to type in the text of a message. Psychologically, human beings were built to accept and transmit voice messages whereas the ability to enter text type messages must be learned as a special skill. More people can use their voice than can correctly write a complete sentence.

Basically, voice mail will be the overall glue that holds together and controls the automated office of the future, because different operators will need to converse with or leave messages for various other operators. Whereas typists will have no problem with electronic mail, nontypist middle managers and executives will rely more on voice mail because the spoken word is faster than the typed word.

Incidentally, voice mail is achieved by digitizing a voice signal and breaking it into a stream of digital bits. Some other systems are now capable of converting a voice signal into a 4800 bit per second or even a 2400 bit per second digital bit stream. This 2400 bit per second digital stream would enable us to have four simultaneous voice conversations on a 9600 bit per second data communication circuit.

Some important application features of voice mail that should be considered are:

- System users:

  Closed system, which subscribers can use and which entail controlled passwords.

  Open systems to be used by nonsubscribers or anyone else.
- Type of hardware to be used such as 12-key touchtone telephones or video telephones.
- Training aids such as booklets, audio prompting, or, if you have a video display telephone, a help key that provides pictorial representations of how to use the system.
- Key templates that might be placed over the 12-key pad to enhance explanations of how to use the system.
- System designed to direct messages to any telephone, to only subscriber telephones, or directly to a user's address message queue without regard to the telephone.
- Other message addressing schemes such as to individuals, to a unique telephone, to groups of individuals, to preorganized numeric codes, to system directories, or custom methods used by an individual organization.
- Type of system data that might accompany the message, such as date of call and time, address of sender and receiver, time of message, or other system data, and so on.
- Provision of priorities and/or different message categories as might be utilized within a large corporation.
- Scheduling of messages for future delivery.
- Whether the system will make repeated attempts to call the recipient.
- How long messages are retained.

- Notifying the sender of the message that the message was or was not received.
- Possible length of a recorded message.
- Adjustable volume and/or the speed at which the message is played back.

Finally, the number one advantage for voice mail is that everyone who currently has a touchtone telephone already has the terminal required to utilize this system.

Electronic mail is similar to voice mail except that the user must key in (or type) the message so that the recipient will receive a textual style printed message. Electronic mail will be very important in the office of the future, although voice mail will be used more. It also will become a major factor in the private home networks that were mentioned in an earlier section on configurations. One of the primary problems with electronic mail is that there is no standard protocol to install as there is with the current switched public telephone network (dial up).

Electronic mail, like voice mail, can be used to send messages to people when a two-way dialogue is not an immediate necessity. Thus, you can schedule future meetings, send messages to individuals or groups of people, overcome the problem of different time zones, and not worry if your recipient is unavailable when an electronic mail message is sent to an electronic mailbox. Even the post office is starting to send letters electronically from one post office to another. These are printed at the receiving post office and delivered in paper form to a home or business. The U.S. Postal Service offers ECOM, and Western Union offers a service called Mailgram.

Any user who has a portable terminal or microprocessor can connect to an electronic mail network. In order to review the features of an electronic mail system, reread some of the features that were listed for voice mail; they are the same.

As more competition arises in the public telephone networks, you will be able to transmit both electronic mail and voice mail over your home telephone. All that is required for the electronic mail portion is a small interactive terminal and an acoustic coupler modem.

In the business office it might be too expensive and too time consuming to give every employee an electronic mail terminal, but most employees today have a telephone and could use voice mail. Also, electronic mail requires more training. Furthermore, human resistance to major change, such as going from voice conversations to typewritten conversations, makes implementation more difficult.

In summary, both electronic mail and voice mail will be major parts of the future automated office. Voice mail will become a major feature of the world's public telephone systems; electronic mail will become part of these telephone systems only when the user has an interactive terminal or microprocessor. In less developed countries it may be difficult enough to teach a person how to use a voice telephone, much less an interactive keyboard-driven terminal.

COMMUNICATIONS IN BANKING/FINANCE

One of the major applications for data communications is in the banking industry. The increased pace and mobility of our society have placed a tremendous demand on our financial institutions. Customers are now demanding around-the-clock service, multiple banking locations, and virtually instantaneous response to their transactions. To meet these demands, financial institutions have implemented massive data communication networks, automated teller machines (ATM), and automated funds management systems for business organizations and government agencies.

Once a bank or group of banks develops a data communication network and interconnects it with one or more host computer processing locations, then any or all of these services can be offered. Some of the financial services that are offered over such a network include:

- Automated teller machines (ATM) where a user can deposit or withdraw money, make transfers between different bank accounts such as savings to checking, pay various bills, and the like.
- At-home banking and telephone bill paying. These are being tested for use with the public telephone system and cable television.
- Point-of-sale terminals (POS) tied directly to a customer's bank account from the store in which the POS terminal is located.
- Automated clearing houses (ACH) that facilitate the paperless dispensing and collection of thousands of financial transaction on a real-time basis.
- National and international funds transfer networks, check verification, check guarantees, credit authorizations, and the like.

Many of the networks developed by banks are private networks that combine point-to-point leased circuits, dial-up circuits, multidrop, multiplex, packet switching and, as we move into the private home, that will include any of the combinations of private home networks. It should be noted that banking networks that used to be within a city are now between various states and countries.

Many banks now participate in shared networks: a bank network system in one area of the country interconnects itself to another bank network in another area of the country.

In closing, it should be noted that banks also have to talk between themselves. There are some networks available for this type of communication:

- SWIFT   (Society for Worldwide Interbank Financial Telecommunications)
- CHIPS   (Clearing House Interbank Payments System)
- MINTS   (Mutual Institution National Transfer System)

- FED WIRE   (Federal Reserve Bank Telecommunications Network)
- BANK WIRE   (Interbank Network)

These networks are used for transfer of information between banks. Banks also create private networks and interconnect them for use with the general public. You might view the preceding list of networks as involving *wholesale* electronic funds transfer between banks and, by contrast, the private bank networks as involving *retail* electronic funds transfer for use by the general public and/or customers of each individual bank. A bank's network for its checking/saving branch offices, ATMs, POS terminals, home banking, and the like would be their retail network. Figure 5–21 shows a typical bank's retail network and its interconnection into the SWIFT wholesale network. If you would like to trace the flow of a message from the SWIFT "USA" Operating Center around the world, look ahead at Figure 9–4.

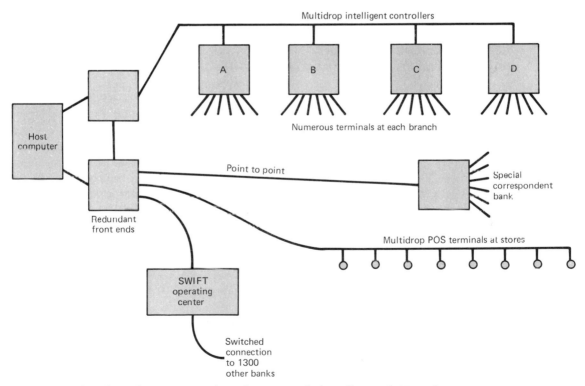

The terminals at branches A–D may be video terminals for tellers and ATMs for customers.
FIGURE 5–21 Bank networks.

COMMUNICATIONS FOR THE AIRLINES

Almost everyone is familiar with the use of data communications in the airline industry. American Airline's SABRE system was the first automated airline reservation system.

Today the major air carriers have developed individual communication networks for reservations, flight planning, inventory, control, flight scheduling, and all of the other applications that are required to run a major airline. The major systems have been developed by American Airlines, Continental Airlines, Trans World Airlines, and United Airlines. There also is an interlinking network system (used by all airlines) to transfer messages among these various systems. The interlinking network is maintained by Aeronautical Radio Incorporated (ARINC). American Airlines and United Airlines have the largest base of terminals installed in travel agents' offices as well as for their own use.

Airline networks are built using various point-to-point lease circuits and combining multiplexing and multidropping in order to get an efficient network layout. In today's world the application of data communications to the airline business is not so much in the network itself, but more in the installed base of terminals located in travel agents' offices. This is because there is a correlation between the number of airline booking terminals in travel agent's offices and the number of flights that are booked on that airline.

COMMUNICATIONS IN THE RENTAL CAR INDUSTRY

Another major application for large national and international communication networks is in the rental of automobiles. Avis Rent-A-Car was the first major company to develop an on-line real-time rental car network. This network interconnects the major locations where automobiles might be rented, picked up, and returned.

The networks used by rental car companies involve multiplexed circuits. These circuits are leased from the major telephone companies and/or special common carriers who are tariffed to sell communication circuits.

A rental car network is quite similar to an airline network in that it is used to keep track of dates when cars have to be returned to their lessor, make rental agreements with people who rent the cars, calculate the cost and mileages utilized, and perform other general accounting functions.

QUESTIONS

1. If the annual budget for voice telephones at the local university is $106,000, what would be a good estimate for their cost of data communications?

2. If there is a dial-up computer system at your organization, compare the steps for dialing in and gaining access to this computer system with the steps that were listed in this chapter.

3. Can you configure a system that combines local intelligent devices, multi-drop, and multiplex configurations?

4. Define the difference between circuit switching, message switching, and packet switching.

5. Can you identify six different packet networks?

6. Identify types of terminals and their locations for a local network that is utilized by your organization, in your department, or at your university or business.

7. Look at Figure 5–11. Determine how many messages per second the local network can handle if it is designed for 100% utilization (technically this is impossible because there would be too many collisions).

8. Using Figure 5–7 as a guide, draw a multidrop configuration.

6
PROTOCOLS AND SOFTWARE

This chapter traces the flow of a message through the critical software packages. It defines the differences between protocols, software, and network architectures. The seven levels of the OSI model are explained in a simplified manner. X.25, teleprocessing monitors, and other key protocol/software are described, as well as several vendor network architectures.

MESSAGE/SOFTWARE FLOW

As you may recall, in Chapter 4 we traced the flow of a data signal through a data communication network and discussed the various technical concepts. In this chapter we will relate the various software packages to that data signal as it is received at the front end communication processor and passed on the host computer.

Looking at Figure 6–1, you will note the various software packages. For example, at the *remote* end of the data communication network there may be half a dozen software packages for switching functions, software for intelligent controllers, statistical TDM (time division multiplexer) software, concentrators may

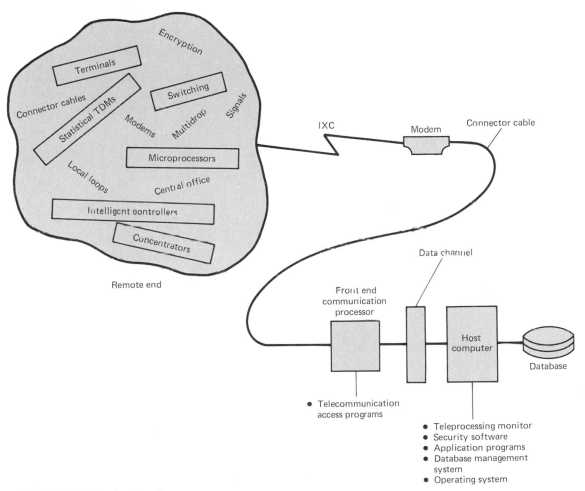

FIGURE 6–1 The basic software concepts.

have software, intelligent terminals use software, and microprocessors have various software packages.

These software programs that are located out in the network maybe scaled-down subsets of the software located at the front end communication processor or in the host computer. From your previous reading you already know that the front end/host can perform switching, multiplexing, control functions, and, of course, any of the application programs that a microprocessor can perform.

Now look at the six software packages (Figure 6–1) listed below the host computer/front end. You should recall that the message would not have been permitted to leave the remote terminal unless the *telecommunication access program*, located in the front end communication processor, polled the remote terminal and gave it permission to send the message. Incidentally, you should note that the telecommunication access programs can also be located in the host computer. Only the older systems' architectures use the host, however. It is more efficient and it distributes the processing loads better if these programs are located in the front end communication processor.

IBM is a good example of a dual location because they have telecommunication access programs in the host computer (Telecommunications Access Methods—TCAM) and they also have telecommunication access programs in the front end communication processor (Network Control Programs—NCP).

After the telecommunication access programs bring the message in from the network, log it, check for errors, and perform any other functions that this software is designed to do, the message is passed from the front end through the data channel and on to the host computer.

The *teleprocessing monitor* software package (located in the host) now takes control of the message. The detailed workings of the telecommunication access programs and the teleprocessing monitor are described later in the section on Software.

The teleprocessing monitor software handles functions such as queuing messages into an input processing queue or an output processing queue, file management, and sometimes database management, depending upon the database management system. Its other functions are restart/recovery for problems, recording statistics for accounting purposes, keeping track of statistics for performance evaluation, handling communication tasks within the host computer such as the handling/processing of requests and interfacing with application programs/database management systems, checkpointing, error handling within the host, security checking, and performing various utility functions such as message switching or checking various tables with regard to control.

Once the teleprocessing monitor has placed a message into the host computer's input processing queue, the input message may go through a security software package for review. A variety of security packages are available to check functions such as the validity of passwords, whether the password is a valid one for the terminal that originated it, whether the password allows access to the requested data, whether the terminal that originated the request is allowed to make the request that was entered, whether both the password and terminal that sent the request are valid at the specific time of day the message entered the system, whether the terminal or password are restricted to specific application programs, files/records/data items, and the like.

Assuming that a request passes all the security checks, the next step is for a specific application program to be called out because of the request that has been made.

At this point the application program begins to process the request. Let us assume that a database management system lookup is required to process this request.

Your database lookup would proceed something like this. The database management system software takes the "program subschema" from the application program and matches it against the database schema. *Subschemas* delineate the specific logical data that the program is allowed to retrieve, while *schemas* delineate the specific logical data layout in the overall database.

Once the location of your data has been pinpointed on the disk, the database management system asks the computer operating system to physically read the disk. Most database management systems do not physically read disks. When the data has been read from the disk and placed into a buffer in the computer memory, the database management system again takes control from the computer operating system and sorts out the specific data that your program subschema allows you to retrieve. The data are sent back to the application program, and processing of the message continues.

When the message processing has been completed, the results are passed back to the teleprocessing monitor and the response is put in the host computer's output queue.

Next, the telecommunication access programs in the front end take the message. They select the remote terminal that originated the request to determine if it can receive the response. If the answer is yes, the response is transmitted to the terminal. At this time, the round-trip transaction has been completed.

Notice that this simple input transaction might have involved any of the five software packages (host/front end) and possibly an application program in the host computer. Furthermore, on the left side of Figure 6–1 (the remote terminal end), it may have involved any of the six software packages listed. For example, the transaction might have gone through 50,000 lines of program code as it passed through the remote software packages, telecommunication access programs, teleprocessing monitor, security package, application program, database management system, and the host computer operating system. At this point you might find it useful to compare Figure 4–1 (the basic technical concepts) with Figure 6–1 (the basic software concepts).

PROTOCOL/SOFTWARE/ARCHITECTURE

Before discussing the details of these three areas, we want to define their differences. It is a somewhat blurred distinction because protocols, software, and network architecture all work together. In fact, a full data communication network cannot operate without the interaction of these three.

Protocol This is a strict set of rules or procedures that are required to initiate and maintain communications. Protocols exist at many different layers in the network, such as link by link, end to end, subscriber to switch/packet switching networks. These layers will become evident in the next section, on Hierarchy of Protocols. Certainly the most famous protocols are the international packet switching protocol X.25 and IBM's protocols such as Synchronous Data Link Control (SDLC) and Binary Synchronous Communications (BSC).

Software These are programs that are located at various points in the network. For example, the host computer might have such software packages as the operating system, the teleprocessing monitor, database management systems, security packages, and application programs. At the front end communication processor there probably are telecommunication access programs for network control. At a remote switch/statistical TDM/concentrator there can be switching software, store and forward software, and control software packages that perform a subset of the telecommunication access program functions that were located at the front end communication processor. Further out in the network there might be software packages located at intelligent terminal controllers or within intelligent terminals themselves, and of course the whole range of software that might be available in a remote microprocessor.

Microprocessor chips have blurred the definition of software, which originally referred to computer programs. The term *firmware* is used increasingly to refer to that halfway point between hardware and software. Firmware is a microcircuit chip that contains the program functions to be performed. When the program functions are placed into an electronic chip, they can operate much faster and they are more secure from unauthorized change or modification.

Also, you can find protocols within various software, such as within the telecommunications access programs. This further blurs the separation between protocols and software.

Architecture System network architectures attempt to facilitate the operation, maintenance, and growth of the communication and processing environment by isolating the user and the application programs from the details of the network. Network architectures use both protocols and software in their operation. The architectures package the software and protocols together into a usable network architecture system. Some of the better known network architectures are System

Network Architecture (SNA) from IBM, Digital Network Architecture (DNA) from Digital Equipment Corporation, Burroughs Network Architecture (BNA) from the Burroughs Corporation, Distributed Communications Architecture (DCA) from Sperry Univac, and local area network architectures such as Ethernet from the Xerox Corporation.

HIERARCHY OF PROTOCOLS (SEVEN-LAYER OSI MODEL)

Most network architectures have little in common because they were designed around very different machines from different vendors. In 1977 the International Organization for Standardization (ISO) created the Open Systems Interconnection Subcommittee to develop an Open System Interconnection (OSI) reference model that could serve as a framework around which a series of standard protocols could be defined. The ISO is an international standards group whose membership comprises the national standards organization of each member country. The ISO and the CCITT (Consultative Committee on International Telephone and Telegraph) usually try to cooperate on issues of telecommunication standards. In July 1979 the seven-layer OSI model was born. It should be noted at this point that the CCITT is moving toward independently developing a separate seven-layer architecture as well. There are some functional differences between these two seven-layer architectures. The ISO standard seems to be the one that is most widely used.

The seven layers of the OSI model (sometimes called ISO standard) are in Figure 6–2. Notice how this figure maps the seven layers with the terms, protocol, software programs, application programs, and system network architecture.

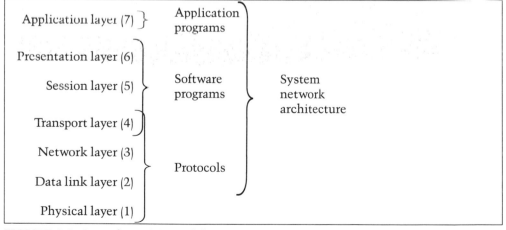

FIGURE 6–2 Seven-layer OSI model.

If you have two machines carrying on a dialogue, each layer in a machine carries on a conversation with an equal layer on another machine. In reality, the data are not transferred directly from, let's say, layer 4 to layer 4. Actually it is passed down through the layers, from 4 to 3 to 2 and on to layer 1, where it finally is transmitted over the physical communication circuits or links. There is physical communication between two machines only at the lowest layer (layer 1). Between each adjacent layer there is an interface that allows the data/information to be passed up and down through the seven layers (see Figure 6–3).

The purpose of these seven layers is to define the various functions that must be carried out when two machines communicate. For example, suppose you want to start a conversation with a friend. You probably would *not* carry out the dialogue by saying, "Hi, do you want to talk with me?" then wait for your friend's answer and, if it is yes, start a conversation. In a real-life situation the protocols you use involve looking at the person, using body language, or just beginning to speak. These protocols cannot be used in a data communication situation because the communication circuit has to be set up.

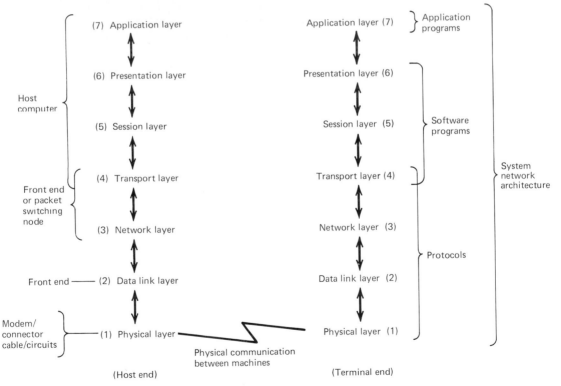

FIGURE 6–3 Communications using the seven–layer OSI model.

Recall how you use your telephone; you must first lift up the receiver, listen for a dial tone, dial the number, and listen for a response such as "Hello." Notice that, without even thinking about it, you went through a set of protocols to establish a communication link using a telephone. The reason you are unaware of all this with human interactions is that the establishment of a face-to-face communication link usually is done by unspoken body language or actions.

In order to understand the seven layers, and the functions performed by each layer, look at Figure 6–3 as you read the following sections. On the left side of this figure we show the piece of hardware that *usually* performs the functions from the host end. The right side shows the protocol/software layers from Figure 6–2. Remember that each layer of protocol/software performs its own specific functions and passes its portion of the message up or down, depending upon whether the message is heading toward layer 7 (the application layer) or layer 1 (physical layer).

Layer 1: Physical Layer   This layer is concerned primarily with transmitting data bits (0s or 1s) over a communication circuit. The main purpose of this layer is to define the rules by which one side (host) sends a 1 bit so that the other side (terminal) defines it as a 1 bit when it is received. This is the physical raw communication circuit layer. At this layer we are concerned with very basic things like voltages of electricity, timing factors such as 1200 bits per second's being equal to 833 microseconds per bit, full duplex or half duplex transmission, rules for establishing the initial connection, how to disconnect when the transmission is complete, and connector cable standards such as RS232C, RS449, or X.21.

At this layer we are concerned with how the physical, electrical, and functional interchange takes place that establishes, maintains, and disconnects the physical link between DTEs (Data Terminal Equipment) and DCEs (Data Circuit-Terminating Equipment).

Remember that layer 1 is the basic link over which all data pass. Communication between layers 2–7 at a host and layers 2–7 at a terminal are only virtual (appearing to exist) communications. In reality, the messages must be passed down to layer 1 (physical layer) for the actual movement of the message between the host computer and a remote terminal.

Layer 2: Data Link Layer   This layer manages the basic transmission circuit that was established in layer 1 and, it is hoped, transforms it into a circuit or link that is free of transmission errors. This error-free transmission link interfaces closely with layer 3, the network layer. The data link layer accomplishes its tasks by breaking up the input data into data frames, transmitting these frames, and processing acknowledgment frames that are sent back to acknowledge the received data. Figure 6–4 shows a typical frame format.

Flag 01111110	Address	Control	Message	Frame check sequence	Flag 01111110

FIGURE 6–4 X.25 frame.

Because layer 1 accepts and transmits only a serial stream of bits without any regard to meaning or structure, it is up to the data link layer to create and recognize frame boundaries and check for errors during transmission. As you can see, in Figure 6–3 the layer 1 protocol is handled in conjunction with the modem, connector cable standards, and communication circuits.

Layer 2 requires some intelligence and therefore probably is located in the front end communication processor at the host end of the circuit and in a remote intelligent terminal control device at the remote end of the circuit. It is up to this layer to solve the problems caused by damaged, lost, or duplicate frames so that the next layer (layer 3) will be working with error-free messages.

Another issue at layer 2 is how to keep a fast transmitter from drowning a slow receiver. Some mechanism or procedure must be employed to let the terminal that is transmitting know that the available buffer space at the receiver is filling to a critical level. This procedure and the error handling usually are integrated, although the problem of a terminal overrunning another terminal is also handled in some of the layers above layer 2.

Finally, some of the other functions that were discussed in regard to the physical layer also have to be acted upon in the data link layer, such as whether transmission is in full duplex or half duplex. Some of the typical data link level protocols are X.25, High-level Data Link Control (HDLC), Synchronous Data Link Control (SDLC), and of course, Binary Synchronous Control (BSC).

Layer 3: Network Layer This layer provides for the functions of internal network operations such as addressing and routing. In other words, it provides services that transport data through the network to its destination node/terminal. The network layer actually controls the operation of the combined layers 1, 2, and 3. This sometimes is called the *subnetwork* and/or the *packet switching network* function.

The control messages created by the network layer include line connection and termination requests, as well as message confirmations. Notice that (see Figure 6–3) a message confirmation might be created here, passed down to layer 2 where it is put into a frame, and then passed down to layer 1 where it is physically moved over the communication circuit. Packets are created in this layer from messages routed through the network. If a packet destined for another node is received, this layer reroutes the packet back down through layers 2 and 1 for transmission.

X.25 is the first internationally recognized data communication protocol and, as such, has been accepted as the network layer standard.

One major issue here is concerned with front end communication processors, packet switching nodes, and host computers. Which of these three physical hardware devices should ensure that all packets are received correctly at their destination and in the proper order? Even though the software takes care of this function, the problem lies in the fact that the software can be physically located in any one of these three devices.

Because this is the packet switching layer, the software accepts messages from a host computer, converts them to packets, and then sees to it that the packets get addressed and directed toward their destination. Routing of the packets is another task. There may be, at this layer, a database of routing tables used to keep track of the various routes a packet can take and to determine how many different circuits there are between any two individual packet switching nodes. Packet routing also involves load leveling the volume of transmission on any given circuit, as well as knowing whether a circuit has failed.

There can even be an accounting function built into layer 3. An example is one that keeps track of how many messages are transmitted for each organization so each group can be billed correctly. However, usually accounting functions are built into one of the higher layers of the seven-layer protocol scheme.

It should be noted that X.25 only defines the requirements of a connection between data terminal equipment (DTE) and a public data network; it is not a standard connection between sets of data communication equipment from different vendors. Each equipment manufacturer must implement X.25 (through software) on its own hardware.

Layer 4: Transport Layer   This layer often is called the *host-to-host layer*. The transport layer provides the facilities that allow end users to pass messages between themselves across *several* intervening stations/nodes. The lower layers (1–3) of protocol or other architectural features of a vendor's network architecture are transparent (invisible) to the user. Layer 4 includes facilities to do all user addressing, data assurance (control), and flow control of messages from source to destination (end to end) across simple or complex networks.

At this layer we may have moved out of the message protocols and into other software programs, peer-to-peer protocols between layers, and the vendor's network architectures. Probably the best known one here is IBM's Systems Network Architecture (SNA).

This layer might include the specifications for broadcast messages, datagram type services, accounting information collection, message priorities, security, response times, and a recovery strategy in case of failure.

The transport layer is a source-to-destination or end-to-end layer because a program at the source machine can carry on a virtual conversation with a similar program on a destination machine using message headers and control messages. The physical path still goes down to layer 1 and across to the destination machine though.

It should be noted that at the lower layers (layers 1–3) the protocols are carried out by each machine and its immediate neighbors, not by the ultimate source and destination machines. These source/destination machines always are separated by many other pieces of hardware such as front end communication processors, concentrators, multiplexers, message switches, modems, and the like. Layers 1 to 3 are chained together in a sequential fashion, whereas layers 4 through 7 are end to end or computer to remote terminal. The message actually is passed down from layer 7 to layer 1 and then transmitted.

The transport layer also can multiplex several streams of messages onto one physical circuit by creating multiple connections that enter and leave each host computer. The *transport header* delineates which message belongs to which connection. Also at this level there is a mechanism that regulates the flow of information so a very fast host cannot overrun a slower terminal or especially an overburdened host. The flow control at the transport layer is a little different from the flow control at the lower layers. The transport layer prevents one host from overrunning another by controlling the movement of messages, whereas in the lower layers the physical flow (speed) of packets or frames is controlled. In fact, a lower layer can hold back data that are sent out by layer 4. Today layer 4 is physically in the host computer, although it may be drifting slowly out toward the front end. Layer 4 functions usually are carried out by software, but they are fast becoming part of the firmware as we get standards that are more clearly defined.

Layer 5: Session Layer The session layer is responsible for initiating, maintaining, and terminating each logical session between end users. In order to understand the session layer, think of your telephone. When you lift the receiver, listen for a dial tone, and dial a number, you create a physical connection (layer 1). When you start speaking with the appropriate person at the other end of the telephone circuit, you are engaged in a session. In other words, the session is the dialogue that the two of you carry out.

In addition, this layer is responsible for managing and structuring all session-requested data transport actions. Session initiation must arrange for all the desired and required services between session participants. Required services could be logging on to circuit equipment, transferring files between equipment, use of various terminal types or features, security authenticators, the software tasks for half duplex or full duplex, and the like.

This layer provides for session termination, which is an orderly way to terminate the session. It also provides the facilities to abort a session prematurely through items such as a break key. The session layer also might keep track of various accounting functions so the correct party receives the bill later. It may have some redundancy built in to recover from a broken transport (layer 4) connection in case of a failure. This is the layer that ensures that, when a database management system is used, a transaction against the database is never aborted. This is a necessary precaution because an abort would leave the database in an inconsistent state; concurrence and deadlock may preclude further use of the database unless that specific transaction is carried to its completion. If there is an abort, software at the session layer has to restore the data to its original condition (before the database transaction) in case there is a break at the session level.

The session layer is very close to the transport layer, although it has more application-oriented functions than does the transport layer. Because the session layer usually is handled by the host computer operating system supervisors, it would be easy for the session and transport layers to be merged into a single layer.

Layer 6: Presentation Layer The presentation layer performs a selectable set of message transformations and formatting in order to present data to the end users. This layer has items such as video screen formatting, peripheral device coding, other formatting, encryption, and compaction.

Basically, any function (except those in layers 1–5) that is requested sufficiently often to warrant finding a general solution for it is placed in the presentation layer. Even though some of these functions can be performed by library routines, placing them in the library routines might overwork the host computer operating system, thus slowing throughput. More generally, different computers have different file formats, so a file conversion option might be useful as well as protocol conversion between incompatible hardware. Other items such as number of printed lines per screen, characters per line, cursor addressing, and the like can be handled by this layer.

Layer 7: Application Layer The application layer is the end user's access to the network and therefore is developed by the individual user organization. Each user program determines the set of messages and any actions it might take upon receipt of a message. There are other considerations at the application layer, such as network management statistics, remote system initiation and termination, network monitoring, application diagnostics, making the network transparent to users, simple processor sharing between host computers, use of distributed databases, and industry-specific protocols such as you might have in banking.

SUMMARY OF SEVEN-LAYER MODEL

Layer 1 involves the circuit interfaces and the standard connector cables. These interfaces allow a wide range of transmission rates on synchronous or asynchronous communication circuits. Layer 2 specifies the facilities for frame transfer between two network nodes/terminals. Layer 3 is the packet layer. It creates packets which go to layer 2 for framing and onto layer 1 for transmission.

Layers 4–6 are end-to-end functions which account for the virtual passing of messages between application programs, initiating, maintaining, and terminating sessions, and special functions that are requested sufficiently often to warrant finding a general solution for them. Finally, level 7 accounts for your own user application programs at your organization.

X.25

The most popular bit-oriented protocol is the international standard X.25. It defines the structure, contents, and sequencing procedures for the transmission of data among DTEs, DCEs, and a public data network. It also defines the techniques used for error detection and recovery.

If you look back in the section that described the seven layers of the OSI model, you should note that the X.25 protocol is involved only in layers 1–3. Layers 4–6 are more concerned with other software and network architecture, while layer 7 involves the user application programs.

Figure 6–4 shows a typical frame from the X.25 protocol. Each frame begins and ends with a special bit pattern (01111110). This is known as the *beginning* and *ending flag*. The beginning flag references the position of the *address* and *control* frame elements and initiates error-checking procedures. The ending flag terminates the error-checking procedures; when you have contiguous frames, it may also be the beginning flag for the next frame.

The *address field* is used to identify one of the terminals. For point-to-point circuits it is sometimes used to distinguish commands from responses or to address a specific terminal device on multifunction terminals. This eight-bit field might contain a station address, a group address for several terminals, or a broadcast address to all terminals.

The *control field* identifies the kind of frame that is being transmitted, such as information, supervisory, or unnumbered. The *information* frame is used for the transfer and reception of messages, frame numbering of contiguous frames, and the like. The *supervisory* frame is used to transmit acknowledgments, such as to indicate the next expected frame, to indicate that a transmission error has been detected, to acknowledge that all frames received are correct, to stop sending, and to call for the retransmission of specified frames. The *unnumbered* frame is used for other purposes, such as to provide a command "disconnect" that allows a terminal to announce that it is going down, or to indicate that a frame with a correct checksum has impossible semantics. Since frames used for control purposes may be lost or damaged, just like information frames, they must be acknowledged also. A special control acknowledgment frame is provided for this purpose; it is called an *unnumbered acknowledgment*.

The *message field* is of variable length. This is the user's message or request (data packet). This field may include a general format identifier, logical channel group numbers, logical channel numbers, packet type identifiers, internal message DTE addresses (calling DTE and called DTE), and, of course, the message that is being transmitted.

Notice that we have *two* types of sequence numbering here. In the previously mentioned control field we sequence number the individual contiguous frames. In the message field we sequence number the individual packets when the system breaks the message into multiple packets.

The *frame check sequence field* provides a 16-bit Cyclic Redundancy Checking (CRC) calculation that is placed within the field by the transmitting station. Upon receipt, the receiving station recalculates the CRC values and matches them in order to determine if there have been any errors during data transmission. If the CRC values do not match, a request for retransmission is sent to the sending terminal station/node.

In summary, X.25 handles call setup and termination, identifies the logical channel number chosen by the call initiator, addresses the called party, counts packets for charging methods, starts the transmission over a virtual circuit (remember that the transmission goes down to layer 1, the physical circuit), breaks the message into various packets/frames, prevents network overload by controlling the flow of packets, determines the path over which the packet is transferred, and provides various security features. These security features for packet switching are discussed in Chapter 8, Security and Control.

IBM's System Network Architecture with its protocol (Synchronous Data Link Control—SDLC) is compatible with the international standard X.25. The frames of SDLC are virtually equal to the frames of X.25.

BINARY SYNCHRONOUS COMMUNICATION (BSC)

The binary synchronous protocol is still in wide use because many organizations have not had the time or money to upgrade to the newer bit-oriented protocols such as X.25. BSC is a byte-oriented protocol because it takes an entire 8-bit byte to send a command signal to the receiving station. With a bit-oriented protocol (X.25), the changing of a single bit within a frame's control byte sends a command to the receiving station. BSC communication takes place in a two-way, half duplex transmission mode with 8-bit byte commands.

Figure 6–5 shows a BSC message format. The (SOH) start of heading (first character of message) is a fixed 8-bit character. The SOH is 10000001 in ASCII coding. It is followed by other header control characters, which are used by the system for terminal addressing or other control purposes. The start of text (STX—01000001) is another special 8-bit character, as is the end of text (ETX—11000000). These two characters sandwich the message text characters so they can be identified

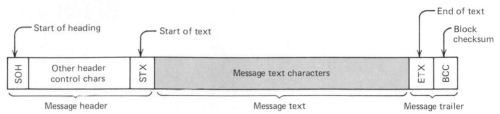

FIGURE 6–5 BSC message format.

easily. Finally, there is a block check character (BCC), a checksum at the end which is similar to the Cyclical Redundancy Check of X.25. The purpose of this BCC is to detect any errors that occurred during data transmission. The list of special control characters used in the binary synchronous protocol can be seen in Figure 3–1, ASCII code structure.

Notice that the BSC protocol does not put messages into tightly defined frames, nor does it handle any packet switching. It is an inefficient protocol because it is designed for half duplex circuits. BSC does not support loop circuits and mixing of terminal device types. Different code structures cause serious problems. For example, the BSC protocol was designed to be used with Extended Binary Coded Decimal Interchange Code (EBCDIC), American Standard Code For Information Interchange (ASCII), or a 6-bit, 64-character code. BSC does not support hub go-ahead polling or some of the fast select polling scenarios.

Under BSC, the initiation of an error recovery procedure at one terminal station can cause line unavailability to other terminal stations until the recovery procedure is complete.

BSC half duplex operation works on the basis of sending one message block to a distant terminal. The terminal then must acknowledge the successful or unsuccessful receipt of that block before another message block can be transmitted.

BSC does not provide as complete an error checking for all control and information transfer messages as does X.25. It does not provide any system-assigned block sequence numbering except for odd/even acknowledgments. BSC cannot be used on satellite circuits because the propagation delay time affects the stop and wait ARQ (see next section). Messages are delayed too long on satellite circuit transmission when you have to send a message block and receive an acknowledgment before you can send the second message block. This is because there is an approximate 0.5 to 0.6 second round-trip delay for propagation time on a satellite circuit.

ARQ (AUTOMATIC REPEAT REQUEST)

A system that detects an error in data and has it retransmitted automatically is called an ARQ (automatic repeat request) system. ARQ systems are of two types: *stop and wait* ARQ and *continuous* ARQ.

With *stop and wait ARQ*, after sending a block, the transmitting terminal waits for a positive acknowledgment (ACK) or a negative acknowledgment (NAK). If it is an ACK, the terminal sends the next block; if it is a NAK, the terminal resends the previous block. Another possible response is a WAK, which means positive acknowledgment but do not transmit any more at this time.

With *continuous ARQ*, the transmitting terminal does not wait for an acknowledgment after sending a block; it immediately sends the next block. While the blocks are being transmitted, the stream of returning acknowledgments is examined by the transmitting terminal. If a NAK is received, the transmitting terminal usually (depending upon the specific vendor's protocol programs) retransmits all the blocks from the one that was in error to the end of the stream of blocks sent. The terminal also may be able to retransmit only the block that was in error, although this method requires more logic and buffering at the terminals. Continuous ARQ requires a full duplex circuit or a reverse channel modem.

Usually the BSC protocol uses the stop and wait ARQ. The newer bit-oriented protocols, like X.25, use the continuous ARQ error recovery procedures.

SOFTWARE

Figure 6–1 listed the various types of software. Now we can discuss a couple of these packages that are directly related to data communications. These are the telecommunication access programs and the teleprocessing monitors.

The *telecommunication access programs*, which used to reside in the host computer, now reside primarily in the front end communication processors, the switching nodes (SN) of a packet network, the remote intelligent controllers, and/or microprocessors. This software package provides for some of the following capabilities:

- Polling and selecting of terminals in a central control network.
- Automatic dial-up and answering of calls.
- Code conversion.
- Message switching and store and forward (although this might be in the host computer's teleprocessing monitor).
- Circuit switching and port contention.
- Logging of all inbound and outbound messages.
- Error detection and the ordering of retransmission when an error is detected (ARQ systems).

The list of functions for the front end communication processor (enumerated in Chapter 2) can provide other ideas as to the type of software functions that are performed in many of today's front end communication processors.

As has been stated, some functions overlap between the telecommuncation access programs of front ends and the teleprocessing monitors within the host computer. IBM is a good example of this. The IBM software package Telecommunication Access Methods—TCAM (this is a telecommunication access program) and the IBM package Customer Information Control System—CICS (this is a teleprocessing monitor) both reside in the host computer. IBM's front end has its own set of programs called Network Control Programs— NCP (telecommunication access programs). NCP and TCAM overlap; therefore, functions such as polling/selecting can be performed from either the host computer or the front end communication processor.

Teleprocessing monitors are software programs that directly relieve the operating system of many of the tasks involved in handling message traffic between the host and the front end or the host and other internal CPU software packages (such as the database management system within the host). Generally speaking, teleprocessing monitors can perform functions such as line handling, access methods, task scheduling, system recovery, and the like.

The telecommunication access programs need to access and move data into and out of the host computer. The teleprocessing monitor must interface with the telecommunication access programs on one side, but also must interface with all of the host computer's software on the other side. Teleprocessing monitors must interface with various network architectures, operating systems, computer architectures, database management systems, security software packages, and application programs.

Typical teleprocessing monitors offer features such as ability to interface with X.25 and other protocols, ability to interface with various operating systems and hardware, and interfaces to database management systems and telecommunication access programs.

Teleprocessing monitors should offer some sort of security, logging, accounting procedures, failure recovery procedures, ability to interface with multiple front ends, terminals, and various communication codes. The monitor provides input/output job task queue management, various methods of instituting priorities for certain transactions or jobs, file/database management, application program management, task and resource control, restart and recovery procedures in case of a failure, special utilities that will perform tasks that are done often enough to warrant setting them up as a utility feature (OSI model layer 6) or in a program library. It will keep track of various accounting features and operating statistics, and it can isolate various programs or various parts of the system from other programs or other parts of the system. In other words, a teleprocessing monitor can be considered a "mini" operating system with data communication interfaces.

The teleprocessing monitor will increase in importance and almost equal the operating system for performing job tasks as we move toward more on-line data communication-oriented networks, and especially as we move into the world of distributed systems and databases. Because of rapid growth in this area, there is an increasing need for guidance on how to evaluate a teleprocessing monitor. For this reason, a step-by-step plan for evaluating teleprocessing monitors is in Appendix 6.

EXAMPLES OF TELECOMMUNICATION ACCESS PROGRAMS

The most familiar communication software packages are the telecommunication access programs of the IBM 360/370 series, some of which have been in use since the late 1960s. They cover a range of capabilities. There are four such packages, which are presented as examples of the range of products with which the designer can work:

- Basic telecommunication access method (BTAM).
- Queued telecommunication access method (QTAM).
- Telecommunication access method (TCAM).
- Virtual telecommunication access method (VTAM).

The *basic telecommunication access method* (BTAM) provides the basic functions needed for controlling data communication circuits in IBM 360/370 systems. It supports asynchronous terminals, binary synchronous communications, and audio response units. BTAM is a set of basic modules that may be used to construct communication programs. It is recommended for use where there are ten or fewer circuits to support or when a specialized communication control program is required. BTAM requires a knowledge of the terminal's operation, link discipline, and a basic knowledge of programming. A BTAM user must write routines for the scheduling and allocation of facilities. The basic flow control and data administration routines are also the responsibility of the BTAM user. It is the least sophisticated of the four data communication software programs listed above, but it does contribute the lowest system overhead. BTAM provides facilities for polling terminals, transmitting and receiving messages, detecting errors, automatically retransmitting erroneous messages, translating code, dialing and answering calls, logging transmission errrors, allocating blocks of buffer storage (OS360/370 only), and performing on-line diagnostics to facilitate the testing of terminal equipment. BTAM resides in the central computer and is the interface between the front end communication processor and the user-written application programs.

The *Queued Telecommunication Access Method* (QTAM) is an extension of BTAM and includes all the BTAM facilities except that it does not support binary synchronous communications; therefore, it supports only asynchronous terminals. QTAM provides a high level and flexible macrolanguage for the control and processing of communication data, including message editing, queuing, routing, and logging. It can schedule and allocate facilities, poll terminals, perform error-checking routines, reroute messages, cancel messages, and the like. QTAM is not utilized much any more and has largely been replaced with TCAM.

The *Telecommunication Access Method* (TCAM) replaces and extends the older QTAM. The most significant features of TCAM are those for network control and system recovery. An operator control facility is also provided for network supervision and modification. It supports asynchronous terminals, binary synchronous communications, and audio response units. TCAM performs all the functions of BTAM and QTAM and handles the data communications in a system that utilizes a high degree of multiprogramming. Unlike the prior basic data communication software, TCAM has its own control program that takes charge and schedules the traffic handling operations. In some cases it can handle an incoming message by itself without passing it to an application program—for example, routing a message to another terminal in a message switching system. TCAM also provides status reporting on terminals, lines, and queues. It has significant recovery and serviceability features to increase the security and the availability of the data communication system. The checkpoint and restart facilities are much more capable than those of QTAM. TCAM has prewritten routines for checkpointing, logging, date and time stamping, sequence numbering and checking, message interception and rerouting, and error message transmission, and it supports a separate master terminal for the data communication system operator. TCAM can manage a network structured on System Network Architecture (SNA). ACF/TCAM also accommodates SNA/SDLC terminals.

The *Virtual Telecommunication Access Method* (VTAM) is the data communication software package that complements IBM's advanced hardware and software. VTAM manages a network structured on SNA principles. It directs the transmission of data between the application programs in the central computer and the components of the data communication network. It operates with front end communication processors. The basic services performed by VTAM include establishing, controlling, and terminating access between the application programs and the terminals. It moves data between application programs and terminals and permits application programs to share communication circuits, communication controllers, and terminals. VTAM controls the configuration of the telecommunication network and permits the network to be monitored and altered in addition to performing all the basic functions of the other three data communication software packages.

In summary, BTAM is the least sophisticated access method and, to be effective, requires the most effort on the part of a user. QTAM has been replaced with TCAM. TCAM provides the facilities for a complete communication system. It supports a wide range of terminals, provides network control and significant recovery facilities, and it now supports SNA. VTAM is the most advanced data communication software package and is intended to support the virtual computer systems using SNA.

SELECTED NETWORK ARCHITECTURES

You will recall that network architectures are involved at layers 1 through 7 of the OSI model while individual message protocols are mainly at Levels 1 through 3 (see Figure 6–2). System network architectures attempt to facilitate operation, maintenance, and growth of the communication and processing environment by isolating the user and the application programs from the details of the network. An ideal general purpose data communication network architecture should have the following characteristics:

- It should be adaptable for multivendor network use.
- It should not make any restrictions on the network configurations (topology) and it should allow for easy changes in configuration.
- The end user should not be required to know how network functions are implemented.
- It should be both program and device location independent.
- It should allow for interprocess or task-to-task communications.
- It should detect all errors and recover from them whenever possible.
- It should be able to recover from hardware/software failures.
- Users should not have to perform any complicated control procedures.
- It should be developed in a systematic layered manner; strict interfaces and protocols should define the interaction of any given layer with its adjacent corresponding layer (seven-layer model).

The selected architectures that will be discussed below are Digital Network Architecture (DNA), Distributed Systems Environment (DSE), Distributed Communications Architecture (DCA), Burroughs Network Architecture (BNA), Distributed Network Architecture (DNA), Distributed Systems (DS), Public Networks, Local Network Architecture (previously described in Chapter 5), and IBM's System Network Architecture (SNA). Because it was mentioned earlier, the new UNIX operating system is also briefly described.

DIGITAL NETWORK ARCHITECTURE (DNA)

DNA is Digital Equipment Corporation's distributed network architecture. It is called Decnet and has five layers. The physical layer, data link control layer, transport layer, and network services layer correspond almost exactly to the lowest four layers of the OSI model. The fifth layer, the application layer, is a mixture of the OSI model (see Figure 6–2) presentation and application layers. Decnet does not contain a separate session layer.

Decnet, like IBM's SNA, defines a general framework for both data communication networking and distributed data processing. The objective of Decnet is to permit generalized interconnection of different host computers and point-to-point, multipoint, or switched networks in such a way as to permit users to share programs, data files, and remote terminal devices.

Decnet supports the X.25 international protocol standard and has packet switching capabilities. An emulator is offered that permits Digital Equipment Corporation systems to be interconnected with IBM mainframes running in an SNA environment. The digital data communications message protocol (DDCMP) is Decnet's byte-oriented protocol which is similar in structure to IBM's Binary Synchronous Communications protocol (BSC).

DISTRIBUTED SYSTEMS ENVIRONMENT (DSE)

DSE is Honeywell's concept in communication processing/architecture. It is not a rigid framework for network implementation. Any two Honeywell processors remotely connected can be considered a DSE system. This network includes a host, a front end processor, one or more remote host processors, and various terminals. High-level Data Link Control (HDLC) is the standard protocol for interprocessor links. HDLC is a bit-oriented, synchronous protocol and is implemented in accordance with the ISO standard. Because this is a distributed processing architecture, some of the application processing is performed at the user site, thus relieving the host computer of some of its processing burden. In addition, this architecture supports distribution of databases and database management functions.

DISTRIBUTED COMMUNICATIONS ARCHITECTURE (DCA)

DCA is Sperry Univac's network architecture and it embraces a concept of distributed networking similar to Honeywell's DSE philosophy. DCA attempts to relocate communication control logic away from the host computer. One of the distinguishing features of DCA is its ability to support an environment consisting of non-Univac terminals, processors, and networks. DCA is compatible with IBM terminal handlers and has protocol compatibility with X.25 packet switching services. Univac's standard protocol is the Universal Data Link Control (UDLC), which is a bit-oriented, synchronous link protocol based on the ISO standard.

BURROUGHS NETWORK ARCHITECTURE (BNA)

BNA consists of two major layers. The *host services* layer provides logical access to the communication network. The *network services* layer provides physical access to the communication system.

The *host services layer* provides the user with a transparent interface to the network system by making use of an enhanced version of the Master Control Program (MCP). The master control program is the teleprocessing monitor and control interface to the host computer operating system. The *network services layer* has three distinct levels: *port level, router level,* and *station level.* The *port level* contains several system ports with a port level manager to allocate and deallocate the ports as needed. The ports provide individual access to the network services layer and the underlying network through the host services layer. It is this level that is responsible for breaking a message into a series of packets, for coordinating message sequencing, and for chaining a series of messages. The *router level* is responsible for routing as well as storage and forwarding of various packets. The router has a database of active hosts used to determine required routing information. The *station level* manages the physical interface between adjacent hosts. The station module checks packets for errors and requests retransmission of a packet if there is an error.

Since either BDLC (Burroughs Data Link Control) protocol or an X.25 protocol may be used, connection to an X.25-compatible public data network is supported.

DISTRIBUTED NETWORK ARCHITECTURE (DNA)

The NCR Corporation has a network architecture called Distributed Network Architecture (DNA). It operates through two software modules, the *telecommunication access method* and the *data transporting network.*

The *telecommunication access method* software provides all queuing and addressing of information functions.

The *data transporting network* has three separate logical layers: (1) communication system services, which manage all messages passing between systems by breaking them into packets and passing them, in a controlled manner, to the route manager; (2) the route manager, which routes each packet dynamically through the system; and (3) the data link control layer, which then builds transmission packets according to an NCR protocol (Synchronous Data Link Control protocol) or X.25 protocols.

This network architecture supports point-to-point or multipoint connections.

It should be noted that the NCR Corporation acquired Comten, an advanced company that produces IBM-compatible front end communication processors. The DNA philosophy has been altered since then and is now more generally referred to as the Communication Network Architecture (CNA). Many further enhancements can be expected from NCR because the front end manufacturer, Comten, was a very advanced fourth-generation front end manufacturer.

DISTRIBUTED SYSTEMS (DS)

Distributed Systems is Hewlett-Packard's distributed network system. DS allows the flexibility of adding an optional front end communication processor to the communication nodes as a means of removing much of the communication processing load from the host computer. The hardware interfaces are all microcoded (firmware). This microcode software builds the transmission protocol units and forms the lowest level (the communications access method) in the DS network architecture. This layer is accessed by the communication management layer, which provides all local system management routing and queuing functions.

There are two higher-level layers which effectively create a master-to-slave relationship between any message generating and receiving node. In a *master-to-slave* relationship, when a slave terminal wants to talk, it must first ask permission from a master terminal which initiates the session. In a *master-to-master*, any terminal can initiate a session. As IBM moves away from the master-to-slave relationship in its SNA, Hewlett-Packard will also, because the X.25 protocol uses a master-to-master rather than a master-to-slave relationship.

PUBLIC NETWORKS

In many countries both the government and private companies have developed and offer networking services on public packet switching networks. The network is owned by the vendor, which provides communication services for its customers' host computers or data terminals. A public network interface involves an intelligent switching node (SN). Most of the public network vendors have agreed upon the standard protocol X.25. As you already know, the X.25 protocol involves the first three levels of the ISO standard. This guarantees that messages from data terminal equipment (DTE) will get to, and be handled by, the public packet switching network. That of course leaves levels 4 through 7 to be handled by you with your own software or vendor's software such as teleprocessing monitors.

Interfacing with the public packet switcher can be through an international standard known as X.3, packet assembler/disassembler (PAD). The standard X.3 describes the functions of a PAD. A PAD accepts strings of characters from non-X.25 terminals and assembles them into packet format for transmission into the packet network. Packets arriving at a PAD are disassembled and delivered, one character at a time, to a non-X.25 terminal.

Other standards in this area are the following: X.28 defines the interface between a nonintelligent terminal and a PAD; X.29 defines the procedure for governing the exchange of data between a PAD and a packet mode terminal; X.75 governs the interface between public data networks and a gateway (entry point to the network).

Gateways may become one of the big items of hardware/software in the future. For example, satellite gateways to public networks can replace front ends and message switches, as well as concentrators, PBX/CBXs, and protocol converters. The purpose of a *gateway processor* (it is a specialized front end/protocol converter) is to connect two different networks. A typical gateway processor might connect a local network to an X.25 packet switched public network. Another gateway could connect a local network to the appropriate satellite links. Gateways can talk to a terminal network on one side and a host computer on the other. Networks themselves are quite limited in use unless the various devices/hosts can be interconnected. Gateways are another way of interconnecting between two network architectures and/or public packet networks. Figure 5–19 showed an X.75 gateway.

LOCAL NETWORK ARCHITECTURE

Local network architecture was described thoroughly in Chapter 5 in the section on Local Area Networks (LAN).

SYSTEM NETWORK ARCHITECTURE (SNA)

SNA is a network specification for a distributed data processing network. Because it is a systems architecture, it includes the software and protocols necessary to operate as an independent data communication network.

This section will describe IBM's System Network Architecture in a little more detail than the previous architectures because it is the most used network architecture today. Also, as was stated previously, it is now compatible with the international standard protocol X.25.

The appropriate place to begin understanding the concept of SNA is to look at it from the viewpoint of the end user.

The end user (terminal operator) talks to the network through a Logical Unit (LU). These logical units are implemented as program code or microcode (firmware) and they provide the end user with a point of access to the network. The program code or microcode can be built into the terminal or it can be implemented into an intelligent terminal controller/concentrator/remote front end.

Before one end user of an SNA network can communicate with any other end user, each of their respective logical units must be connected in a mutual relationship called a *session*. Because a session joins two logical units, it is called a *LU-LU session*. Figure 6–6 depicts the inner connection of logical units when one end user wants to talk to another.

The exchange of data by end users is subject to a number of procedural rules (protocols) that the logical units specify before beginning the exchange of information. These procedural rules specify how the session is to be conducted, the format of data, the amount of data to be sent by one end user before the other end user replies, actions to be taken if errors occur, and other types of protocols that usually are handled by body language when two human beings converse.

Each logical unit (LU) in a network is assigned a network name. Before a session begins, the SNA network determines the network address that corresponds to each LU network name. This scheme allows one end user (for example, a terminal operator) to establish communication with another end user (for example, an application program) without having to specify where that end user is located in the network. These network names and addresses are used for addressing messages.

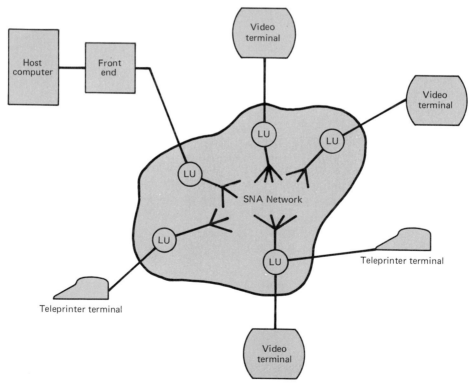

FIGURE 6–6 SNA session.

The flow of data between users actually moves between two logical units in a session. This flow of data moves as a bit sequence (frame) and generally is referred to as a *message unit*. The message unit also contains the network address of the logical unit that originated the message and the logical unit that is to receive the message. These are the basic protocols at work. Refer back to Figure 6–4, which depicted the X.25 frame (this is identical to the IBM Synchronous Data Link Control—SDLC frame).

A session between a pair of logical units is initiated when one of them (the end user) issues a request to send a message. Once a session has been activated between a pair of logical units, they can begin to exchange data. This is where the basic IBM protocol, Synchronous Data Link Control (SDLC) handles the movement of the data in order to have an orderly data flow. SDLC controls such items as the rate at which data flows, whether the sending LU expects to receive a response after every message unit (frame) it sends, sequencing, and the like.

A session between a pair of logical units is deactivated when one of them sends a deactivation request or when some other outside event interrupts the session. This outside event could be intervention by a network operator or the failure of some part of the network.

The logical organization of an SNA network, regardless of its physical configuration, is divided into two broad categories of components. The first category is called *network addressable units* and the second category is called the *path control network*.

Network addressable units are sets of SNA components that provide services which enable end users to send data through the network and help network operators perform network control and management functions. Physically, network addressable units are hardware and programming components within the terminals, intelligent controllers, and front end communication processors. Network addressable units communicate with one another through the path control network.

The network addressable units in SNA contain three kinds of addressable units. The *first* has already been introduced and it is called a *logical unit* (LU). The *second* category of network addressable units is the *physical unit* (PU). This is not truly a physical device; it is a set of SNA components that provide services to control communication links, terminals, intelligent controllers, front end communication processors, and host computers. Each terminal, intelligent controller, front end processor, and the like contains a physical unit which represents the terminal, intelligent controller, and the like to the SNA network. The third kind of network addressable unit is the *System Services Control Point* (SSCP). This also is a set of SNA components, but its duties are broader than those of the physical units and logical units. Physical units and logical units represent machine resources and end users, whereas the SSCP manages the entire SNA network or a significant part of it called a *domain*.

Just as sessions exist between logical units, sessions can exist between other kinds of network addressable units such as SSCP-LU. Even an SSCP-PU session is possible. Figure 6–7 shows the location of LUs, PUs, and SSCPs.

System network architecture defines a *node* as a point within an SNA network that contains SNA components. For example, each terminal, intelligent controller, and front end communication processor that is designed into the SNA specifications can be a node. Each SNA node contains a physical unit (PU) that represents that node and its resources to the System Services Control Point (SSCP). When the SSCP activates a session with a physical unit (SSCP-PU session), it makes the node (terminal, intelligent controller, or front end communication processor) that contains that physical unit an active part of the SNA network. It is convenient to think of an SNA node as being a terminal, intelligent controller, or front end communication processor within the network. Certain more powerful nodes also can have an SSCP.

The *path control network* provides for routing and flow control; transmitting data over individual links is the major service provided by the data link control

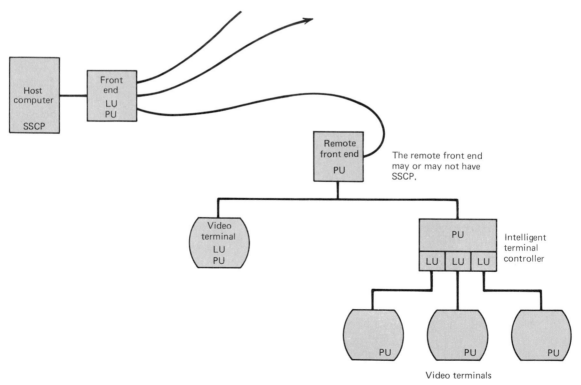

FIGURE 6–7 SNA SSCP/LU/PUs.

layer within the path control network. In the path control network, logical units must establish a path before a LU-LU session can begin. Eash SSCP, PU, and LU has a different network address, which identifies it to other network addressable units and also to the path control network. Path control provides for the following:

- Virtual routing so all sessions send their messages by different routes.
- Transmission priorities.
- Multiple links to maximize throughput.
- Message pacing to keep a fast transmitter from drowning a slow receiver.
- The ability to detect and recover from errors as they occur.
- Facilities to handle disruption because of a circuit failure.
- Facilities to inform network operators when there is a disruption in the network.

The path control network has two layers (*path control layer* and *data link control layer*). Routing and flow control are the major services provided by the path control layer, whereas transmitting data over individual links is the major service provided by the data link control layer. It is in this area that the Synchronous Data Link Control (SDLC) is used by links for serial bit-by-bit transmission of frames. SDLC is a discipline for the management of information transfer over data communication links. The SDLC function includes the following activities (layer 2 of the OSI model):

- Synchronizing or getting the transmitter in step with the receiver
- Detecting and recovering from transmission errors.
- Controlling the sending and receiving stations.
- Reporting improper data link control procedures.

The SDLC procedures take each message and sandwich it into a frame for transmission. In the SDLC concept, the frame is the vehicle for every command and response and for all information that is transmitted using SNA. Figure 6–4 depicted the X.25 frame, which is identical to the SDLC frame. All messages are put into this frame format for transmission from one node to another node (PU-PU). The error checking for each message is incorporated in the frame check sequence portion of the SDLC frame.

In summary, SNA describes an integrated structure that provides for all modes of data communication and upon which new data communication networks can be planned and implemented. It is a network architecture that incorporates protocols and various software packages. SNA encompasses distributed functions in which many network responsibilities can be moved from the host computer to other network components such as the front end communication processor or remote intelligent control units or terminals. SNA describes paths between the end users (LU-LU). This allows multiple network configurations. SNA uses the principle of device independence which permits an application program to communicate with an input/output device without regard to any unique device requirements. This also allows application programs and communication equipment to be added or changed without affecting other elements of the communication network. SNA uses standardized functions and protocols (now compatible with the international standard X.25) between logical units and physical units for the communication of information between any two points. This means that there can be one architecture for general purpose network systems which can use terminals of many varieties interconnected through one network protocol.

A data communication network built on SNA concepts can be considered to consist of the following (see Figure 6–8):

- A host computer.
- A front end communication processor (intermediate node).

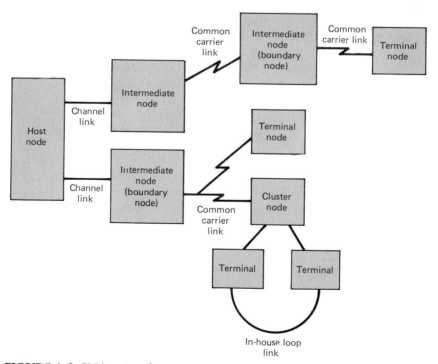

FIGURE 6–8 SNA network.

- Remote intelligent controller (intermediate node or boundary node).
- A variety of general purpose and industry-oriented terminals (terminal node or cluster node).

Readers who would like to obtain a more detailed explanation of Systems Network Architecture should order IBM Manual GC30-3072, entitled "Systems Network Architecture: Concepts and Products."

UNIX

UNIX is a general purpose, interactive, timesharing *operating system* which was developed by Bell Laboratories. The UNIX operating system has become more popular as the power of minicomputers and microprocessors increased and their cost decreased. Five years ago there were more than 600 installations running the UNIX system. In November 1981, Bell Labs released System III as the first "commercial" UNIX system. UNIX may be the operating system of the future, with expectations that it will become the standard operating system for 16-bit and 32-bit microcomputers. During 1982 it was reported that approximately 90 percent of all university computer science departments license the UNIX system.

The UNIX operating system is a candidate to control local network applications of the future. With regard to office automation, local networks allow different functions in the automated office to communicate with one another and to perform in an integrated manner. The UNIX operating system is well suited to support office automation because it can be transported easily from host computer to minicomptuer and on to microprocessors. In addition, UNIX can control many kinds of hardware, including terminals, copying machines, facsimile devices, and typesetting machines.

Office automation can be viewed as having four major aspects: data processing, administration of databases, automation of office administrative functions, and data communications. In an integrated office system these four areas must be combined to allow efficient performance.

With UNIX, the end user can control multiple operations on different hardware/terminals. It allows communication between many different brands of computer hardware ranging in size from microprocessors to host computer mainframes.

The UNIX system contains the networking and telecommunications software necessary to tie many brands and types of computers together and it allows the prospective end user to deal with a common language, regardless of the size of the computer.

Also of interest to the data communication user is the ability to use the output of an executing program as the input to another program. This information is transferred through software channels called PIPES. A special kind of PIPE, called a TEE, can be used to allow output data to be transferred simultaneously to two other programs. A FILTER, which is an extension of PIPEs and TEEs, is a feature that can selectively extract information from one file, modify it, and send the result to another file or into a PIPE. A typical use of a FILTER is to help a data communication user monitor data from several nodes in a local network.

UNIX users need to become familiar with only six basic groups of commands: User Access Control, File Manipulation, Manipulation of Directories and File Names, Running Programs, Status Inquires and Communication. Here we will discuss only communication.

All of the commands in the Communication group are of importance to the prospective data communication-oriented UNIX user. The three most commonly used commands are: MAIL, WRITE, and MESG. The MAIL command is used to mail a message to one or more users. It also can be used to read and dispose of incoming mail. At log-in the user is advised that there is mail. The WRITE command allows the user to establish direct terminal communication with another user. The MESG command will inhibit receipt of messages from the WRITE command when invoked by the user.

The most powerful commands in the Communication group are the CU and UUCP commands. The CU command is used to communicate with another timesharing system, even though the remote computer system may not be running under UNIX; the user interface is transparent. This command allows the user to transmit ASCII files to the remote machine at standard bit rates (the default is 300 bps) or to engage in interactive conversation. In addition the CU command allows the user to take remote input from a file on the local system or to place remote output into a file on the local system. The UUCP command is used for file transfer between two UNIX systems. It provides automatic queuing until a circuit is available and the remote machine is ready to receive data. The syntax of the command is quite complex, and the available options may vary depending upon which UNIX version is being used and how much it has been customized.

QUESTIONS

1. Identify the software packages that were referred to as being located at the remote end of a data communication circuit.

2. In a modern day communication network, is there any difference between the programs in the host computer and those in the front end communication processor?

3. Can you define the difference between a protocol, software, and network architectures?

4. Describe the seven layers of the OSI model.
5. Summarize the basic functions of the X.25 protocol.
6. What is the biggest drawback of the binary synchronous communication (BSC) protocol?
7. Which protocols use a continuous automatic repeat request (ARQ)?
8. When system network architecture (SNA) is used, can a remote front end communication processor have all three addressable units such as LU, PU, and SSCP?
9. In UNIX there are specific communication commands. Can you name them?

PART TWO

NETWORK MANAGEMENT AND SECURITY

Part Two of this book is devoted to the human and organizational aspects of managing networks as well as security requirements and the controls that are necessary.

7

MANAGEMENT OF NETWORKS

This chapter discusses the basic management skills required to be a successful network manager. It also describes departmental functions, how to manage the department, required reports, error testing, and test equipment. Network control and troubleshooting are the overall themes of this chapter.

THE DATA COMMUNICATION FUNCTION

In both government and private business organizations, one of the major growth sectors within the organization has been the internal "service function." The service functions have grown rapidly and the management techniques need to be strengthened. These service functions include: staff assistance, research departments, planning groups, coordinators, data processing departments, data communication networks, and the like. Such functions are organized to support manufacturing, sales, or the specific product for which the organization was originally conceived.

The data communication function has the primary responsibility for moving and conveying data/information that has been developed internally. This transfer of information takes place within or between departments of an organization. Remember that data are nothing more than meaningless characters, whereas information takes these meaningless characters and assembles them into a fact or idea that can be used for decision making by managers. Information presupposes adequate communication because information is useless if it is not available when needed.

The manager of a data communication function should always remember that data or information transmitted over any network must CATER to the overall needs of its users. CATER is an acronym that stands for *consistent, accurate, timely, economically feasible,* and *relevant.* While the manager of the data communication function may not have direct responsibility for *consistency* or *relevance* (that is the responsibility of the data or information owner/gatherer/developer), the data communication manager is responsible for ensuring *accuracy, timeliness,* and *economic feasibility.*

Information activities present a special organizational problem because they have to be both centralized and decentralized. The original developers/gatherers of data may be either centralized or decentralized, but the total organization is best served if the data communication function is centralized. This may seem to be a dichotomy in today's world, because we are moving quickly toward distributed data processing and distributed databases. Nevertheless, the data communication function should be centralized. The data communication function can be likened to the nervous system of the human body because it controls the paths over which all control messages and data/information flow. When it is viewed in this manner, it is obvious why centralized control is necessary in order to interconnect all of the various terminals/CPUs/databases, and so on.

The individual manager who is responsible for the data communication function must be able to adeptly perform the five key management tasks of *planning, organizing, directing, controlling,* and *staffing.* These functions require the following expertise:

- Planning activities require . . .
  Forecasting
  Establishing objectives
  Scheduling
  Budgeting
  Developing policies
- Organizing activities require . . .
  Developing organization structure
  Delegating
  Establishing relationships
  Establishing procedures
- Directing activities require . . .
  Initiating activities
  Decision making
  Communicating
  Motivating
- Controlling activities require . . .
  Establishing performance standards
  Measuring performance
  Evaluating performance
  Correcting performance
- Staffing activities require . . .
  Interviewing people
  Selecting people
  Developing people

NETWORK ORGANIZATION

Network organization, as discussed here, focuses upon the management and organization of the *people* running the network rather than upon the physical organization of the network communication circuits. Management must define a central control philosophy with regard to the overall network functions. This means that there is a "single" control source for all emergency problems, testing, and future planning. The details on how to implement this policy are discussed in the next section, Network Management.

The data communication network organization should have a written charter that defines its mandate, operational philosophy, and long-range goals. These goals must conform both to the parent organization's information processing goals and to its overall organizational goals. Along with its long-term policies, the organization must develop the individual procedures that implement the policies. These policies and procedures provide the structure that guides the day-to-day job tasks of people working in the data communication function.

The ultimate objective of the data communication function is to move data from one location to another in a timely fashion, and to provide and make available the resources that allow this transfer of data. This major objective, however, is sacrificed all too often to the immediacy of problems generated by factors thought to be outside the control of management. Such factors might be problems caused by unexpected circuit failures, pressure from end users to meet critical schedules, unavailability of certain equipment/circuits, or insufficient information (on a day-to-day basis) to ensure that the network is providing adequate service to all users. In reality, network managers must develop their own decision-making information in order to perform such essential tasks as measuring network performance, identifying problem areas, isolating the exact nature of problems, restoring the network (how to do this is discussed in the next section), and predicting future problems.

Too many managers spend too much time on the management function of *control* because they must contend daily with a series of breakdowns and immediate problems. These managers do not spend enough time on the management functions of *planning* and *organizing* in order to develop a proper information base so they can foresee problems and thereby reduce the need to drop everything to fix a breakdown (sometimes called firefighting).

Another major organizational problem is the area of voice communications versus data communications. Characteristically, voice communications are handled by a manager who has always handled the telephone system, whereas data communications are handled by people within the data processing function. This separation can cause enormous political problems within an organization because neither manager wants to relinquish his or her position or have his or her staff merged into another department. We cannot present a fixed solution to this problem because it must be studied individually within each organization. We can state unequivocably that an organization that avoids combining these two departments will have inefficient communication systems, lower employee productivity, and the associated higher costs of operation. While you might quickly gloss over this statement, we predict that by 1990 the total cost of both voice and data communication will equal or exceed the total cost of data processing. We are moving from an era where the computer is "king" to one in which the communication network is "king."

We have alluded to two separate problems that directly affect organization charts: whether data communications is placed in the data processing function, and whether voice and data communications are combined. If data communications is placed in the data processing department, the manager of the data communication function should report directly to the highest-ranking data processing director or manager. When this function is placed under another section of the data processing department, such as computer operations, it has neither the budgetary or management control authority needed to provide communication networks that will CATER to the needs of the total organization.

The second area involves how to handle the management and control of the total communication functions when voice and data communications are combined. The choice might be either to have this communications manager report to the highest-ranking data processing director or to move the function entirely out of data processing and create a new high-level position, such as vice-president of voice and data communications.

As communication costs increase between now and 1990, more consideration should be given to the idea of having a communications department that is totally separate, not only from data processing but from all other departments as well. As microprocessors proliferate throughout the organization, the major centralized thread of continuity will be the combined voice and data communication networks. It is conceivable that in the next few years anywhere from 10 to 25 percent of an organization's employees might work at home several days a week instead of coming into the office. In such a situation, the organization relinquishes some control over its computers (microprocessors), but it must maintain control over the communication functions.

One other issue, with regard to combining voice and data communication into a new and totally separate organization, involves cost. Today, approximately three-quarters of an organization's communication costs are voice related, while only one-quarter are data related. Many organizations, by combining these two functions, could eliminate at least 20 percent of the combined costs of these two functions by sharing management personnel, circuits (voice over data), hardware (PBX/CBX), and software.

A major topic that should concern management is security of the network (the details of security and control are discussed in Chapter 8). An outside organization, one that is separate from the voice and data communication functions, should develop security standards that must be implemented and observed. The internal audit function should review the security standards and ensure compliance.

For example, some specific security standards that need to be addressed include (Chapter 8 expands on these):

- Encryption of data/information
- Network disaster and contingency planning
- Documentation for network-related items
- Access controls for the network
- Communication-related hardware/software controls
- Network program-and-data backup/recovery policy
- Physical security for network components
- Security training for personnel

NETWORK MANAGEMENT

Figure 7–1 is a typical organization chart for a data communication function that is organized within the Information Systems Department. Notice that the network manager reports directly to the director of information systems. You could combine the voice and data communications into one function by removing the network manager from the Information Systems Department, adding the job responsibilities for the effective management of voice communications, and moving the combined function to an independent location within the organization.

With regard to the job functions listed under network manager (Figure 7–1), each of the four functions involves a set of specific job tasks that require the utmost in management expertise and control of personnel, as well as in-depth technical knowledge. Let us examine each of these four sets of job tasks.

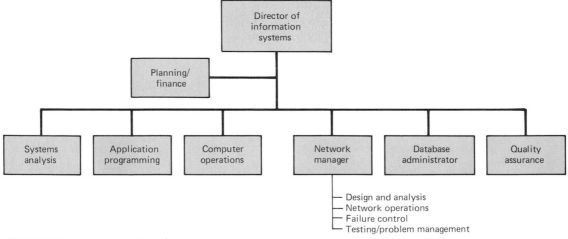

FIGURE 7–1 Organization chart.

Design and Analysis This function involves planning overall network design and continual analysis of the network. Management requires, for example, ongoing statistics with regard to network performance and feedback with regard to user satisfaction. Obviously, these statistics require close interaction with the other three job functions.

This function should be responsible for developing operations procedures and standards for the networking personnel. The network design people use measurement tools such as network models, simulators, statistical measurements, daily data collection routines, and other tools in order to manage the ongoing network throughout its life. Network design should not be viewed as something that is done when the network is created and then never again reviewed. Network design and analysis is a continuous "redesign" of the hardware, circuits, and software of the data communication network.

Network Operations This function is responsible for the network's day-to-day operations. This group maintains the ongoing services for the organization. One major function is the master network monitoring and daily gathering of statistics that are used by the network designers and the network manager. This group sometimes may interface with irate users as well as with the various hardware, software, and circuit vendors.

Failure Control This is the "central control group" that receives telephone calls when problems occur and records the incidents of problems. This task may be performed by the network operations personnel. Basically, it is a "help desk" that is called when anything goes wrong anywhere in the system. This group has appropriate customer service representatives to record problems, report them to the testing/problem management people, follow up, and generally ensure that the network is back into operation as soon as possible. This group also might be responsible for change scheduling, coordination, and follow up on any changes, whether they involve hardware, software, or circuits. In other words, this is the "user's interface" whenever there is a problem of any kind.

Failure control involves developing a central control philosophy for problem reporting and other user interfaces. This group should maintain a central telephone number to call when there is any problem in the network. As a central troubleshooting function, only this group or its designee should have the authority or responsibility to call any hardware vendor, software vendor, or common carrier.

A bound notebook with two carbon copies should be developed to record problem incidents that are reported. Each page (original and 2 carbons) should be prenumbered to avoid missing incident reports. One bound page should always be kept at the "trouble log desk." Two carbon copies may not be required, but it may be desirable to give one copy of the original trouble call to the vendor and one to internal testing/problem management personnel.

When a problem incident is reported, personnel at the "trouble log desk" should record the following to the best of their ability:

- Who reported the incident
- The problem reporter's telephone number
- The time and date of the problem (not the time of the call) as closely as it can be identified
- Location of the problem
- The nature of the problem
- When the problem was identified
- Why the problem happened (probably unable to identify this in most cases)
- How the problem occurred (probably not able to identify this in most cases)

Once this information is recorded, the failure control personnel should use an electric time-and-date stamp machine to indicate when they received the incident report.

The purpose of this procedure is to mandate central control of all problems and totally eliminate unnecessary service requests to vendors. Remember that there may be many hardware or software vendors for a data communication network. In fact, a typical network might have different vendors for the following equipment: terminals, cable connectors, modems, multiplexers, circuits (sometimes the local loops and the IXC circuits have different vendors), front end communication processors, host computer, and probably two or three different vendors for the various software packages.

The purpose of the bound trouble log volume is to record problems on paper so people will correct them and follow up, as well as to keep track of statistics with regard to problem incidents. For example, after a period of time utilizing a centralized failure control group, the organization might learn that there were 37 calls for software problems (3 for one package, 4 for another package, and 30 for a third software package), 26 calls for modems evenly distributed among the two vendors, 49 calls for terminals, and 85 calls to the common carrier that provides the network circuits. Data of this type are valuable when the design and analysis group begins redesigning the network to meet future requirements. Also, hard statistics like this enable you to put more pressure on the vendor who supplied the software package with the high number of problems.

Testing/Problem Management   This group establishes test and validity criteria and coordinates the various testing functions. It maintains the complex testing equipment that is needed to diagnose problems quickly and sometimes fix them in house. Probably its single most important function is to interface with the failure control group. This is because the failure control group, when it becomes aware of a problem, immediately passes it to the testing group so they can diagnose the problem and identify what needs to be corrected. Depending on the severity of the problem, the operations group may be notified as well. The complete problem handling procedure may involve either fixing the problem in house or notifying the appropriate vendor so that corrections can be made and the system can be operating again.

As soon as a problem is reported, the failure control group should immediately send a copy of the trouble log incident report to the testing/problem management group so they can diagnose the problem and possibly fix it themselves. The testing/problem management group should report back to the failure control group as soon as they have diagnosed the problem. This is so the time required to diagnose the problem can be recorded. In other words, the organization should keep track of Mean Time To Diagnose (MTTD), which is an indicator of the efficiency of testing/problem management personnel. This is the first of three different *times* that should be kept for future statistics.

Assume that a vendor is contacted in order to have a problem corrected. Either testing or failure control personnel should keep track of the time the vendor takes to respond. In other words, the Mean Time To Respond (MTTR) is identified. This is a valuable statistic because it indicates how well vendors respond to emergencies. A collection of these figures over a period of time can lead to a change of vendors or, at the minimum, can put severe pressure on vendors who do not respond promptly to problems.

Finally, after the vendor arrives on the premises, the last statistic to record is the Mean Time To Fix (MTTF). This figure tells how quickly the vendor is able to correct the problem. A very long time to fix in comparison with the time of other vendors may be indicative of faulty equipment design, inadequately trained customer service electronic technicians, or even the fact that inexperienced personnel are repeatedly sent to fix problems.

Some organizations combine the Mean Time To Respond (MTTR) and the Mean Time To Fix (MTTF); this is called Mean Time To Repair (MTTR).

One other statistic should be gathered. It usually is developed by the equipment vendor and it is called Mean Time Between Failures (MTBF). The Mean Time Between Failures of vendor-supplied network interface equipment should be very high. Usually the figure is greater than 30,000 hours.

With the various mean times, a formula is then worked out for calculating network availability (see Figure 7–2). Remember that MTBF can be influenced by the original selection of vendor-supplied equipment. The MTTD is related directly to the ability of in-house personnel to isolate and diagnose failure by hardware, software, and circuit. This means that test personnel need adequate training. The MTTR can be influenced by showing the vendor how good or bad their response has been in the past. The MTTF can be influenced by the use of redundant inter-

$$\text{Network availability} = \frac{\text{Uptime}}{\text{Uptime} + \text{Downtime}}$$

$$\text{Network availability} = \sum_{J=1}^{N} \frac{\text{MTBF}_J}{\text{MTBF}_J + (\text{MTTD}_J + \text{MTTR}_J + \text{MTTF}_J)}$$

$$N = \text{Total number of network elements.}$$

FIGURE 7–2 Network availability.

face equipment, alternate circuit paths, adequate recovery/fallback procedures to earlier versions of software, and the technical expertise of internal staff. Since all four of these mean time statistics are used to calculate network availability, their collection is vital if network performance is to be measured accurately and if performance is to be improved.

Another set of statistics that should be gathered are those collected on a daily basis by the network operations group. These statistics record the normal operation of the system, such as the number of errors (retransmissions) per communication circuit, per terminal, or whatever else is appropriate. Statistics also should be collected on a daily volume of transmissions (characters per hour) for each communication link or circuit, each terminal, or whatever else is appropriate for the network. These data can identify terminal stations/nodes that have a higher-than-average error rate as well as communication circuits that have higher-than-average error rates, and they can be used for predicting future growth patterns and failures.

This can be accomplished by setting up a simple quality control chart similar to those utilized in manufacturing processes. Such programs use an upper control limit and a lower control limit with regard to the number of blocks in error per day or per week (see Figure 7–3). Notice how this figure identifies when the common carrier moved a circuit from one microwave channel to another (circuit B), or how a deteriorating circuit can be located and fixed before it goes through the upper control limit (circuit A) and causes problems for the users.

NETWORK REPORTING

Poor network management leads to an organization that is overburdened with today's problems (firefighting) and does not have time to address future needs. Management requires adequate reports if it is to address future needs. Information for these reports can be gathered from host computers, front end communication processors, network monitors, the network management group, vendors, test equipment, and the like.

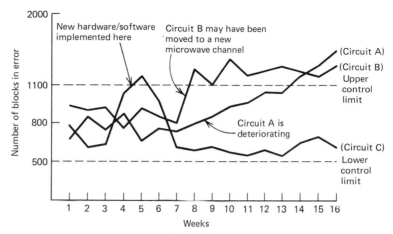

FIGURE 7-3 Quality control for circuits.

Technical reports that are helpful to management should contain some or all of the following:

- Cumulative network downtime.
- Detail of any subnetwork downtime.
- Circuit utilization.
- Response time analysis per circuit or per terminal.
- Usage by various types of terminal stations, such as interactive versus remote job entry.
- Voice versus data usage per circuit.
- Queue-length descriptions, whether in the host computer, in the front end communication processor, or at remote terminal sites.
- Histograms of daily/weekly/monthly usage, number of errors, or whatever is appropriate to the network.
- Failure rates for circuits, pieces of hardware, or software programs.
- Utilization rate of critical pieces of hardware, such as host computer or front end processor.
- File activity rates for database systems.
- Local device activity.
- Network gateway failure statistics and daily activity.
- Distribution of packet volume (for packet networks), or distribution of character volume per circuit link.
- Statistical profile of network traffic.

- Distribution of traffic by time of day and location.
- Peak volume statistics as well as average volume statistics per circuit.
- Correlation of activity between today and a similar previous period.
- Correlation of queue-length statistics by time and volume to a similar previous period.

In the area of management reporting, network documentation is a mandatory requirement for control of the network. *Network documentation* can consist of the following:

- Network maps:
    Worldwide
    Within a single country
    Within a state or province
    Within a city
    Within a specific building or facility
- Circuit layout records.
- Vendor maintenance records (MTBF, MTTD, MTTR, MTTF) with hardware/software/circuit cross-references.
- Software listings by hardware.
- Software listings by network tasks performed.
- All user site telephone numbers and individual contacts.
- Hardware maintenance history logs located at each user site.
- Circuit control telephone contact index and log (whenever possible establish a national account with the common carrier rather than dealing with individual common carriers in separate states and provinces).
- Serial number inventory (property control) of all network components.
- Network switching criteria and redundancy locations.
- Vendor contractual agreements.
- Legal requirements to comply with local or federal laws, control, or regulatory bodies; also have legal requirements for other countries (international).
- Operations manuals for network operations personnel.
- Vendor-supplied hardware operation manuals.
- Software documentation manuals.
- Escalation levels (where to go when the problem cannot be resolved).

- Preventive maintenance guidelines.
- Record of user site tests required by network security monitor.
- Disaster plan/recovery techniques.
- Diagnostic techniques by hardware component or type of trouble.

NETWORK STATUS

As part of the network organization and control, the present network status should be continually monitored and assessed by the design and analysis group. The data collected from a status review can be used both for future planning and to validate the performance of the network manager. Some questions that can be used to review the network status are as follows:

- Is the voice and data communication system combined?
- Does the network manager report at a high enough level in the management hierarchy?
- Is the network manager within the Information Systems Department?
- How many independent data networks are used in the organization?
- Are the networks application dependent or independent?
- What was the system availability for yesterday, the last five days, the last month, and so on?
- Are any of the reports mentioned in the above two lists available?
- Are the security and control aspects, described in Chapter 8, available?
- What is the data communications system's annual budget?
- What is the monthly communications cost or the total cost last year? For voice? For data?
- Is the network critical to the organization's revenue-stream management, expense-stream management, cash-flow management, etc?
- Is network operation erratic or difficult to monitor?
- Can trouble areas be pinpointed quickly (fault diagnosis)?
- Are 95 percent of all response times less than or equal to 3 seconds for on-line real-time traffic?
- Are management reports timely and do they contain the latest up-to-date statistics?
- How many hours per day or days per week are utilized for network operation versus network management functions?

- What is the inventory of the current network configuration for all of the pieces of hardware?
- What is the network configuration for all circuits?
- What is the inventory of software and where is it located in the network?
- Does a formal network management organization exist with mandated goals, policies, procedures, and the like?

TEST EQUIPMENT

Ten years ago many data communication facility managers did not have test equipment, and simple analog test equipment was all that seemed to be in network control rooms. Today, analog and digital test equipment is a necessity. Tomorrow the network manager will have microprocessor-based, software-driven, automated test equipment. Some of today's line monitors and network analyzers already are based on microprocessors, although software is still being developed for the automated monitoring, diagnosing, and possible automatic redundant fallback switching.

Techniques and devices for monitoring network performance can range from comparatively simple analog and digital test equipment to highly sophisticated network controllers and overall management systems. The following is a short description of some of the various types of test equipment.

Analog Test Equipment   This simple equipment performs tests for envelope delay distortion, bias distortion, phase jitter distortion, and the like. Tests such as these usually are performed by the common carrier, but it can be advantageous for the network manager to determine whether the communication circuit is functioning properly and at the error rate that is considered minimal for the type of circuit. For example, one bit error in 10^{-5} or 10^{-6} may be normal for a copper wire voice grade circuit.

Breakout Box   This hand-held device represents the next higher level in data communication monitoring and test equipment. It can be plugged into a modem's digital side to determine the voltage values for the RS232C connector cable interface (25 pin cable).

Bit-Error Rate Tester (BERT)   This piece of digital test equipment is somewhat more sophisticated than a breakout box because it sends a known pseudorandom pattern over the communication circuit. When this pattern is reflected back, the BERT compares the pattern transmitted with the pattern received and calculates the number of bit errors that occurred on the communication circuit. Various test patterns are used and common pattern lengths are 63, 511, 2047, and 63511 bit patterns. The odd numbers allow simple circuitry in this test equipment.

Bit-Error Rate (BER) measurements can be made with this type of equipment. A BER is the number of bits received in error divided by the total number of bits received. BER measurements are used by service personnel to tune the communication circuit and to make a subjective evaluation as to the quality of a specific circuit or channel. BER cannot be related directly to throughput because error distribution is not taken into account. Assume that 1000 one-bit errors occur during a time interval of 1000 seconds. If the errors are distributed evenly (one per second) the effect on throughput would be disastrous; however, if all the errors occur in a single second, the effect is minimal.

Block-Error Rate (BKER) is the number of blocks received that contain at least one bit error divided by the number of blocks received. BKER is more closely related to throughput. Assume a BKER measurement has been made and the BKER value is 10^2 (1/100). This means that out of every 100 blocks received, one contained an error. Therefore, you would expect to see one retry for every 100 blocks transmitted (1 percent error rate).

Another error rate parameter (only for digital networks) is Error Free Seconds (EFS). EFS is similar to BKER except that it indicates the probability of success rather than the probability of failure, and the block size is the number of bits transmitted in a one-second time period. For example, for a 4800 bit-per-second channel, the one-second block would contain 4800 bits.

On digital communication channels, AT&T Communications guarantees that 99.5 percent of all seconds of data transmission will be error-free seconds.

Self-Testing Modems   If a self-testing feature is in a modem, a test pattern is generated inside the modem (after appropriate buttons are pushed) as close as possible to the normal digital input (which is, of course, disconnected). The test pattern travels through 90 percent of the modem's circuitry, passes through an artificial telephone circuit, and is returned to its point of origin. The artificial telephone circuit acts as a local analog loop. The returning pattern is then compared with the transmitted pattern and the operator is advised of discrepancies via an indicator lamp.

Some modems also have digital or analog remote loopback testing, whereby the signal actually is sent over the communication circuit and is looped back to the originating modem by the remote modem. Then a comparison of the signal is made and the operator is advised of any discrepancies.

Finally, some modems have internal circuit diagnostic checks whereby they can diagnose their own failures in case of circuit or chip failure. Self-diagnostics are made possible by the use of firmware and microprocessor chips.

Newer modems contain some of the features of network analyzers. They actually keep track of poll times and other types of network analysis information.

Response Time Analyzer At this point our discussion of test equipment begins to be less clear-cut because there is so much overlap between the different types of equipment and the names used by various vendors. Basically, a response time analyzer checks the operation of the communication protocols. When the equipment is operating in a polled network, the elapsed time from the initiation of polling until receipt of the response varies. Response time analyzers measure the responses of all hardware in the system. This equipment can determine whether or not the network equipment is meeting specifications.

A typical piece of this equipment might measure poll-to-poll time, the time from the detection of the poll being sent to the terminal until the time that poll is again sent to the same terminal. This measurement is updated continuously as the polls are generated by the host computer.

Another typical measurement is poll/response time. This measurement starts at the second SYN character of the poll and is terminated by the second SYN character of the response (you might review the section on Data Signaling/Synchronization in Chapter 4). Clear definitions are impossible because some data line monitors also perform response time analysis.

Data Line Monitors As is noted above, some data line monitors perform response time analysis in addition to many other tasks. Therefore, they are sometimes known as *protocol monitors*. They check the actual data (both control and data characters) on the communication circuit. You can use this device to check the interaction of software and hardware by looking at all the data passing in both directions on a communication circuit. It is mandatory with the new bit-oriented protocols because line control is no longer dependent upon an entire byte to transmit the control message, but on a single bit within the 8-bit byte sequence.

Users can capture data in an external tape storage or internal memory, as well as freeze the most current data on the video screen. What makes the line monitor unique is that usually it is the only test equipment that displays all the "control characters" which are not usually seen on the video screen. A line monitor can show when a carriage return or a line feed occurs, as well as when a communication control code is transmitted. The technician can count the number of SYN characters, identify the 8-bits of each field/character within a frame, and the like.

There are two basic categories of data line monitors, *active* and *passive*. The *active monitors* can generate data, are interactive on the circuit, and can emulate various terminals (they are programmable). *Passive units* merely monitor and collect data to be examined later. Active data line monitors contain all the features of a passive data line monitor, and more. It should be noted that data line monitors are a security risk because of their ability to generate data, interactively place it on a communication circuit, and do this while emulating another terminal located somewhere else.

Automated Test Equipment (ATE)   Future generations of automated test equipment will be a consolidation of all of the above pieces of test equipment but with built-in microprocessors and programmable testing features. With this type of a programmed environment, technicians instead of electrical engineers will be able to perform testing/problem management in networks.

The equipment will become more digital as communication circuits go to digital and as voice and data are combined into one digital transmission stream.

This automated testing equipment will include functions such as diagnostics, polling, and statistics gathering. It also will be integrated into all of the equipment in a network. For example, a centralized network control system will be talking constantly to automated test equipment that has been built into intelligent multiplexers, concentrators, modems, terminal control units, and the terminals themselves. Equipment like this will be capable of protocol emulation, measurements as to whether the bandwidth (capacity) is being used efficiently, self-diagnostic features for every piece of hardware and software packages, analog circuit testing, digital circuit testing, automatic switching to back up hardware/software/circuits, and the like.

QUESTIONS

1. Data or information transmitted over any network must CATER to the overall needs of the network users. Can you define the acronym CATER?

2. What are the five key management tasks that must be performed by all managers?

3. If you were going to initiate a data communication network control department, what would be some of the major job tasks or organizations that would be set up in this department?

4. Identify some of the items that should be reported at the trouble log desk.

5. Define the following and give a description of each: MTBF, MTTD, MTTR, and MTTF.

6. If you were using a bit-error rate tester, would you use a BKER test for asynchronous transmission?

8

SECURITY AND CONTROL

This chapter identifies the 17 network control points that must be addressed for security and control. Specific hardware/software/ protocol controls are reviewed. Other control areas that are reviewed are: management controls, error control, recovery/back up/disaster, and the use of a matrix to identify, document, and evaluate security and control in a data communication network. Hundreds of specific controls that relate to the matrix approach are listed in Appendix 2.

NETWORK SECURITY

In recent years organizations have become increasingly dependent upon data communication networks for their daily business communications, database information retrieval, and distributed data processing. This commitment to data communications/teleprocessing has changed the potential vulnerability of the organization's assets. This change has come about because the traditional security, control, and audit mechanisms take on a new and different form in data communication based systems. Increased reliance on data communications, the consolidation of many previously manual operations into computerized systems, use of database management systems, and the fact that on-line real-time systems cut across many lines of responsibility have increased management concern about the adequacy of current control and security mechanisms used in a data communication environment.

There also has been an increased emphasis upon computer network security because of numerous legal actions involving officers and directors of organizations, because of pronouncements by government regulatory agencies, and because the losses associated with computerized frauds are many magnitudes larger, per incident, than those from noncomputerized frauds. These factors have led to an increased vigilance with regard to protecting the organization's information assets from many potential hazards such as fraud, errors, lost data, breaches of privacy, and disastrous events that can occur in a data communication network.

With regard to data communication networks, the organization must be able to implement adequate control and security mechanisms within its facilities, including building facilities, terminals, local area networks, local loops, interexchange channel circuits, switching centers, network interface units (gateways), packet networks, hardware (modems, multiplexers, encryption devices, and the like), network protocols, network architecture software, test equipment, and network management control.

As an example, Figure 8-1 depicts a network typical of one that an organization might develop.

In such a network all of the areas mentioned above would require a positive decision (policies and procedures) as to security and control. It should be noted that with this kind of network the organization is vulnerable to many points of entry from an unwanted intruder. In fact, every terminal in the network is a potential entry point for an unauthorized intruder.

The rest of this chapter will be a discussion on each of the major portions of a data communication network, such as hardware and software, and a description of the various controls that relate to that specific area.

Finally, a matrix methodology for identifying network controls, documenting them, and evaluating their effectiveness will be presented. This is to provide the network manager with a good picture of the current controls, their effectiveness, and adequate documentation.

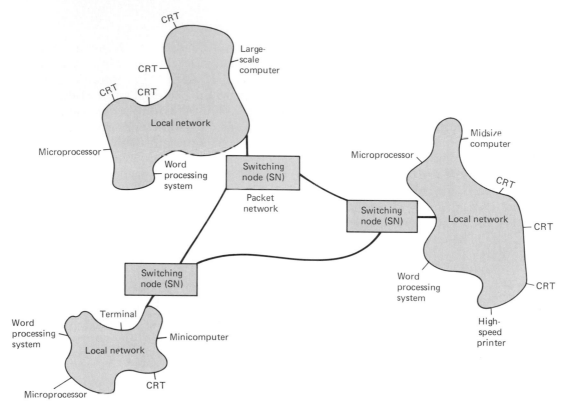

FIGURE 8–1 Future network.

NETWORK CONTROL POINTS

The 17 control points, or areas where control and security mechanisms must be implemented, are depicted in Figure 8–2. The network manager, quality assurance personnel, security officer, or the organization's EDP auditor should examine these areas to ensure that proper controls are implemented and are functioning properly. The numbers on Figure 8-2 are described in the following list.

1. Physical security of the building or buildings that house any of the hardware, software, or communication circuits must be evaluated. Both local and remote physical facilities should be secured adequately and have proper controls.

2. Operator and other personnel security involves implementation of proper access controls so that only authorized personnel can enter the closed areas where network equipment is located or can access the network itself. Proper security education, background checks, and the implementation of error/fraud controls fall into this area.

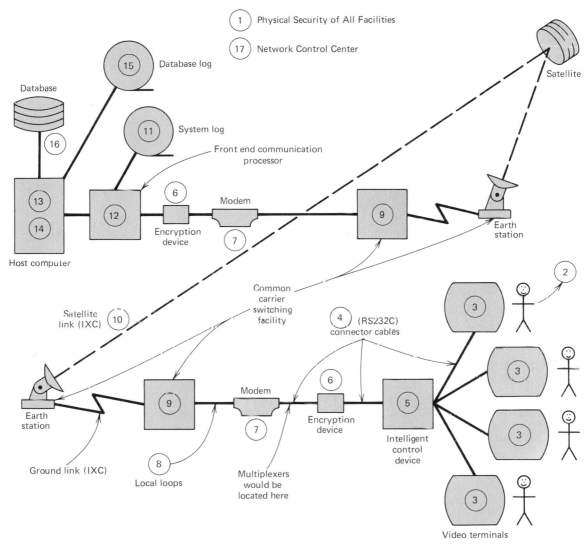

FIGURE 8-2 Network control points.

3. Terminals are a primary area where both physical and logical types of security controls must be enforced.

4. Local connector cables and wire pairs that are installed throughout the organization's facilities must be reviewed for physical security.

5. Local intelligent control devices that control groups of terminals should be reviewed for both physical and logical programmed controls.

6. Hardware encryption is a primary control point, especially with regard to security of messages.

7. The modems and multiplexer hardware should be reviewed with regard to control and security at this point in the network.

8. Local loops that go from the organization to the common carrier's switching facility should be reviewed.

9. The physical security and backup of the common carrier switching facility (telephone company central office) should be evaluated. If this facility were destroyed, all the circuits would be lost. This review may include both central offices in a city and earth stations for satellite transmission.

10. Review the security/control mechanisms in place with regard to the interexchange channel (IXC) circuits.

11. A major control point is the system log that logs all incoming and outgoing messages.

12. The front end communication processor is another major control point to review. At this point there may be a packet switching node (SN) that must be reviewed for security and control.

13. Within the host computer, any controls that are built into the software should be reviewed.

14. Also within the host computer, review for any controls that are designed into the host computer's hardware mechanisms/architecture.

15. Another major control point, only in database systems, is the database before-image/after-image logging tape. This should be reviewed for any controls that may be in existence at this point. Many other security/control items of data are logged at this point.

16. With regard to database-oriented systems, another control point is the database management system (DBMS) itself. The database management system software may have some controls that help with regard to security of the data communication network and the control of data/information flow.

17. The last control point is the network control center itself. This area has controls that relate to management and operation, test equipment utilized, reports, documentation, and the like (see Chapter 7).

These 17 control points are the specific areas where control features can be implemented and maintained within a data communication network. Now let's review the specific controls that can be used to secure your data communication network.

ENCRYPTION

Encryption (introduced in Chapter 2) is the process of disguising information by the use of many possible mathematical rules known as *algorithms*. Actually, *cryption* is the more general and proper term. *En*cryption is the process of disguising information, *de*cryption is the process of restoring it to readable form. Of course, it makes no sense to have one process without the other. When information is in readable form, it is called *clear* or *plaintext*. When in encrypted form, it is called *ciphertext*.

The art of cryptography reaches far into the past and until recently has almost always been used for military and political applications. By today's exacting standards, such ciphers are insecure and therefore obsolete. They were usually alphabetic ciphers (rules for scrambling the *letters* in a message) and were designed for manual processing. Today's world of binary numbers and the speed of computers has given birth to a new class of cryption algorithms.

The acceleration of new research began during the Second World War and has continued into the present time for four reasons:

- Recognition of the necessity of encrypting communications for military purposes.
- The advent of high-speed computational electronics (computers).
- A growing interest in cryptography within academic circles.
- An interest on the part of private corporations, as well as governments, in protecting their proprietary information.

Interest in cryptographic protection runs highest in the world of communications. Of all the routes and resting places of information, communicated information is the most vulnerable to disclosure. Data stored on magnetic tapes, on disks, and in computer memory can be protected to a large extent by physical security, passwords, and other software access control systems.

Modern data communications takes advantage of existing public telephone circuits, microwave transmissions, and satellite relays. As a result, communicated information is highly exposed in a variety of forms. It can be captured at minimum expense and risk to the data thief, and at maximum loss to the organization.

A striking example of this exposure is the daily Electronic Funds Transfer (EFT) of billions of dollars between domestic and foreign banks over public links. The covert alteration of bank account numbers, amount of funds, and the like can have disastrous results.

There are always two parts to an encryption system. First there is the algorithm itself. This is the set of rules for transforming information. Second, there is always a *key*. The key personalizes the use of the algorithm by making the transformation of your data unique. Two pieces of identical information encrypted with the same algorithm but with *different keys* produce completely different cipher-texts. When using most encryption systems, it is necessary for communicating parties to share this key. If the algorithm is adequate and the key is kept secret, acquisition of the ciphertext by unauthorized personnel is of no consequence to the communicating parties.

The key is a relatively small numeric value (in number of bits) that should be easily transportable from one communicating node to another (see item 6 on Figure 8-2). The key is as it sounds. It is something that is small, portable, and with the aid of a good lock, the algorithm, it keeps valuables where they belong.

Good encryption systems do not depend on keeping the algorithm secret. Only the keys need to be kept secret. The algorithm should be able to accept a very large number of keys, each producing different ciphertexts from the same clear-text. This large "key space" protects ciphertext against those who would try to break it by trying every possible key. There should be a large enough number of possible keys that an exhaustive computer search would take an inordinate amount of time or would cost more than the value of the encrypted information.

Almost every modern encryption algorithm transforms digital information. Scrambling systems have been devised for analog voice signals, but it generally is agreed that their algorithms are not as strong as those used for digital signals made up of binary bits. The most recent advances in analog signal protection have not been in newer and better algorithms. Instead, they have been in the technology of high-speed conversion of analog signals to digital information bits in preparation for encrypting them with digital algorithms. In any case, the vast majority of today's proprietary information is digital. For this reason we will discusss only digital techniques.

Encryption algorithms may be implemented in software or hardware. The former has some advantages in protecting stored data files and data in the host computer's memory. However, hardware implementations have the advantage of much greater processing speed, independence from communication protocols, ability to be implemented on "dumb" devices (terminals, TELEX, facsimile machines, etc.), and greater protection of the "key" because it is physically locked in the encryption box. Unauthorized tampering with the box causes erasure of the keys and related information. Hardware implementations have been reduced to the chip level because they are simply specialized microprocessors housed in small hardware boxes.

By far the most widely used encryption algorithm is the Data Encryption Standard (DES). It was developed in the mid-1970s by the U.S. government in conjunction with IBM. DES is maintained by the National Bureau of Standards (NBS) and is often referred to as NBSDES or DEA (Data Encryption Algorithm). The U.S. government recommends that DES be used for the encryption of commercial and *un*classified military data. The American Banking Association has endorsed its use for the commercial banking industry.

This combination of credentials makes DES the technique of choice by private institutions. This concept of "choice" is somewhat misleading. DES is the *only* algorithm endorsed by the government. The academic literature is full of alternatives, but practical reasons, such as obtaining insurance against third party fraud, and the lack of mathematical sophistication on the part of encryption system users, presently leave little choice.

DES is classified as a *block cipher*. In its simplest form the algorithm encrypts data in independent 64-bit blocks. Encryption is under the control of a 64-bit key. DES expects a full 64-bit key but it uses only 56 of the bits (every eighth bit may be set for parity). Therefore, the total number of possible keys is 2^{56} or over 72 quadrillion combinations.

DES ciphertext is composed of blocks containing highly randomized bit sequences. The algorithm is so thorough in its randomizing of any 64-bit block (almost without regard to the cleartext of the key) that ciphertext almost always passes standard tests for randomness. The random quality of ciphertext is a crucial factor in the design of communication networks that will convey ciphertext. Communication control characters (for message routing or error detection) cannot be mixed with ciphertext because there is always some probability that DES will generate one of these control characters and thwart the communication system.

As a result, DES hardware usually is employed as shown in Figure 8-3. Communication protocols, parity, and checksums are in place with the message *before* it enters the originating DES hardware device. As is shown, this information may originate from a terminal, a front end, or a variety of communicating devices. The hardware encryption boxes usually are utilized on a link-to-link basis as depicted in Figure 8-3.

Placing the DES device between the modems can present a number of problems. First, most DES boxes are digital devices. They usually do not accept the analog signals output by modems. Second, in asynchronous communications at least the start bit must be sent in the clear. Encryption can, and usually does, begin with the first data bit and end with the last. Similar problems can occur if synchronous timing signals are encrypted.

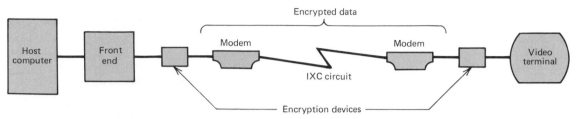

FIGURE 8–3 Encryption device location.

The randomized information is now transmitted to a network switch, computer, terminal, or other receiving device. The receiving DES hardware, which must be loaded with the *same* key as the originating DES hardware, then decrypts the information before it enters the receiving terminal device. Any communication protocols are verified *after* the decryption.

DES in some ways provides better error detection than standard parity or checksum techniques. If a single bit of any 64-bit ciphertext block is flipped during transmission, on decryption of that block the result will be 64 bits of random nonsense. This "error propagation" virtually ensures that parity and checksum will fail after decryption.

A more serious problem occurs if a bit is picked up or dropped during communication. The message loses 64-bit block "synchronization" at the point of the dropped or added bit, and the message decrypts into nonsense. The result can be the loss of an entire message.

This magnification of communication errors is not without its price. Since the *minimum* loss of information is usually 64 bits, a retransmission almost always is required if there is a single bit communication error.

DES is a member of a class of algorithms known as "symmetric." This means that the key used to decrypt a particular bit stream must be the *same* as that used to encrypt it. Using any other key produces cleartext that appears as random as the ciphertext. This can cause some problems in the complex area of key management; keys must be dispersed and stored with great care. Since the DES algorithm is publicly known, the disclosure of a key can mean total compromise of encrypted messages. Therefore, in order for two nodes in a network to establish communication of ciphertext, it is first necessary to define and communicate a common key over a secure channel or send it by personal courier.

Alternatives to DES have been proposed by a number of academic cryptologists. These fall under the category of "asymmetric" or "public key" algorithms. In these systems the key needed to decrypt a message is *different* from the one used to encrypt it. The two keys are related distantly in a mathematical sense. The security of asymmetric systems depends on the extreme difficulty (analytical impossibility or computational unfeasibility) of deriving one key from the other.

Asymmetric algorithms can greatly reduce the key management problem. Each receiving node has its publicly available key (hence the name "public key") that is used to encrypt messages sent by any network member to that node. These public keys may be listed in a telephone book type directory. In addition, each user has a *private key* that decrypts only the messages that were encrypted by its *public key*. The net result is that if two parties wish to communicate with each other, there is no need to exchange keys beforehand. Each knows the other's public key from the public directory and can communicate encrypted information immediately. The key management problem may be reduced to each user's being concerned only with the on-site protection of its private key.

It is expected that the National Bureau of Standards will endorse a public key algorithm by 1985.

In order to visualize how a public key algorithm works, look at Figure 8-4. At the top of this figure there is a public directory which contains all of the public keys for each organization utilizing public key encryption. Our public directory contains five different banks.

In order to use the public key encryption methodology, a bank also has a secret key known as a private key; therefore, there are two separate keys, the private (secret) key and the public key. In this case, the bank places its public key into the public directory and carefully secures its own copy of the private key.

The middle of Figure 8-4 shows what an encrypted message would look like. When bank 4 wants to send a message to bank 1, it encrypts the message with the bank 1 public key, which is obtained by bank 4 from the public directory. This represents a straightforward encryption of a message between bank 4 and bank 1. Obviously, when the message is received at bank 1, it decrypts the message using its secret private key.

For more complex encryption, bank 4 can include its signature so bank 1 also can verify the signature or, in other words, be sure that the message originated from bank 4.

In order to perform a signature verification (see the bottom message of Figure 8-4), bank 4 first encrypts its ID (signature) plus some of the "key-contents"[1] of the message, using the bank 4 private key. This is its own private key and is known only to bank 4.

Next, bank 4 encrypts both the message contents and the already encrypted bank ID using the bank 1 public key from the public directory. This means that the bank 4 ID has been double encrypted, first using the bank 4 private key and then a second time using the bank 1 public key. The message is then transmitted to bank 1.

Upon receipt of the message, bank 1 uses its private key to decrypt the entire message. At this point, bank 1 is able to read the contents of the message, except

---

[1]The term *key-contents* means unique information from the message such as date, time, or dollar amount. It does not refer to the public/private keys.

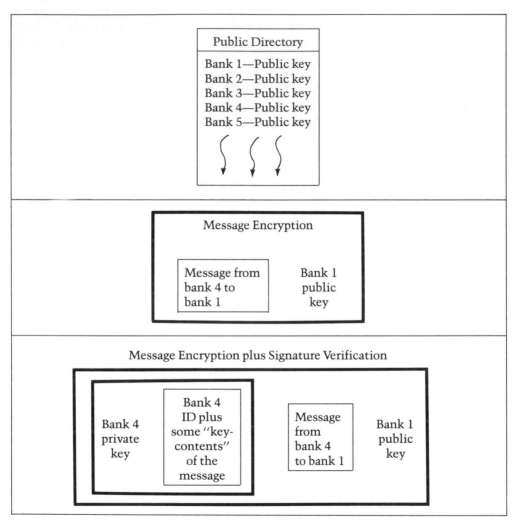

FIGURE 8–4 Public encryption.

for a block of data that is still encrypted (unidentifiable). Because the message was received from bank 4, bank 1 assumes that bank 4 secretly encrypted its ID plus some "key-contents" of the message for signature verification purposes. At this point, bank 1 takes the Bank 4 public key (from the public directory) and decrypts the trailing block of data that contains the bank 4 ID plus some "key-contents" of the message.

In this way, the public key system encrypts messages and also offers electronic signature identification without an exchange of keys among all the thousands of banks throughout the world. The public directory need only be updated as often as necessary. You can encrypt with a public key and decrypt with a private key or you can encrypt with a private key and decrypt with a public key.

The algorithms for public key systems are very different from symmetric algorithms like DES, but in practice their ciphertexts are similar when viewed from the data communication standpoint. Each produces ciphertext consisting of randomized bit patterns. There is almost always some degree of inherent error propagation. Therefore, the practicalities of handling the communication of both types of ciphertexts are for the most part identical.

The world of cryptology is full of controversy and debate. This is caused in part by tension between governments and independent academic cryptologists. National security is always the issue. The underlying cause of this tension, however, is the fact that cryptology is an art rather than a science. Except for a few noteworthy exceptions, it is impossible to mathematically *prove* whether an encryption/decryption algorithm can be broken. This means that the only route to breaking a cipher is an artful (and perhaps time-consuming and expensive) trial and error approach. Debates about the security of ciphers often end with mathematical generalizations and seat-of-the-pants type expressions.

This combination of embryonic but exciting mathematics and a dramatic increase of interest in encryption by corporations and governments means that it indeed will be an exciting arena in the years to come.

HARDWARE CONTROLS

This section describes the hardware controls found in a network. The pieces of network hardware are discussed in terms of the controls that relate to them. We will review controls that relate to front end processors, packet switching controllers, modems, multiplexers, remote intelligent controllers, and terminals.

Front End Processors   The front end processor that controls a centrally controlled data communication network can be one of the single most important areas for security and control. It is only a piece of hardware, but within it are software programs/protocols that control the access methods for data flow.

Some specific controls that might be housed within the front end communication processor are:

- Polling of the terminals to ensure that only authorized terminals are on the network.
- Logging of all inbound and outbound messages (systems log) for historical purposes and for immediate recovery should the system fail.

- Error detection and retransmission for messages that arrive in error (ARQ—see Chapter 6).
- Message switching that reduces the possibility of lost messages (there also can be circuit switching or packet switching).
- Store and forward techniques help avoid lost messages (although store and forward opens up the possibility of a network programmer's copying messages from the storage disk).
- Serial numbers for all messages between all nodes.
- Automatic call-back on dial-up facilities to prevent the host computer from being connected to an unauthorized dial-up terminal.
- Systems editing such as rerouting of messages, triggering of remote alarms if certain parameters are exceeded or if there is an abnormal occurrence.
- Collection of network traffic statistics for long-term control of the total network.

Also review the list in Chapter 2 of functions for the front end.

Packet Switching Controllers   A packet switching controller or switching node (SN) is similar to a front end communication processor, but it has some specialized features that pertain to the operation of a packet network. It is possible for a packet switching controller to perform any of the control functions previously mentioned for front ends. In addition, it performs other specific control functions such as the following:

- It keeps track of messages between different nodes of the network.
- It controls the numbering of each packet to avoid lost packets, messages, or illegal insertions.
- It routes all messages. It may send different packets, containing parts of the same message, on different circuits (unknown circuit path). This may prevent an unauthorized user/perpetrator from receiving all parts of a sensitive message.
- It contains global and/or local databases that contain addresses and other sensitive data pertaining to each node. These databases can be cross-referenced with other written documentation when network nodes are reviewed for security.
- On dial-up packet networks, it keeps track of the sender of each message that is delivered.
- It can either restrict the users to dial-up or allow use of leased circuits into the packet network.

Modems   The modem may be an interface unit either for broadband (analog) communication circuits or for baseband (digital) communication circuits. It does not matter which because these hardware units can perform any of the controls listed below, depending upon the features installed by each manufacturer.

Modems can offer loopback features that allow the network manager to isolate problems and identify where they are occuring in the network. Some modems contain automatic equalization microprocessor circuits to compensate for electronic instabilities on transmission lines, thereby reducing transmission errors. Some modems have built-in diagnostic routines for checking their own circuits. Mean Time Between Failure (MTBF) statistics should be collected for modems because low MTBF indicates that downtime is excessive.

Some dial-up modem controls include changing the modem telephone numbers periodically, keeping telephone numbers confidential at both user sites and the central data center, possibly disallowing automatic call receipt at the data center (using people to intercept), removing telephone numbers from both local and remote dial-up modems, and requiring the use of terminals that have an electronic identification circuit for all dial-up ports. Finally, it may be desirable to utilize a *dial-out-only facility*, whereby the act of dialing into the network and entering a password automatically triggers a disconnect; the front end or host computer then dials the "approved" telephone number that matches the password used during the original dial-in. In other words, dial-in triggers a dial-out.

Multiplexers   Because many multiplexer sites are at remote locations, a primary control is to prevent physical access to the multiplexer. Another consideration is whether the multiplexer should have dual circuitry and/or backup electrical power since loss of a large multiplexer site can knock out several hundred terminals. Because time division statistical multiplexers have internal memory space, and some have disk storage, special precautions must be taken. Memories and disk storage make illegal copying of messages easier. Other controls include logging all messages at the remote multiplexer site before transmission to the host computer and manually logging all vendor service call visits.

Remote Intelligent Controllers   A remote intelligent controller can be a special form of multiplexer or a remote front end communication processor that is located several hundred miles from the host computer. These devices usually control large groups of terminals. All of the controls that were mentioned for multiplexers also apply to remote intelligent controllers.

A review of software controls that can be programmed into this device is suggested. For example, daily downline loading of programs can help ensure that only authorized programs are in this device. Another control is the periodic counting of bits in the software memory space. This identifies a minor program change so that a new one can be downline loaded immediately. Each controller should have its unique address on a memory chip (instead of software) to thwart anyone who wants to change hardware addresses. Remote logging of each inbound/outbound message should be considered seriously. If hardware encryption boxes are located in the same facility as the remote intelligent controllers, then access to these devices should be controlled by implementation of strict physical control procedures and locked doors.

Terminals There are two basic areas that must be considered with regard to the control of terminals in a data communication network. The first is human error prevention controls and the second is security controls. Each is listed below:

Human Error Prevention Controls

- Ensure adequate operator training with regard to self-teaching operator manuals and the periodic updating of these manuals.
- Keep dialogue simple between the operator and the application system (menu selection might be utilized).
- Terminals should be easy to use and have functional keyboards.
- Consider preprinted forms for printing terminals and a fill-in-the-blank format (preprinted forms on a video screen) for video terminals.
- Instructions should be preprogrammed and available for recall when an operator needs help. Secured systems, where assistance should be more difficult to obtain, may be an exception.
- Operators should have restart procedures that can be used for error recovery during a transaction.
- Work area extremes in light, noise, temperature, and so on must be minimized if operators are to reduce errors to a minimum.
- Reasonably fast response times reduce errors because longer response times produce error-causing frustration in operators. Long response times also reduce productivity.
- Intelligent terminals can edit for logical business errors and verify data before transmission.
- When video terminals are used, they should have the largest dot matrix screen to reduce operator eyestrain (10 times 14 is easier to read than 5 times 7), screens should have an anti-glare surface, characters should not jitter on the screen, and the cursor should be visible at 8 feet. Reverse video (black on white as compared to white on black) provides a choice for individual operators. Yellow/green screens are easiest to see, and terminals that have brightness/focus/contrast adjustments are preferred by terminal operators.

Security Controls

- Terminals can have a unique electronic chip built in that provides positive identification. With chips, the front end or host can identify each terminal electronically.
- Physically lock terminal on/off switches or have locks that disable the screen and keyboard.
- Keep terminals in a physically secure location.
- Lock off all of the communication circuits after hours (positively disable the communication circuits).
- Each system user should have an individual password.
- Each user could have a plastic identification card that runs through an identification card reader. Such cards replace the need for individual passwords.
- Utilize special log-in numbers that can be entered only by a key person in the department.
- Consider using one of the newer types of personal identification such as signature identification, fingerprint identification, voice identification, or hand image identification.
- Transaction code each terminal. This prevents any transaction that is not related to the work area in which the terminal is located. In other words, the terminal is made transaction specific.
- Develop a security profile of the types of data being entered and the user log-in procedures. If a violation occurs, the terminal that was used can be shut down automatically. In addition, a terminal security report should be delivered the next day to the manager of the user work area.
- Restrict terminals to read-only functions.
- Sequence number, time stamp, and date all messages.
- Passwords should not print when they are typed.
- Ensure proper disposal of hard copy terminal output.
- Allow intelligent terminals to perform editing on transactions before they are transmitted.
- When looking at the control and security of dial-up terminals, review the controls for dial-up modems that were listed previously in the section on modems.

CIRCUIT CONTROLS

Some of the communication circuits that must be reviewed are the wire pairs and cables that are placed throughout the user facility, the local loops that go between the user facility and the common carrier (telephone company), and the interexchange channel (IXC) circuits between cities.

The wire pairs and cables within the user facility should be made as physically secure as possible, because this is where anyone wanting to tap the system would enter. It is 100 times easier to tap a local loop than it is to tap an interexchange channel. Ensure that the lines are secured behind walls and above ceilings, and that the telephone equipment and switching rooms are locked and the doors alarmed.

With regard to local loops, there is not much that can be done except to visit the common carrier switching facility. This provides some idea as to the physical security, fire protection, and disaster prevention controls implemented by the common carrier. If these are inadequate, about the only thing that can be done is to split local loops among your facility and two or three different common carrier switching facilities (telephone company end offices).

For security on interexchange channels, encryption of messages is the only dependable method. If the data/information is so sensitive that a breach of privacy or the insertion/modification of a message cannot be allowed, then encryption must be considered.

With regard to internal cables within your user facility, the use of fiber optics might be considered. Fiber optic cable uses light-emitting diodes or laser light to transmit pulses of light through hair-thin strands of plastic or glass. These devices offer security through their immunity to electrically generated noise, their resistance to taps, their isolation, and their small size. They also have some special benefits in an environmental sense.

Because optical fibers are immune to electrically generated noise such as radio interference, they offer a bit error rate of approximately 10^{-9} as compared with 10^{-6} for metallic connectors.

Fiber optic cable is an attractive security measure because it is almost totally immune to unauthorized access by tapping. Taps can be made only by breaking the cable, polishing it off, and inserting a splice or nicking into the core, to detect the light. The first method, using a T-splice adapter, does give a detectable power loss in the optic fiber system, so it can be detected easily. The second method, nicking, might be possible with a step index cable, which has a silica core and a plastic cladding around it. The plastic cladding could be nicked so that the light could leak out, although if too much light leaked out the signal would be lost. The nicking technique is almost impossible to accomplish in a graded index fiber, however, because the core and the cladding are one piece of silica. A graded index fiber is made of silica and the cladding is also silica, but it has a different index of refraction and therefore reflects light down the cable. Because these two are melted together during manufacture, nicking would be almost impossible (the glass would crack).

With regard to isolation, optical fiber cables provide complete isolation between transmitters and receivers, thus eliminating the need for a common ground. This structure provides electrical isolation from hardware and eliminates problems such as ground loops within an installation. It also reduces the amount of electrical noise that produces errors on data communication circuits. For communications in a dangerous atmosphere, such as a petroleum refinery or a paint factory, it has another advantage because static spark is eliminated.

The small size and light weight of fiber cables offer users better opportunities to secure this medium physically. Because fiber optic cable is nonconductive, it is free from electromagnetic noise radiation and therefore is resistant to conventional passive tapping techniques.

Finally, in most cases fiber optic cable is less restricted under harsh environmental conditions than its metallic counterparts. It is not as fragile or brittle as might be expected and it is more corrosion resistant than copper. The only chemical that affects optical fiber is hydrofluoric acid. In case of fire, an optical fiber can withstand greater temperatures than copper wire. Even when the outside jacket surrounding the fiber has melted, a graded index fiber optic system can still be operational in an emergency signaling system. One word of warning, however: care must be taken when pulling these cables through the building so the cable is not separated because its tensile strength is exceeded.

One more caution with regard to control and security of the connector cables against surreptitious taps. The maximum 50-foot cable length of the RS232C or the 4,000-foot cable length of the RS449 could be prime targets. The RS449 offers extra control features such as special circuits for moving from a primary private line service to a packet switched service when backup is needed or simply to access another database that is not normally used. This eliminates manual patching, switching keys, and so on. The RS449 can invoke tests to isolate problems with either the local or remote data circuit-terminating equipment (DCE) or the communication circuit itself. You might review the section in Chapter 4 on RS232C/RS449/X.21.

PROTOCOL CONTROLS

Protocols are simply the rules by which two machines talk to each other. The word "protocol" comes from the greek *protokollon*, which is the first sheet glued to a papyrus roll; it was the table of contents.

The International Organization for Standardization (ISO) has developed a seven-layer (OSI model) protocol. These layers and some ideas for their control are as follows (if you are not familiar with the seven-layer OSI model, it is described in Chapter 6):

- **Layer 1: The Physical Link Control**   The physical layer is concerned with transmitting raw bits over a communication channel. It describes the physical, electrical and functional interchange that establishes, maintains, and disconnects the physical link between data terminal equipment (DTE) and data circuit-terminating equipment (DCE).

  At this layer controls are needed to physically protect the connector cable. An example might be that an RS449 cable offers more control pins than an RS232C cable (see Figures 4-4 and 4-5). The goal at this layer is to control physical access by employees and vendors and to try to identify breaches of security and/or restrict entry to the system at this physical layer, as well as each of the following six layers.

- **Layer 2: Data Link Control (DLC)**   Data link control contains the functions that transfer data over the link established by layer 1. The task of the data link layer is to take a raw transmission facility and transform it into a circuit that appears free of transmission errors to the network layer (layer 3). It accomplishes this task by breaking the input data into *data frames*, transmitting the frames sequentially, and processing the acknowledgment frames back to the original sender.

  At this layer the protocol should contain controls such as sequence counting of frames, error detection and retransmission capabilities, identification of lost frames, reduction of possible duplicate transmissions to zero. It should solve problems caused by damaged/lost/duplicate frames, prevent a fast transmitter from drowning a slow receiver in data, provide limited restart capabilities in case of abnormal termination situations, ensure that some of the transmitted data are not misinterpreted as line control characters, increase flow control efficiency to ensure that the maximum number of frames can be sent without requiring an acknowledgment, properly terminate a session, and the like.

- **Layer 3: Network Control**   Network control provides for the functions of internal network operations such as addressing and routing. This is probably the software located in the terminal or intelligent controller at the remote end and the front end communication controller at the host end, although it may be in the packet switching node (SN). Layer 3 determines the chief characteristics of how packets (the units of information) are exchanged and routed within the network. The major issue here, which is confusing, is the division of labor between the host computer and the front end processor.

  Some of the controls to be questioned in this layer involve who should ensure that all packets are received correctly at their destinations and in the proper order. This layer of protocol should accept messages from the host, convert them to packets, and see to it that the packets get directed toward their destination. Packet routing should be controlled here; there also might be some global or local databases at this layer that should be kept secure. Control of congestion, such as too many packets on one channel, should be controlled by this layer. Also, this layer can contain billing routines for charging users and should be reviewed for possible problems such as error, theft of time, or improper message charges.

- **Layer 4: Transport Control**   Transport control provides transport services to the users for network independent interfacing from source to destination (end-to-end) across the network. At this layer we are out of the area of message protocols and into the area of software and network architectures. Layers 4 through 7 involve network architectures, whereas layers 1 through 3 involve basic message protocols. Layer 4 is unique because it can be either protocol or network architecture software. The basic function of the transport layer (also known as the host-host layer) is to accept entire messages from the session layer (layer 5), split it into smaller units, pass these to the network layer (layer 3), and ensure that all the pieces arrive correctly at the other end.

   Some of the controls that should be checked at layer 4 are related to network connections because the transport layer might have to create multiple network connections in order to get the required number of circuit paths. At this layer multiplexing might be invoked, so multiplexing controls should be reviewed. At this layer also a program on a source machine carries on a conversation with a similar program on the destination machine using headers and control messages; therefore, some of the controls might be in the application programs. At the lower layers (layers 1–3) the protocols are carried out by each machine and its immediate neighbors rather than the ultimate source and destination machines, which may be separated by many hardware devices and circuit links. Another needed control is one that determines if the software at this level can tell which machine belongs to which connection. Other controls that are performed at this level, even though they may be performed elsewhere as well, are source/destination machine addressing and flow control (here it is flow of messages rather than flow of packets) so one machine cannot overrun another.

- **Layer 5: Session Control**   Session control supports the dialogue within a session. Operating system supervisors traditionally support this function. The session layer is the user's interface into a network. It is at this layer that the user negotiates to establish a connection with a process on the other machines.

   Controls that should be examined at this layer are the typical controls that relate to a terminal (dedicated or dial-up), such as passwords, log-in procedures, terminal addressing procedures, authentication of terminals and/or users, correct delivery of the bill, and so on. Another control occurs when the transport control (layer 4) connections are unreliable; the session layer may be required to attempt to recover from broken transport connections. As another example, in database management systems it is crucial that a complicated transaction against the database never be aborted halfway through the routine because this leaves the database in an inconsistent state. The session layer often provides a facility by which a group of messages can be set aside so that none of them is delivered to the remote user until all of them have been completed. This mechanism ensures that a hardware or software failure within the subnetwork can never cause a transaction to be aborted halfway through. The session layer also can provide for sequencing of messages when the transport layer does not.

- **Layer 6: Presentation Control**   Presentation control provides for the transformation or conversion of data formats. Examples are compaction, encryption, peripheral device coding, and formatting. The presentation layer performs functions that are requested sufficiently often as to warrant finding a general solution for them.

  At this layer there are controls such as software encryption, text compression, text compaction, and conversion of incompatible file formats/file conversions so two systems can talk to one another. Also this layer can take incompatible terminals and modify line and screen length, end-of-line conventions, scroll versus page mode, character sets, and cursor addressing to make them compatible. Simple errors at the remote terminal might be caused by the software at this layer.

- **Layer 7: Application Control**   The application control layer performs the application programs and system activities in support of the business functions. The content of the application layer is up to the individual user/organization.

  This layer has controls that are related logically to the business system under review. Controls in the application layer are the typical day-to-day logical controls that are built into business systems.

The most common protocol that you might have to review is the international standard for packet switching networks. It is called X.25 and it is described in Chapter 6.

NETWORK ARCHITECTURE/SOFTWARE CONTROLS

Controls that relate to network architecture/software typically are associated with layers 4 through 7 of the OSI model. Additional architecture/software controls relate to computer operating systems, teleprocessing monitors, telecommunication access programs, databases, and security software packages.

Teleprocessing monitors are programs that relieve the operating system of many of the tasks involved in handling message traffic between the host and remote terminals, such as line handling, access methods, task scheduling, and system recovery. System throughput is increased by offloading these data communication functions from the operating system to the teleprocessing monitor.

Some of the controls that should be reviewed for teleprocessing monitors are access controls, who can sign onto a terminal, and who can access program routines (sometimes called *exits*). With regard to these exits, the code of each exit routine should be checked for correctness and security. These exit program modules should be placed in software-controlled libraries.

The vendor's "system generation" manuals or the teleprocessing monitor should be reviewed to determine if any security controls were built into the monitor by the vendor. See Appendix 6, Evaluating Teleprocessing Monitors.

Telecommunication access programs are vendor-supplied programs that control the transmission of data to and from the host computer and various data communication devices. The telecommunication access programs are more likely to reside in a front end communication processor but they can reside in the host computer.

Some controls for the telecommunication access program may be documented in the vendor's "system generation" manuals. The controls that were built in by the software vendor should be evaluated. As with teleprocessing monitors, user program routines (exits) and access methods need to be examined. Review the list of front end functions that were described in Chapter 2.

Another area that interfaces with the teleprocessing monitor is the very sensitive network control database. For example, there might be a *system database* containing global information about addresses and logical names of all peripheral devices, locations of system files, system timing parameters, task locations and priorities, peripheral device control tables, and system supply command lists. Because of its importance to the security of the entire system, data in the system database must be protected adequately from copying or destruction.

Another database is the *network database*. It contains data such as the number of stations, polling/selecting lists, current station identifier, communication control port addresses, logical terminal identifiers, terminal device list, terminal poll/call sequences, dial-up numbers, message and process information, and the like. The network database also must be protected from unauthorized copying.

The network database can be used to cross-check against manual documentation. Because it contains information such as the number of stations and logical terminal identifiers, a copy of the database can be matched against the written network documentation as a means of verifying the currency of the documentation.

The impact of any security software packages that restrict or control access to files, records, or data items should be reviewed. These packages are independent of the data communications software.

Finally, software should be protected in case of a disastrous situation such as a power failure. Restart recovery routines should be available, and the system should only have one master input terminal for entering sensitive or critical commands. All default options should be identified, and the impact of default options that do not operate properly should be assessed to determine whether adequate software maintenance is available. Also all sensitive tables (passwords) should be protected in the memory.

ERROR CONTROL IN DATA COMMUNICATIONS

There are two categories of errors. The first category involves corrupted (changed) data, and the second involves lost data. With regard to selecting an error control system, some of the following factors should be considered:

- The extent and pattern of error-inducing conditions on the type of circuit used.
- The effects of no error control, or error detection/retransmission (type of ARQ, see Chapter 6), and of automatic error detection and correction (forward error correction).
- The maximum error rate that can be tolerated.
- Comparison of the cost of increased accuracy with the present cost of correcting errors.
- Comparison of different application systems as to the overall transmission accuracy currently being achieved.
- Cost of errors remaining in the received station to flag them and reenter them.

Data Communication Errors Errors are a fact of life in today's data communication networks. Depending on the type of circuit/line, they may occur every few minutes or every few seconds or even more frequently. They occur because of noise on the lines (types of line noise are discussed in the next section). No data communication system can prevent all these errors from occurring, but most of them can be detected and many corrected by proper design. Common carriers that lease data transmission lines to users provide statistical measures specifying typical error rates and the pattern of errors that can be expected on the different types they lease.

Normally, errors appear in bursts. In a burst error more than one data bit is changed by the error-causing condition. This is another way of saying that 1-bit errors are not uniformly distributed in time. However, common carriers usually list their error rates as the number of bits in error divided by the number of bits transmitted, without reference to their nonuniform distribution. For example, the error rate might be given as 1 in 500,000 when transmitting on a public voice grade telephone circuit at 1,200 bps.

The fact that errors tend to be clustered in bursts rather than evenly dispersed has both positive and negative aspects. If the errors were not clustered (but instead were evenly distributed throughout the day), with an error rate of 1 bit in 500,000 it would be rare for two erroneous bits to occur in the same character, and consequently some simple character checking scheme would be effective. But this is not the case, because bursts of errors are the rule rather than the exception. They sometimes go on for time periods that may obliterate 50 to 100 or more bits. The positive aspect is that, between bursts, there may be rather long periods of error-free transmission. Therefore no errors at all may occur during data transmission in a large proportion of messages. For exa#ple, when errors are #ore or less evenly distrib#ted, it is not di#ficult to gras# the me#ning even when the error #ate is high, as it is in this #entence (1 charac#er in 20). On the other hand, if errors are concentrated in bursts, it becomes more difficult to recover the meaning and much more reliance must be placed on knowledge of message #######[2] or on special logical/numerical error detection and correction methods.

It is possible to develop data transmission methodologies that give very high error detection and correction performance. The only way to do the detection and correction is to send along extra data. The more extra data that are sent, the more the error protection that can be achieved. However, as this protection is increased, the throughput of useful data is reduced. Therefore, the efficiency of data throughput varies inversely as the desired amount of error detection and correction is increased. Errors will even have an effect on the length of the block of data to be transmitted when synchronous transmission is used. The shorter the message blocks used, the less likelihood there is of needing retransmission for any one block. But the shorter the message block, the less efficient is the transmission methodology as far as throughput is concerned. If the message blocks are long, a higher proportion may have an error and have to be resent.

In transmissions over the dial-up switched network, a considerable variation in the error rate is found from one time of the day to another. The error rate is usually higher during the periods of high traffic (the normal business day). In some cases the only alternative open to the user of these facilities is to transmit the data at a slower speed because higher transmission speeds are more error prone. Dial-up lines are more prone to errors because they have less stable transmission parameters than private leased lines, and, because different calls use different circuits, they usually experience different transmission conditions. Thus, a bad line is not necessarily a serious problem in dial-up transmission; a new call may result

---

[2]In case you could not guess, the word is "context."

in getting a better line. Line conditioning, a service that is not available on dial-up lines, but only on private leased lines, consists of special electrical balancing of the circuit to ensure the most error-free transmission.

Line Noise and Distortion   Line noise and distortion can cause data communication errors. In this context we define noise as undesirable electrical signals. It is introduced by equipment or natural disturbances and it degrades the performance of a communication line. If noise occurs, the errors are manifested as extra or missing bits, or bits whose states have been "flipped," with the result that the message content is degraded. Line noise and distortion can be classified into roughly 11 categories: white noise, impulse noise, cross talk, echoes, intermodulation noise, amplitude changes, line outages, attenuation, attenuation distortion, delay distortion and jitter.

White or Gaussian noise is the familiar background hiss or static on radio and telephones. It is noise caused by the thermal agitation of electrons and because of this, it is inescapable. Even if the equipment utilized were perfect and the wires were perfectly insulated from any and all external interference, there would still be some white noise. White noise is usually not a problem unless its level becomes so high that it obliterates the data transmission. Sometimes noise from other sources such as power line induction, cross modulation from adjacent lines, and a conglomeration of random signals resembles white noise and is labeled as such even though it is not caused by thermal electrons.

Impulse noise (sometimes called *spikes*) is the primary source of errors in data communications. An impulse of noise can last as long as 1/100th of a second. An impulse of this duration would be heard as a click or a crackling noise during voice communications. This click would not affect voice communications but it might obliterate a group of data bits, causing a burst error on a data communication line. At 150 bps, 1 or 2 bits would be changed by a spike of 1/100th of a second, whereas at 4,800 bps, 48 bits would be changed. Some of the sources of impulse noise are voltage changes in adjacent lines or circuitry surrounding the data communication line, telephone switching equipment at the telephone exchange branch offices, arcing of the relays at older telephone exchange offices, tones used by network signaling, maintenance equipment during line testing, lightning flashes during thunderstorms, and intermittent electrical connections in the data communication equipment.

Cross talk occurs when one line picks up some of the signal traveling down another line. It occurs between line pairs that are carrying separate signals, in multiplexed links carrying many discrete signals, in microwave links where one antenna picks up a minute reflected portion of the signal from another antenna on the same tower, and in any hard-wire telephone circuits that run parallel to each other, are too close to each other, and are not electrically balanced. You are experiencing cross talk during voice communication on the public switched network when you hear other conversations in the background. Cross talk between lines will increase with increased communications distance, increased proximity of the two wires, increased signal strength, and higher frequency signals. Cross talk, like white noise, has such a low signal strength that it is normally not bothersome on data communication networks.

Echoes and echo suppression can be a cause of errors (echo suppressors were discussed in Chapter 3). An echo suppressor causes a change in the electrical balance of a line and this change causes a signal to be reflected so it travels back down the line at reduced signal strength. Whenever the echo suppressors are disabled, as in data transmission, this echo returns to the transmitting equipment. If the signal strength of the echo is high enough to be detected by the communication equipment, it will cause errors. Echoes, like cross talk and white noise, have such a low signal strength that they are normally not bothersome.

Intermodulation noise is a special type of cross talk. The signals from two independent lines intermodulate and form a product that falls into a frequency band differing from both inputs. This resultant frequency may fall into a frequency band that is reserved for another signal. This type of noise is similar to harmonics in music. On a multiplexed line, many different signals are amplified together and slight variations in the adjustment of the equipment can cause intermodulation noise. A maladjusted modem may transmit a strong frequency tone when not transmitting data, thus yielding this type of noise.

Amplitude noise involves a sudden change in the level of power. The effect of this noise depends on the type of modulation being used by the modem. For example, amplitude noise does not affect frequency modulation techniques; this is because the transmitting and receiving equipment interprets frequency information and disregards the amplitude information. Some of the causes of amplitude noise may be faulty amplifiers, dirty contacts with variable resistances, sudden added loads by new circuits being switched on during the day, maintenance work in progress, and switching to different transmission lines.

Line outages are a catastrophic cause of errors and incomplete transmission. Occasionally a communication circuit fails for a brief period of time. This type of failure may be caused by faulty telephone branch office exchange equipment, storms, loss of the carrier signal, and any other failure that causes an open line or short circuit.

Attenuation is the loss of power that the signal suffers as it travels from the transmitting device to the receiving device. It results from the power that is absorbed by the transmission medium or is lost before it reaches the receiver. As the transmission medium absorbs this power, the signal gets weaker, and the receiving equipment has less and less chance of correctly interpreting the data. To avoid this, telephone lines have repeater/amplifiers spaced through their length. The distance between them depends upon the amount of power lost per unit length of the transmission line. This power loss is a function of the transmission method and circuit medium. Also, attentuation increases as frequency increases or as the diameter of the wire decreases.

Attenuation distortion refers to high frequencies losing power more rapidly than low frequencies during transmission. The received signal can thus be distorted by unequal loss of its component frequencies.

Delay distortion can cause errors in data transmission. Delay distortion occurs when a signal is delayed more at some frequencies than at others. If the method of data transmission involves data transmitted at two different frequencies, then the bits being transmitted at one frequency may travel slightly faster than the bits transmitted at a different frequency. A piece of equipment, called an *equalizer*, compensates for both attenuation distortion and delay distortion.

Jitter may affect the accuracy of the data being transmitted. The generation of a pure carrier signal is impossible. Minute variations in amplitude, phase, and frequency always occur. Signal impairment may be caused by continuously and rapidly changing gain and/or phase changes. This jitter may be random or periodic.

Approaches to Error Control   Error control implies (1) techniques of design and manufacture of data communication transmission links and equipment to reduce the occurence of errors (an area that is outside the scope of this book), and (2) methodologies to detect and correct the errors that are introduced during transmission of the data. In the sense of the second meaning of error control, the methodologies fall into three categories, and possibly four if you consider the option of ignoring the errors.

- Loop or echo checking
- Error detection with retransmission
- Forward error correction (FEC)

Each of these three approaches will be discussed in this section.

Loop or Echo Checking   Loop or echo checking does not use a special code. Instead, each character or other small unit of the message, as it is received, is transmitted back to the transmitter, which checks to determine whether the character is the same as the one just sent. If it is not correct, then the character is transmitted a second time. This method of error detection is wasteful of transmission capacity because each message (in pieces) is transmitted at least twice and there is no guarantee that some messages might not be transmitted three or four times. Also, some of this retransmission of characters for a second or third time might not be necessary since the error could have occurred on the return trip of the character. This would require the transmitter to retransmit the character even though it was in reality received correctly the first time. Loop or echo checking is usually utilized on hard-wire, short lines, with low-speed terminals. This type of error checking does give a high degree of protection but it is not as efficient as other methods. It is sometimes confused with full duplex transmission.

Error Detection with Retransmission   Error detection and retransmission schemes are built into data transmitting and receiving devices, front end computers, modems, and software. These schemes include detection of an error and immediate retransmission, detection of an error and retransmission at a later time, or detection of an error and retransmission for up to, say, three tries and then retransmission at a later time, or the like. Error detection and retransmission is the simplest, and if properly handled, the most effective and least expensive method to reduce errors in data transmission. It requires the simplest logic, needs relatively little storage, is best understood by terminal operators, and is most frequently used. Retransmission of the message in error is straightforward. It is usually called for by the failure of the transmitter to receive a positive acknowledgment within a preset time. Various methods are used to determine that the message that has just been received has, in fact, an error imbedded in it. Some of the common error detection methods are parity checking, constant ratio codes, and polynomial checking. ARQ (stop-and-wait/continuous) was described in Chapter 6.

Parity checking: If you examine a character from the ASCII coding structure, it soon becomes apparent that 1 of the 8 bits encoding each character is redundant—i.e., its value is solely determined by the values of the other 7 and is therefore unnecessary. Since this eighth bit cannot transmit any new information, its purpose is the confirmation of old information. The most common rule for fixing the value of the redundant bit uses the "parity" (evenness or oddness) of the number of 1s in the code. Thus, for an even parity code system using ASCII:

- Letter "V" is encoded 0110101. Since the number of 1s is 4, already an even number, a zero is added in the parity (eighth) position, yielding V = 01101010.
- Letter "W" is encoded 0001101, which has an odd number of 1s. Therefore a 1 is added in the parity position to make the number of 1s even, yielding W = 00011011.

A little thought will convince you that any single error (a switch of a 1 to a 0 or vice versa) will be detected by a parity check but that nothing can be deduced about which bit was in error. If the states of *two* bits are switched, the parity check may not sense any error. Of course, it may be possible to sense such an error because the resulting code, although correct as far as parity is concerned, is a code that is "forbidden," e.g., undefined or inappropriate in its context. Such detection, of course, requires more circuitry or software.

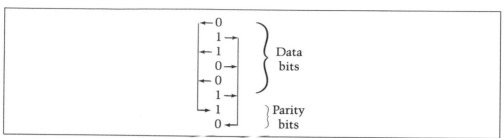

FIGURE 8–5 Cyclical parity check for a 6-bit code.

Another parity checking technique is the *cyclical parity check* (sometimes called *interlaced parity*). This method requires two parity bits per character. Assuming a 6-data bit code structure, the first parity bit would provide parity for the first, third, and fifth bits, and the second parity bit would provide parity for the second, fourth, and sixth bits. Figure 8–5 shows an even parity cyclical parity check on a 6-bit code.

Constant ratio codes: N out of M codes are special data communication codes that have a constant ratio of the number of 1 bits to the number of 0 bits. The most common one is IBM's 4-of-8 code that was discussed in Chapter 3. N out of M codes detect an error whenever the number of 1 bits and 0 bits are not in their proper ratio. For example, in the 4-of-8 code there are always supposed to be four 1 bits and four 0 bits in the received bit configuration of the character. Whenever this ratio is out of balance the receiving equipment knows that an error has occurred. N out of M codes are not widely utilized because they are inefficient. As an example of their inefficiency, consider that the 4-of-8 code has 70 valid character combinations while a 7-bit ASCII code has 128 valid character combinations $(2^7 = 128)$.

Polynomial checking: Polynomial checks (also called *Cyclical Redundancy Check* or *CRC*) on blocks of data are often performed for synchronous data transmission. In this type of message checking, all the bits of the message are checked by application of a mathematical algorithm. For example, all the 1 bits in a message are counted and then divided by a prime number (such as 17) and the remainder of that division is transmitted to the receiving equipment. The receiving equipment performs the same mathematical computations and matches the remainder that it calculated against the remainder that was transmitted with the message. If the two are equal, the entire message block is assumed to have been received correctly. In actual practice, much more complex algorithms are utilized.

One of the most popular of the polynomial error checking schemes is *Cyclical Redundancy Check* (CRC). It consists of adding bits (about 10 to 25) to the entire block. In CRC checking the data block can be thought of as one long binary polynomial, P. Before transmission, equipment in the terminal divides P by a fixed binary polynomial, G, resulting in a whole polynomial, Q, and a remainder, R/G.

$$\frac{P}{G} = Q + \frac{R}{G}$$

The remainder, R, is appended to the block before transmission, as a check sequence k bits long. The receiving hardware divides the received data block by the same G, which generates an R. The receiving hardware checks to ascertain if the received R agrees with the locally generated R. If it does not, the data block is assumed to be in error and retransmission is requested. In ARQ systems, a 25-bit CRC code added to a 1000-bit block allows only three bits in 100 million to go undetected. That is, for a 2.5% redundancy, the error rate is 3×10^{-8}.

Forward Error Correction This approach uses codes that contain sufficient redundancy to permit errors to be detected and corrected at the receiving equipment without retransmission of the original message. The redundancy, or extra bits required, varies with different schemes. It ranges from a small percentage of extra bits to 100% redundancy, with the number of error-detecting bits roughly equaling the number of data bits. One of the characteristics of many error-correcting codes is that there must be a minimum number of error-free bits between bursts of errors. For example, one such code, called a *Hagelbarger Code*, will correct up to six consecutive bit errors provided that the 6-bit error group is followed by at least 19 valid bits before further error bits are encountered. Bell Telephone engineers have developed an error correcting code that uses 12 check bits for each 48 data bits, or 25% redundancy. Still another code is the *Bose-Chaudhuri Code*, which, in one of its forms, is capable of correcting double errors and can detect up to four errors.

To show how such a code works, consider this example of a forward error-checking code, called a *Hamming Code*, after its inventor, R. W. Hamming.[3] This code associates even parity bits with unique combinations of data bits. Using a 4-data-bit code as an example, a character might be represented by the data bit configuration 1010. Three parity bits P_1, P_2, and P_4 are added, resulting in a 7-bit code, shown in the upper half of Figure 8–6. Notice that the data bits (D_3, D_5, D_6, D_7) are 1010 and the parity bits (P_1, P_2, P_4) are 101.

As depicted in the upper half of Figure 8–6, parity bit P_1 applies to data bits D_3, D_5, and D_7. Parity bit P_2 applies to data bits D_3, D_6, and D_7. Parity bit P_4 applies to data bits D_5, D_6, and D_7. For the example, in which $D_3, D_5, D_6, D_7 = 1010$, P_1 must equal 1 since there is but one 1 among D_3, D_5, and D_7 and parity must be

[3]William P. Davenport, *Modern Data Communication-Concepts, Language, and Media* (New York: Hayden Book Company, Inc., 1971), p. 96.

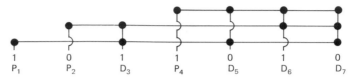

Checking Relations Between Parity Bits (P) and Data Bits (D)

0 = Corresponding parity check is correct 1 = Corresponding parity check fails			Determines in which bit the error occurred
P_4	P_2	P_1	
0	0	0	→ no error
0	0	1	→ P_1
0	1	0	→ P_2
0	1	1	→ D_3
1	0	0	→ P_4
1	0	1	→ D_5
1	1	0	→ D_6
1	1	1	→ D_7

Interpreting Parity Bit Patterns

FIGURE 8–6 Hamming Code for forward error correction.

even. Similarly P_2 must be 0 since D_3 and D_6 are 1s. P_4 is 1 since D_6 is the only 1 among D_5, D_6, D_7.

Now, assume that during the transmission, data bit D_7 is changed from a 0 to a 1 by line noise. Because this data bit is being checked by P_1, P_2, and P_4, all three parity bits will now show odd parity instead of the correct even parity. (D_7 is the only data bit that is monitored by all three parity bits, therefore whenever D_7 is in error, all three parity bits will show an incorrect parity.) In this way, the receiving equipment can determine which bit was in error and reverse its state, thus correcting the error without retransmission.

The bottom half of Figure 8–6 is a table that determines the location of the bit in error. A 1 in the table means that the corresponding parity bit indicates a parity error. Conversely, a 0 means the parity check is correct. These 0s and 1s form a binary number that indicates the numerical location of the erroneous bit. In the example above, P_1, P_2, and P_4 checks all failed, yielding 111, or a decimal 7, the subscript of the erroneous bit.

Error detection and correction methodologies come in many varieties. The data communication network designer must give careful consideration to all facets of the system being designed and make appropriate use of the various methodologies to control and correct errors.

MANAGEMENT CONTROLS

Network management involves setting up a central control philosophy with regard to the overall network functions. The network manager should be independent of the other managers in the data processing environment or even independent of the information systems department itself. Some of the general responsibilities of this job function would be design and analysis, network operations, failure control, and testing/problem management (see Chapter 7).

Some network management controls include the following. The network management team should have a national account with the common carrier when possible. There should be a central call number to log all problems in regard to who, what, where, when, why, the telephone number, date, and time of a problem. The failure control group should compile statistics for their hardware such as Mean Time Between Failures (MTBF). The network hardware vendor often can supply these data. The network management people also should maintain statistics on the time from failure to recovery. In its most detailed form this is comprised of the Mean Time To Diagnose (MTTD) plus the Mean Time To Respond (MTTR) plus the Mean Time To Fix (MTTF). Chapter 7 presents the detailed usage for MTBF, MTTD, MTTR, and MTTF.

In addition to the central control for problem reporting on the network, the network management group should maintain other statistics such as the cumulative network downtime, subnetwork downtime, circuit utilization reports, response time analysis, queue length descriptions, histograms of daily usage (such as number of characters transmitted per day per circuit), failure rates of the circuits such as the number of retransmitted messages, local host and file activity statistics, local device error activities, network gateway failures, distribution of character and packet volume, distribution of traffic by time of day and location, peak volumes, and a statistical profile of all time-related network traffic. These reports, or similar ones, should be available for managing the network.

With regard to network documentation (some of this can be cross-checked for currency in the global or local network databases), a good network management team should have some or all of the following: circuit layout record, network maps, hardware/software cross-references, all network vendor maintenance records, software listings by network task and component, all user site telephone numbers and names of individuals to contact, an interface component maintenance history log, circuit controlled telephone contact index, maintenance history by component, inventory by serial number of network components, network redundancy locations and switching criteria, vendor contracts and vendor contacts, and a current list of personnel working in the network center.

A control review of network management should ascertain the existence of appropriate operational manuals and a comprehensive description on how the network operates. There should be adequate recovery procedures, backup procedures and disaster plans.

With regard to communication test equipment, not much can be done to control it because it is in continuous use. Network management must recognize that misuse can allow breaches of privacy or the insertion of illegal messages.

A network monitor is mandatory when bit-oriented protocols such as X.25 are used. The only control that can be put on such a device is a keylock for its switch. The keylock should be turned off and the key removed when the equipment is not being used. This prevents people from browsing over data as they pass through the data communication network.

Loopback test equipment should be used to diagnose the location and cause of problems.

Microprocessor based network analyzers permit checks for poll-to-poll or poll-to-response times. Such checks aid network management in assessing polling efficiency.

Other, smaller hand-held test devices, such as break-out boxes, allow test personnel to send test patterns of data bits to a modem through the RS232C or RS449 cable.

The primary control for test equipment is to ensure that only qualified people use this equipment and that they use it only when necessary.

RECOVERY/BACKUP/DISASTER CONTROLS

Recovery and backup controls within the data communication network encompass many areas. The person who is reviewing these controls may start at either end of the network (remote terminals or central host computer). The object is to check for recovery procedures and backup hardware throughout the network. Perhaps the most important question to ask is this: Is it cost effective to back up each piece of hardware encountered between a remote terminal site and the central host computer? A related question is: Are there software procedures for recovery of data files, network databases, network software, and the like? Use of Figure 8–2 during the review of recovery/backup/disaster controls will help ensure that all network control points are considered.

One important area involves backup of the communication circuits. One option is to lease two separate physical circuits (that have been alternately routed) in order to have one for backup. Another option is to utilize dial-up communication circuits as backup to leased circuits. Of course, another alternative is to have manual procedures that can be used if the circuit is down for a very short period of time, perhaps several hours at the maximum.

There should be recovery and restart capabilities in the event of either a hardware crash or a software crash. Backup facilities should include backup power, possibly at both the local and remote sites.

With regard to a data communication network disaster plan, a separate plan for each of *six* different areas should be developed. These disaster plans are for (1) the data communication network control center, (2) the communication circuits, (3) remote switches/concentrators/intelligent terminal controllers, (4) common carrier (telephone company) facilities, (5) electric power for the data communication facilities and user terminals/lights, and (6) the user application systems.

A data communication network disaster plan should spell out the following details:

- The decision-making manager who is in charge of the disaster recovery operation. A second manager should be indicated in case the first manager is unavailable.
- Availability and training of backup personnel with sufficient data communication knowledge and experience.
- Recovery procedures for the data communication facilities. This is information on the location of circuits, who to contact for backup data circuits and documentation, as well as preestablished priorities as to which data circuits will be reconstructed first.
- How to replace damaged data communication hardware and software that is supplied by vendors. Outline the support that can be expected from vendors, along with the name and telephone number of who to contact.
- Location of alternate data communication facilities and equipment such as connector cables, local loops, IXCs, common carrier switching facilities, and other public networks.
- Action to be taken in case of partial damage or threats such as bomb threat, fire, water, electrical, sabotage, civil disorders, or vendor failures.
- Procedure for imposing extraordinary controls over the network until the system returns to normal.
- Storage of the disaster recovery procedures in a safe area where they will not be destroyed by the catastrophe. This area must be accessible, however, by those who need to use the plans.

MATRIX OF CONTROLS

In order to be sure that the data communication network has all the necessary controls and that these controls offer adequate protection, it is advisable to build a two-dimensional matrix that incorporates all the controls that *currently* are present in the network.

This matrix is constructed by identifying first all threats facing the network and second, all the network's component parts.

- Errors and Omissions—The accidental or intentional transmission of data that is in error, including the accidental or intentional omission of data that should have been entered or transmitted on the on-line system. This type of exposure includes, but is not limited to, inaccurate data, incomplete data, malfunctioning hardware, and the like.
- Message Loss or Change—The loss of messages as they are transmitted throughout the data communication system, or the accidental/intentional changing of messages during transmission.
- Disasters and Disruptions (natural and man-made)—The temporary or long-term disruption of normal data communication capabilities. This exposure renders the organization's normal data communication on-line system inoperative.
- Breach of Privacy—The accidental or intentional release of data about an individual, assuming that the release of this personal information was improper to the normal conduct of the business at the organization.
- Security/Theft—The security or theft of information that should have been kept confidential because of its proprietary nature. In a way, this is a form of privacy, but the information removed from the organization does not pertain to an individual. The information might be inadvertently (accidentally) released, or it might be the subject of an outright theft. This exposure also includes the theft of assets such as might be experienced in embezzlement, fraud, or defalcation.
- Reliability (Uptime)—The reliability of the data communication network and its "uptime." This includes the organization's ability to keep the data communication network operating and the mean time between failures (MTBF) as well as the time to repair equipment when it malfunctions. Reliability of hardware, reliability of software, and the maintenance of these two items are chief concerns here.
- Recovery and Restart—The recovery and restart capabilities of the data communication network, should it fail. In other words, How does the software operate in a failure mode? How long does it take to recover from a failure? This recovery and restart concern also includes backup for key portions of the data communication network and the contingency planning for backup, should there be a failure at any point of the data communication network.
- Error Handling—The methodologies and controls for handling errors at a remote distributed site or at the centralized computer site. This may also involve the error handling procedures of a distributed data processing system (at the distributed site). The object here is to ensure that when errors are discovered they are promptly corrected and reentered into the system for processing.
- Data Validation and Checking—The validation of data either at the time of transmission or during transmission. The validation may take place at a remote site (intelligent terminal), at the central site (front end communication processor), or at a distributed intelligence site (concentrator or remote front end communication processor).

FIGURE 8–7 General threats.

- A *threat* to the data communication network is any potential adverse occurrence that can harm the network, interrupt the systems that use the network, or cause a monetary loss to the organization. For example, lost messages are a potential threat.

- A *component* is one of the individual pieces that, when assembled together, make up the data communication network. A component can be viewed as the item that is being reviewed or the item over which we are attempting to maintain control. Thus, the components are the hardware, software, circuits, and other pieces of the network.

In Figure 8–7 several *general* threats to a data communication network are shown. Figure 8–8 identifies several *general* component parts for a data communication network.

- Host Computer—Most prevalent in the form of a central computer to which the data communication network transmits and from which it receives information. In a distributed system, with equal processing at each distributed node, there might not be an identifiable central computer (just some other equal-sized distributed computer).

- Software—The software programs that operate the data communication network. These programs may reside in the central computer, a distributed-system computer, the front end communication processor, a remote concentrator or statistical multiplexer, and/or a remote intelligent terminal. This software may include the telecommunications access methods, an overall teleprocessing monitor, programs that reside in the front end processors, and/or programs that reside in the intelligent terminals.

- Front End Communication Processor—A hardware device that interconnects all the data communication circuits (lines) to the central computer or distributed computers and performs a subset of the following functions: code and speed conversion, protocol, error detection and correction, format checking, authentication, data validation, statistical data gathering, polling/addressing, insertion/deletion of line control codes, and the like.

- Multiplexer, Concentrator, Switch—Hardware devices that enable the data communication network to operate in the most efficient manner. The *multiplexer* is a device that combines, in one data stream, several simultaneous data signals from independent stations. The *concentrator* performs the same functions as a multiplexer except it is intelligent and therefore can perform some of the functions of a front end communication processor. A *switch* is a device that allows the interconnection between any two circuits (lines) connected to the switch. There might be two distinct types of switch: a switch that performs message switching between stations (terminals) might be located within the data communication network facilities that are owned and operated by the organization; a circuit or line switching switch that interconnects

---

FIGURE 8–8   General components. (*Continued on next page.*)

various circuits might be located at (and owned by) the telephone company central office. For example, organizations perform message switching and the telephone company performs circuit switching.

- Communication Circuits (Lines)—The common carrier facilities used as links (a *link* is the interconnection of any two stations/terminals) to interconnect the organization's stations/terminals. These communication circuits include satellite facilities, public switched dial-up facilities, point-to-point private lines, multiplexed lines, multipoint or loop configured private lines, and many others.

- Local Loop—The communication facility between the customer's premises and the telephone company's central office or the central office of any other special common carrier. The local loop is usually assumed to be metallic pairs of wires.

- Modems—A hardware device used for the conversion of data signals from terminals (digital signal) to an electrical form (analog signal) which is acceptable for transmission over the communication circuits that are owned and maintained by the telephone company or other special common carrier.

- People—The individuals responsible for inputting data, operating and maintaining the data communication network equipment, writing the software programs for the data communications, managing the overall data communication network, and those involved at the remote stations/terminals.

- Terminals/Distributed Intelligence—Any or all of the input or output devices used to interconnect with the on-line data communication network. This resource would specifically include, without excluding other devices, teleprinter terminals, video terminals, remote job entry terminals, transaction terminals, intelligent terminals, and any other devices used with distributed data communication networks. These may include microprocessors or minicomputers when they are input/output devices or if they are used to control portions of the data communication network.

FIGURE 8–8 *Continued*

Identifying and documenting the controls in a network require the task of identifying the *specific* threats and components that relate to whatever network is used by the organization. After identifying the organization's specific threats and components, the individual controls that are in place can be related to these threats and components.

Once the threats and component parts of the network have been identified, the next step is to place a short description of each threat across the top of the matrix. Likewise, a short description of each component is placed down the left vertical axis of the matrix as shown in Figure 8–9.

When the horizontal and vertical axes have been labeled, the next step is to identify all of the specific controls that are being used currently in the data communication network. These "in-place" controls should be described and placed in a numerical list. For example, assume 24 controls have been identified as being in

THREATS

COMPONENTS	Errors and Omissions	Message Loss or Change	Disasters and Disruptions	Breach of Privacy	Security/ Theft	Reliability (Uptime)	Recovery and Restart	Error Handling	Data Validation and Checking
Host Computer or Central System									
Software									
Front End Communication Processor									
Multiplexer, Concentrator, Switch									
Communication Circuits (Lines)									
Local Loop									
Modems									
People									
Terminals/ Distributed Intelligence									

FIGURE 8–9 Blank matrix.

1. Ensure that the system can switch messages destined for a down station/terminal to an alternate station/terminal.

2. Determine whether the system can perform message switching to transmit messages between stations/terminals.

3. In order to avoid lost messages in a message-switching system, provide a store and forward capability. This is where a message destined for a busy station is stored at the central switch and then forwarded at a later time when the station is no longer busy.

4. Review the message or transaction logging capabilities to reduce lost messages, provide for an audit trail, restrict messages, prohibit illegal messages, and the like. These messages might be logged at the remote station (intelligent terminal), they might be logged at a remote concentrator/remote front end processor, or they might be logged at the central front end communication processor/central computer.

5. Transmit messages promptly to reduce risk of loss.

6. Identify each message by the individual user's password, the terminal, and the individual message sequence number.

7. Acknowledge the successful or unsuccessful receipt of all messages.

24. Consider the following special controls on dial-up modems when the data communication network allows incoming dial-up connections: change the telephone numbers at regular intervals; keep the telephone numbers confidential; remove the telephone numbers from the modems in the computer operations area; require that each "dial-up terminal" have an electronic identification circuit chip to transmit its unique identification to the front end communication processor; do not allow automatic call receipt and connection (always have a person intercept the call and make a verbal identification); have the central site call the various terminals that will be allowed connection to the system; utilize dial-out only where an incoming dialed call triggers an automatic dial-back to the caller (in this way the central system controls those telephone numbers to which it will allow connection).

FIGURE 8–10 Control list.

use in the network. Each one is described and they are numbered consecutively 1 through 24. The numbered list of controls has no ranking attached to it: the first control is number 1 just because it is the first control identified. Figure 8–10 shows what a list of in-place controls looks like.

Next, each of the controls that has been identified is placed into the proper cell of the matrix. This is accomplished by reading the description of each control and the control list and then asking the following two questions:

1. Which threat or threats does this control mitigate or stop?

2. Which component or components does this control safeguard or control?

For example, if the description of control 1 is "ensure that the system can switch messages for a down station/terminal to an alternate station/terminal," then the number 1 should be placed in the very first cell in the upper left corner (see Figure 8–11). This is because a control that ensures that the system can switch messages when a station is down helps control errors, and it also is a control that safeguards or resides in the host computer and/or front end. Figure 8–11 also shows control 1 in the cell that intersects between Message Loss or Change and Host Computer. Control 1 also appears in several other cells. The point is that by answering these two questions, you can place each control in the proper cells of the matrix.

The finished matrix with controls (Figure 8–11) shows the interrelationship of each "in-place" control to the threat that it is supposed to mitigate and the component that it safeguards or controls.

For a complete list of all of the controls for Figure 8–11 along with this control matrix, see Appendix 1.[4]

The last step in designing a custom matrix of controls for your specific data communication network involves a personal evaluation as to the adequacy of the controls. This is accomplished by reviewing each subset of controls as it relates to each threat and component area of the matrix. For example, the subset of controls that are listed down a column below a threat are evaluated. The object of this step is to answer the specific question, "Do we have the proper controls and are they adequate with regard to each specific threat?" Using Figure 8–11, look down the column under errors and omissions. The matrix clearly defines the specific subset of controls that relate to the threat area Errors and Omissions. They are 1, 2, 3, 4, 7, 12, 18, and 5.

This type of review also can be performed for various other subsets of controls. For example, individual subsets of controls can be evaluated as they relate to:

- Threats (columns)
- Components (rows)
- Individual cells
- Empty cells

---

[4]This material was taken from Chapter 4 of the textbook, *Internal Controls for Computerized Systems* by Jerry FitzGerald, published by Jerry FitzGerald and Associates, 506 Barkentine Lane, Redwood City, Calif. 94065.

THREATS

COMPONENTS	Errors and Omissions	Message Loss or Change	Disasters and Disruptions	Breach of Privacy	Security/ Theft	Reliability (Uptime)	Recovery and Restart	Error Handling	Data Validation and Checking
Host Computer or Central System	1, 2, 3, 4, 7	1, 2, 3, 4, 5, 7	1, 8, 11, 13, 16	6, 8, 24	6, 8, 24	1, 13, 16			6, 24
Software	1, 2, 3, 4, 7	1, 2, 3, 4, 5, 7	1, 8, 16	6, 8, 24	6, 8, 24	1			6, 24
Front End Communication Processor	1, 2, 3, 4, 7	1, 2, 3, 4, 5, 7	1, 8, 13, 16	6, 8, 24	6, 8, 24	1, 13, 16			6, 24
Multiplexer, Concentrator, Switch	1, 2, 3, 4, 7	1, 2, 3, 4, 5, 7	1, 8, 13, 16	6, 8, 24	6, 8, 24	1, 13, 16			6, 24
Communication Circuits (lines)	12		10, 15, 16, 18			15, 16			
Local Loop	12								
Modems	12, 18	18, 24	8, 9, 10, 11, 13, 14, 15, 16, 18	24	24	9, 10, 11, 13, 14, 15, 16, 17, 18	9, 10, 11, 14, 15	18, 19, 20, 22, 23	
People	5	5, 7		6, 8, 24	6, 8, 24				6
Terminals/ Distributed Intelligence		2		6, 8, 24	6, 8, 24	1			6, 24

FIGURE 8–11 Matrix with controls.

Looking at Figure 8–12, we see a pictorial diagram describing the above four areas that should be reviewed. The matrix approach offers a perfect tool for a detailed microanalysis of the controls in a data communication network. The matrix clearly shows the relationship between various subsets of controls and specific threat areas, component parts, individual cells, and empty cells.

LISTS OF DATA COMMUNICATION CONTROLS

In order to help you construct a matrix of controls that relate to your organization's data communication network, we have supplied 14 specific lists of controls.[5] Appendix 2 contains the 14 control lists for data communication networks. Each list has a definition of that particular area and is followed by its own set of controls. The areas that are addressed are:

- Software controls (data communication)
- Disasters and disruptions (data communication)
- Modems
- Multiplexer, concentrator, switch
- Communication circuits (lines)
- Error handling (data communication)
- Local loop (lines)
- Data entry and validation (data communication)
- Errors and omissions (data communication)
- Restart and recovery (data communication)
- Message loss or change (data communication)
- People controls (data communication)
- Front end communication processor
- Reliability/uptime (data communication)

RISK ANALYSIS FOR NETWORKS

Management sometimes requests that a risk analysis be performed on the network. Risk analysis can be used to show the potential "average annual loss." This loss figure can be used to justify the cost of controls, to formulate insurance re-

[5]These lists of controls were taken from the book, *Designing Controls into Computerized Systems* by Jerry FitzGerald, published by Jerry FitzGerald and Associates, 506 Barkentine Lane, Redwood City, Calif. 94065. The 14 lists of controls in Appendix 2 were extracted from the 101 lists of controls in the above book.

COMPONENTS

Callout annotations within the figure:
- "This shows the subset of controls that mitigate the threat, Errors and Omissions."
- "This shows the subset of controls that control the component, Communication Circuits (Lines)."
- "Empty cells show a lack of control which may be a serious problem."
- "Some individual cells may be more sensitive to your network or your company; therefore, the controls in these cells should be reviewed very closely."

Components	Errors and Omissions			Security/Theft	Reliability (Uptime)	Recovery and Restart	Error Handling	Data Validation and Checking
Host Computer or Central System	1, 2, 3, 4, 7	1, 2, 3, 4, 5, 7	1, 8, 11, 13, 16	6, 8, 24	1, 13, 16			6, 24
Software	1, 2, 3, 4, 7	1, 2, 3, 4, 5, 7	1, 8, 16	6, 8, 24	1			6, 24
Front End Communication Processor	1, 2, 3, 4, 7	1, 2, 3, 4, 5, 7	1, 8, 13, 16	6, 8, 24	1, 13, 16			
Multiplexer, Concentrator, Switch	1, 2, 3, 4, 7	1, 2, 3, 4, 5, 7	1, 16		13, 16			6, 24
Communication Circuits (lines)	12		10, 15, 16, 18		15, 16			
Local Loop	12							
Modems	12, 18	18, 24	8, 9, 10, 11, 13, 14, 15, 16, 18	24	9, 10, 11, 12		18, 19, 20, 22, 23	
People	5	5, 7		6, 8, 24				6
Terminals/Distributed Intelligence		2	6, 8, 24	6, 8, 24				6, 24

FIGURE 8-12 Matrix evaluation of the controls.

quirements, and to visualize potential loss that the organization *might* suffer should one or more of the threats actually occur.

Risk analysis that produces actual "dollar loss" figures involves many estimates because some figures are not available, such as exact loss suffered in a fire, fraud loss, fraud occurrence rate, or other probabilities.

Appendix 4 presents a step-by-step methodology for conducting a risk analysis of your data communication network.

QUESTIONS

1. Look at Figure 8–2. In order to protect your message as it moves from the earth station to the satellite, which of the 17 control points would be the primary one that you would consider?

2. Is it possible to first encrypt with a public key and then decrypt with a private (secret) key, as well as to first encrypt with the private key and then decrypt that message with the public key?

3. Do packet switching controllers have job tasks beyond that of a typical front end communication processor?

4. What do you think are the three most important security controls that can be placed on a terminal?

5. Can you penetrate a system by nicking a graded index fiber?

6. Which layer of the OSI model would perform controls such as error detection and retransmission or sequence counting of contiguous frames for transmission?

7. When the signals from two independent communication circuits join together and form a product that falls into the frequency band that is different from the first two but interferes with a third frequency, what type of noise results?

8. What is the difference between polynomial checking and Cyclical Redundancy Checking?

9. As the manager of a major data communication network, how many types of disaster plans might you consider?

10. Define a threat and a component.

PART THREE

NETWORK DESIGN

Part Three of this book is devoted to describing the available communication services as well as designing data communication networks.

9

COMMUNICATIONS
SERVICES

This chapter defines common carriers and describes tariff structures, although the individual tariffs (prices) are listed in Appendix 5. The private lease services, measured use services, and other special services for the United States are described. Also, selected Canadian data communication services and a number of worldwide services are described. The chapter closes with a discussion on future communications services.

DATA COMMUNICATION FACILITIES

A data communication facility is the physical apparatus that is required to carry data or information (bits), in the form of data signals, from one point to another. These facilities include hardware, software, and the circuit links over which these data bits travel. At this point we will discuss the various communication services that are available.

A company or government agency that wants to develop its own data communication network can select from a variety of leased communication services to build a private network. If an organization chooses not to develop its own private communication network, another option is to use a public packet switched network. With this option, the organization pays only charges based on usage (volume of data transmitted). This type of usage relieves the user organization of network design problems, most network operations, maintenance and troubleshooting, and the other technical operations that are required for private networks.

The many problems encountered when one is trying to select a facility are compounded by the number of vendors that supply communication circuits, hardware, and software. The different rate structures (known as *tariffs*) and the diversity of facilities available demand that the user organization have a high degree of technical expertise. The cost of some of the more important communication services are listed in Appendix 5.

COMMON CARRIERS AND TARIFFS

A *common carrier* is a government-regulated private company that furnishes the general public with data communication services and facilities (primarily circuits). The best examples might be AT&T Communications and Bell Canada.

A *tariff* is the schedule of rates (prices) and the description of services received with regard to the type of communication service that is being purchased or leased. The circuits are leased, but hardware may be either leased or purchased. The best example is the price structure for home telephones and the description of what is provided for the basic monthly fee. A monthly fee allows you to be connected to the dial-up telephone network; you must either buy the telephone or pay a further small monthly fee for rental of the telephone instrument.

Tariffs are filed with the appropriate regulatory agency. There are two classes of regulatory agencies, federal and state. The best-known regulatory agency is the Federal Communications Commission (FCC). This is an independent federal government agency that regulates interstate (between states) and international communications. The FCC has regulatory powers to compel common carriers (companies that wish to sell/lease communications services) to conform to the Communications Act of 1934 and its recent rewrite/revisions. The federal government currently is in the process of deregulating interstate communications in order to give common carriers more independence in the competitive business environment. Every common carrier that engages in interstate or international communications is under the jurisdiction of the FCC and is required to abide by its regulations.

The other regulatory bodies are the Public Utility Commissions (PUC) that are present in each of the fifty states. The individual state Public Utility Commissions are empowered to regulate intrastate (within a state) communications. While it appears that the federal government is beginning a program of deregulation for common carriers, the individual state Public Utility Commissions do not appear to be following the same course at this time.

A common carrier that wants to sell its communication services must have its services approved. In order to obtain this approval, it must file basic information with either the FCC or the local state Public Utilities Commission. Such a filing provides details of its offered services, the charges for these services, justification for the charges, and so on. These documents are called *tariffs* and they form the basis of the contract between the common carrier and the user of that common carrier's communication service.

In all countries of the world either there is some sort of regulatory agency that controls communications, or the government itself becomes the sole supplier (monopoly) of communication services. For example, in Mexico and the countries of South America the federal government is the sole supplier of data communication services (voice and data). Another example is Canada, where the federal government regulates communication services through the Canadian Radio-Television and Telecommunications Commission (CRTC). In England the British Post Office Corporation has responsibility (monopoly) for providing voice and data communication services. Originally the British Post Office was a government department; then it became a nationalized commercial company. The nationalized commercial company that offers data communication services in Britain currently is up for sale. This sale would move the government out of the data communication business and probably into the business of regulating communication services rather than owning them.

Many other countries of the world operate data communications as a government-controlled monopoly. For example, Germany has the Deutschen Bundespost and in France there is the Postes Telephonique et Telegraphique.

Figure 9–1 lists major common carriers for the United States. The common carriers listed under United States Interstate Services also include intrastate services. The United States International Services offer communications outside of the United States.

UNITED STATES INTERSTATE SERVICES

American Satellite Company
1801 Research Boulevard
Rockville, MD 20850

AT&T Communications
32 Avenue of the Americas
New York, NY 10013

Graphnet, Inc.
99 W. Sheffield Avenue
Englewood, NJ 07631

GTE Sprint Communications Corp.
1 Adrian Court
Burlingame, CA 94010

GTE Telenet Communications Corp.
8229 Boone Boulevard
Vienna, VA 22180

MCI Communications Corp.
1150 17th Street, NW
Washington, D.C. 20036

RCA American Communications, Inc.
400 College Road East
Princeton, NJ 08540

Satellite Business Systems
8003 Westpark Drive
P.O. Box 908
McLean, VA 22101

Tymnet, Inc.
20705 Valley Green Drive
Cupertino, CA 95014

United States Transmission Systems,
Inc. (ITT)
67 Broad Street
New York, NY 10004

Western Telecommunications, Inc.
54 Denver Technological Center
Denver, CO 80201

Western Union Telegraph Company
One Lake Street
Upper Saddle River, NJ 07458

UNITED STATES INTERNATIONAL SERVICES

AT&T Information Systems
32 Avenue of the Americas
New York, NY 10013

Communications Satellite Corp.
950 L'Enfant Plaza South, SW
Washington, D.C. 20024

FTC Communications, Inc.
25 Broad Street
New York, NY 10004

ITT World Communications, Inc.
"Via ITT"
67 Broad Street
New York, NY 10004

RCA Global Communications, Inc.
60 Broad Street
New York, NY 10004

TRT Telecommunications Corp.
1747 Pennsylvania Avenue, NW
Washington, D.C. 20006

Western Union International, Inc.
One WUI Plaza
New York, NY 10004

FIGURE 9–1 Common carriers.

COMMUNICATIONS SERVICES OFFERED

The American Telephone and Telegraph Company (AT&T) has been divided into two separate companies. One half is comprised of the 21 operating telephone companies and is called the Bell Operating Companies (see Figure 9–2). The unregulated "half" is comprised of AT&T Information Systems (computer-based communications products), AT&T Communications (formerly Long Lines), and Bell Laboratories (which remains unchanged).

This new structure means that regular telephone and data communication services are offered through the Bell Operating Companies. At the same time, AT&T Information Systems, AT&T Communications, and Bell Laboratories will

1. Pacific Tel. & Telegraph
2. Bell Telephone of Nevada
3. Pacific Northwest Bell Telephone
4. The Mountain States Tel. & Telegraph
5. Northwestern Bell Telephone
6. Southwestern Bell Telephone
7. Wisconsin Telephone
8. Michigan Bell Telephone
9. Illinois Bell Telephone
10. Indiana Bell Telephone
11. South Central Bell Telephone

12. South Central Bell Telephone
13. Southern Bell Tel. & Telegraph
14. Chesapeake & Potomac Tel. Co's.
15. Bell Telephone of Pennsylvania
16. Cincinnati Bell
17. New York Telephone
18. Diamond State Tel. (Sub. Bell PA)
19. New Jersey Bell Telephone
20. New England Tel. & Telegraph
21. Southern New England Telephone

FIGURE 9–2 Telephone companies.

be entering the data processing business, offering long line communications services and data processing services.

The communication services described in the remainder of this chapter are divided into three basic areas, as follows:

- Private circuit (leased) services
- Measured use services
- Other special services

Each of the three sections contain descriptions of selected services (the most prominent services offered). These descriptions will provide an understanding of the types of communication services (circuits) offered and also provide a basis with which to carry out preliminary design analysis for various types of data communication networks. The tariffs (costs) of these services are in Appendix 5. A thorough understanding of the various communication services is a prerequisite for Chapter 10, Network Design Fundamentals.

PRIVATE CIRCUIT (LEASE) SERVICES

The private circuit services are those in which the user organization leases the service from the common carrier. In other words, private circuits are available for use by organizations twenty-four hours per day, seven days per week. In effect, they are for the exclusive use of the leasing organization.

It would be helpful to understand the distinction between a leased circuit and a dial-up (measured time) circuit. If you had a leased circuit from San Francisco to Los Angeles, it would be one continuous unbroken circuit path. In other words, this leased circuit would be wired around any switching equipment at the telephone company central offices. On the other hand, a measured time dial-up circuit would go through all the switching equipment in the telephone company central office; there would not be one continuous unbroken circuit path for your use. Every time a call is placed on a dial-up circuit, a new/different circuit path is used.

Private lease circuits are so much at the disposal of the lessee that it is as if the circuits were owned by the organization, even though they are leased by payment of a monthly lease rate. Some of the more prominent private circuit services are as follows:

Voice Grade Channels These are sometimes known as Series 2000/3000 Voice Grade Circuits. Technically, a Series 2000 Voice Grade Channel is for voice communications, remote operation of radio telephones, connection of private voice interconnecting systems, and interconnecting remote central offices (FX service that is described later).

A Series 3000 channel is primarily for data transmission, remote metering, supervisory control of electronic devices, and miscellaneous signaling. Actually everyone just refers to this as a voice grade equivalent channel. This is the same channel or communication circuit that was described in Chapter 3, with a 4000 hertz outside bandwidth but an internal usable bandwidth of 3000 hertz or less.

When this type of circuit is leased, it may include copper wire pairs, bundles of wire pairs, microwave transmission, coaxial cables, or even optical fibers. Normally users are told if the voice grade equivalent channel is a satellite channel, because network designers must take into account the propagation delay times associated with satellite transmissions. Also, pricing structures are different for a satellite voice grade equivalent channel than for wires, cables, or microwave.

When series 2000/3000 channels are used for data transmission, users have the option of having them conditioned. Conditioning reduces attenuation distortion and envelope delay distortion, which in turn then reduce line impairments and, thus, errors during transmission. AT&T Communications offers two levels of conditioning, type C or type D, at an extra monthly charge.

C Conditioning offers five types of conditioning. C1 and C2 conditioned lines may be ordered for point-to-point, multipoint, and switched configurations. C3 conditioning, which is similar to C2 conditioning, applies only to private switched networks with a maximum of four circuits and two access lines in tandem. C4 conditioning can be ordered in two, three, and four-point circuits only. C5 conditioning can be ordered only in point-to-point circuits; it is similar to C2 conditioning, except that it is intended primarily for overseas circuits. C level conditioning has specific limits on attenuation distortion and on envelope delay distortion. Such limits reduce line impairments so that a data signal can get to its destination with less noise and distortion (fewer retransmissions caused by errors). Bell's Technical Reference Publication 41004 gives the exact specifications.

D Conditioning is available in two types. D1 conditioning is offered for point-to-point circuits and D2 is for two- or three-point circuits. D type conditioning specifically limits noise and harmonic distortion. Again this type of conditioning reduces line impairments and therefore has fewer retransmissions caused by errors.

Modem manufacturers specify whether conditioning is required on voice grade circuits when their modems are used.

In addition to conditioning, some local telephone companies, such as New York Telephone (NYNEX), offer what is called a *straight copper circuit*. This is a pair of copper wires and it is available only when the entire circuit is within the same telephone company central office. In essence, the straight copper circuit is comprised of two local loops connected at the central office. Since there are no repeater/amplifiers or loading coils in the circuit, the user can utilize less expensive limited distance modems (1 to 15 miles). Limited distance modems also may be used on standard series 3000 unconditioned channels, depending on the length of the circuit and the data rate.

Other similar private lease services that are available include Series 1000 channels, which are low-speed signaling and teletypewriter channels that vary between 30 and 150 baud data transmission rates. Because of their slow signaling speed and the fact that they may cost more than a Series 2000/3000 data circuit, Series 1000 channels are falling into disuse and the telephone company is beginning to drop this offering.

Series 4000 channels are private lease circuits for telephoto transmission in the speed range of 1200 to 2600 characters per second.

Several common carriers other than AT&T Communications offer the typical Series 3000 voice grade channel data transmission circuit (see Figure 9–1).

Wideband Services There used to be a set of Series 5000 channels called Telpak. This service is available now only on an intrastate basis, and it is not available in all fifty states. Series 5000 service involves a wider bandwidth because a type 5700 channel is compromised of a bundle of 60 voice grade circuits and a type 5800 channel is comprised of a bundle of 240 voice grade channels. There are no series 6000/7000 channels.

Series 8000 channels are for data transmission or for alternate voice/data transmission. These channels are for high-speed data transmission at up to 56,000 bits per second. They also may be used in conjunction with a 50,000 bit per second switched FX service, a digital data service extension channel, or high-speed facsimile transmission. This type of communication channel may be delivered to the user as a 48,000 hertz bandwidth or as 12 individual voice grade channels (48,000 hertz ÷ 12 = 4000 hertz). This type of channel is used by an organization that needs wider bandwidths in order to transmit greater quantities of data between its facilities.

Digital Services This includes point-to-point and multipoint configurations for the transmission of data in a digital manner (analog modem conversion not required). This type of service operates at 2400, 4800, 9600, 56,000, and 1.544 million bits per second. It spans both the voice grade communication channels (4000 hertz bandwidth) and the wideband communication channels (greater than 4000 hertz). The advantage of using digital transmission is that the digital modems are much less costly (although increased local loop costs may account for the lower modem cost) and the transmission error rate is far less than with analog circuits.

The designer of a network first ascertains whether digital service is available in the area where the organization wants to transmit and then compares the cost of the digital service with the appropriate analog service such as the Series 3000 data transmission channel.

Satellite Services Several common carriers offer satellite channels for voice, data, facsimile, and various wideband applications. Basically, a *satellite channel* is defined as a four-wire voice grade circuit; therefore, the user gets a 4000 hertz bandwidth and a four-wire equivalent circuit.

As was mentioned in the earlier discussion of voice grade circuits, user organizations are notified when transmission is going over a satellite circuit because it may affect their protocols. A Binary Synchronous Communications (BSC) protocol cannot be used successfully with satellite transmission. Instead, one of the newer protocols must be used, such as X.25, SDLC, HDLC, or the like. Satellite channels usually are less costly than voice grade channels that are on the ground such as microwave transmission, wire pairs, coaxial cables, and the like. If a greater capacity is needed, it is possible to lease a bundle or group of voice grade satellite channels in order to increase bandwidth beyond the standard 4000 hertz for a single voice grade channel.

Communications Network Service Communications Network Service is a two-point or multipoint network service that operates exclusively between or among the network access centers located on the customer's premises. These network access centers access a satellite via rooftop satellite antennas installed by the common carrier. In the future, there might be similar services using cable television or cellular radio access methods. In other words, this type of communication service is different in that a common carrier creates an entire private network using their communication facilities, whether they be satellite, cellular radio, ground-based microwave antennas, or the like.

In summary, the above private circuit (lease) services are the basic circuits/ channels that may be used to design and develop a network. They are the basic building blocks to which we connect the various pieces of hardware and the protocols/software programs that ultimately create a working data communication network.

MEASURED USE SERVICES

The measured use services are communication services in which charges are based upon how much the system is used. For example, a call from your home telephone is based on the length (time) of the conversation and how far (miles) the other person is from you. For a packet switched service, charges are based on how many individual packets of information are transmitted without regard to where the other party is located. With Wide Area Telecommunications Service (WATS), a flat fixed monthly fee is charged for a fixed number of hours of circuit usage time. If the fixed time is exceeded, there is another per hour rate that is charged.

In other words, measured use services are those in which payment is based on utilization. Some of the primary services in this group are Direct Distance Dialing (DDD), WATS, packet switching, Telex/TWX, Datagram, and 50 kilobit service.

Direct Distance Dialing (DDD) With direct distance dialing, the normal telephone network is used for data transmission. The user dials the host computer telephone number, receives appropriate control signaling, enters password/authorizations, and connects to the host computer system.

In direct distance dialing an entirely different circuit path between telephone company central offices is used each time a number is dialed. Charges are based on the distance between the two telephones (in miles) and the time the connection is held open (the data transmission). The data communication user pays the same rate as the individual who uses the telephone for voice communication.

When compared with the private leased voice grade Series 3000 channel, dial-up voice grade channels have more noise and distortion because the signals go through the telephone company central office switching equipment.

Conditioning is not available with DDD because each call dialed gets a new circuit path or routing. Also, the rate of transmission may be a little less than could be achieved on a private leased voice grade circuit because of more noise and distortion. Only two-wire connections are available, so if full duplex transmission is required, a special modem is needed to transmit in both directions simultaneously.

Wide Area Telecommunications Service (WATS) WATS is a special bulk rate service that allows direct dial station-to-station telephone calls, although it may be eliminated in the future. It can be used for both voice communications and data transmission. WATS is the (800) area code series in the United States. The 48 contiguous states are divided into 58 different service areas. The geographical coverage of WATS from any one of these service areas is determined by the "band of service" to which the customer subscribes. For example, interstate service from the state of California uses the following bands:

- **Band 1** Arizona, Idaho, Nevada, Oregon, Utah, and Washington
- **Band 2** Colorado, Montana, Nebraska, New Mexico, and Wyoming
- **Band 3** Iowa, Kansas, Minnesota, Missouri, North Dakota, Oklahoma, South Dakota, and Texas
- **Band 4** Alabama, Arkansas, Illinois, Indiana, Kentucky, Louisiana, Michigan, Mississippi, Tennessee, and Wisconsin
- **Band 5** Connecticut, Delaware, Florida, Georgia, Maine, Maryland, Massachusetts, New Hampshire, New Jersey, New York, North Carolina, Ohio,

Pennsylvania, Rhode Island, South Carolina, Vermont, Virginia, Washington, D.C., West Virginia, Puerto Rico, and the U.S. Virgin Islands.

- **BAND 6** Alaska and Hawaii

These are the six service bands from California. The state of California has two service areas, Northern California and Southern California. The list of states served in bands 1 to 5 differs, depending upon the state. As might be guessed, the first five bands out of New York are almost the direct opposite of bands 1 to 5 out of California. Band 5 out of California is similar to band 1 out of New York. When subscribing to a band, such as band 4, service is automatic to all lower bands (in this case bands 1 to 3).

The WATS bands described above for interstate service have no relationship to intrastate WATS service. Using California as an example, there is a northern service area and a southern service area. Therefore, WATS intrastate service can be for northern California only, southern California only, or statewide. As you may have guessed by now, interstate WATS service does not include your home state; therefore, if access is needed to your home state, it is necessary to lease both interstate and intrastate WATS services.

Interstate WATS service is available on the basis of 0–15, 15–40, 40–80, and over 80 hours of usage per month. For either intrastate or interstate, charges are a fixed flat fee for a certain monthly usage, in hours.

If WATS service is used for data communications and the call holding time is less than 60 seconds, billing will be for one minute of usage (one minute average call holding time). WATS service also is limited to one direction only: it is either "outward dialing" or "incoming" calls only. Inward and outward capability cannot be combined into a single WATS circuit. To do both, the user has to subscribe to two circuits.

Public Packet Switched Services Packet switched communication services offer transmission speeds up to 56,000 bits per second. Differences in transmission speed and different protocols between various switching nodes (SN) are compensated for by the network. The packet network also provides code conversion from one code to another. Data are segmented into 128-character (1024-bit) blocks called packets. Users may access the service via private communication circuits, public dial-up circuits, or other packet switched networks. Figure 9–3 pictures a packet switched network that has both ground-based circuits and satellite circuits.

In packet switching, the user may not design or maintain the network. The user may be only a "user of the network." This may be an advantage since it relieves the user of many of the technical burdens of designing and maintaining a private network.

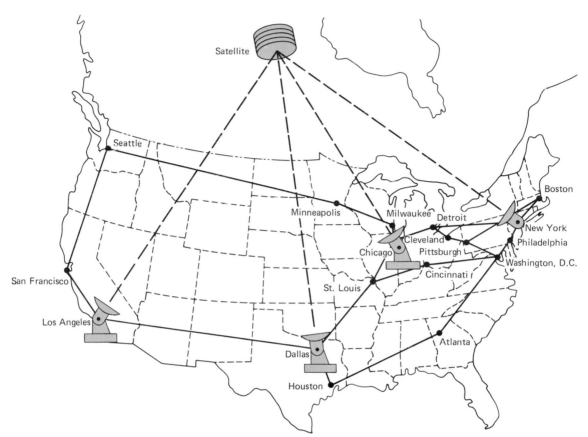

FIGURE 9–3 Packet switched network.

Telex/TWX Teletypewriter exchange service is a data transmission exchange between two terminals (it excludes voice communications). Each subscriber has a terminal and can contact any other subscriber in the telex or TWX network. This is nothing more than an alternative to the voice telephone network, but it offers the ability to transmit hard copy between subscribers. Subscribers pay a monthly fee plus so many cents per minute of connect time. The connect time is charged only for the time it takes to transmit a message from one subscriber's terminal to another. This is a very popular service for businesses and government agencies, especially in the area of purchasing/procurement. Western Union, the common carrier that offers both of these services, is in the process of upgrading the services to increase the speed of transmission. The upgraded version is called Teletex and is described in the section on Future Communication Services at the end of this chapter.

Electronic Computer Originated Mail (ECOM)   This is a new service offered by the U.S. Postal Service. Available 24 hours per day, seven days per week, it enables a user to transmit messages, in an electronic computer form, to designated post offices. At the post office the mail pieces are printed, placed in an envelope, and put into first class local mail for delivery. Messages received at a designated post office before midnight are delivered within two days if the delivery is within the service area of that post office. There are three types of messages: single address messages, common text messages (identical letters to multiple addresses), and text insertion message addressee where unique information is inserted into a master letter.

This service at present is limited to users who can generate at least 200 messages per transmission per designated post office. This is probably the forerunner to a complete electronic mail system, whereby the mail is delivered electronically to your home rather than your local post office. Current charges for ECOM are $50 to establish an account, 26 cents for the first page of a letter, and 5 cents for the second page of the same letter. When the cost of only a postage stamp is 20 cents, 26 cents is a very good price considering that the post office supplies the paper and the envelope, stuffs the envelope, and includes the stamping process. The service is available in 25 major areas including Atlanta, Boston, Dallas, Detroit, Los Angeles, New York, Philadelphia, Pittsburgh, St. Louis, San Francisco, Seattle, and Washington, D.C.

Fifty Kilobit Service   Fifty kilobit switched service is an interstate dial-up service for two-point high-speed data communication and facsimile transmission. In addition to this high-speed data transmission service, a voice channel also is provided for coordination purposes. The bandwidth of the channel provided on the dial-up service is 48,000 cycles per second (48 KHertz). Intrastate 50 kilobit switched service is available in some states on a limited basis. Currently the interstate 50 kilobit switched service is available only between Chicago, Los Angeles, New York, San Francisco, and Washington, D.C. It is being displaced by 56,000 bit per second digital service.

OTHER SPECIAL SERVICES

Flocom is a service of Western Union that provides a computer-controlled store and forward switching service. Multidrop configurations in a polled environment are offered to provide the user with the equivalent of a private line network. Traffic can be transmitted between any of the subscriber's stations/nodes. Flocom is available in only a limited number of cities, including Boston, Los Angeles, Pittsburgh, Chicago, New York, San Francisco, Kansas City, Philadelphia, and Washington, D.C.

Discount Voice Services are offered by companies such as Sprint, ITT, and MCI. They offer voice telephone transmission at a discount over the same services as the Bell System. When the service is used, a local access number is dialed (you must have tone dialing), a billing password is entered, and then the telephone number is dialed. The call is routed over the Bell System local loops to the Sprint, ITT, or MCI computer. From there it is routed on Sprint, ITT, or MCI long-distance (IXC) circuits to the other city. From there, it is routed again over the Bell System local loops to the party that was called.

This service allows voice or data telephone calls to average 10 to 40 percent less in cost than if dialed over the Bell System DDD network. Because of the necessity of entering a billing password, these services are limited to users with touch tone telephones.

Hotline is a service that directly connects two telephones in distant cities. When either of the two receivers is lifted, the telephone rings at the other end of the connection. This is a point-to-point service, available only in selected cities.

Common Control Switching Arrangement (CCSA) is a private long-distance dialed network. A switching arrangement is provided to allow interconnection of channels terminated in the common carrier-provided switching equipment. The service is offered for large corporations and government agencies to interconnect several or hundreds of business operations, thus saving on telephone costs. A flat fee is paid for which the common carrier establishes a private telephone system interconnection for voice telephones, data hardware, and PBXs/CBXs.

The largest CCSA is probably the U.S. government's Federal Telecommunications System (FTS). Many states also have set up private telephone systems that are designed, implemented, and maintained by the common carrier that provides the CCSA service.

Mailgram is Western Union's service for overnight telegraph messages with delivery made via normal postal service at the city of designation. For messages received at Western Union offices between 8:00 A.M. and 6:00 P.M., delivery is made the following day by the U.S. Postal Service. This is similar to ECOM.

Datel is a measured use overseas data transmission or voice service via a U.S. international service carrier (see Figure 9–1) and the public telephone network at the overseas point (country to which the message is transmitted). The average rate of transmission is 1200 bits per second, but it is being increased to 2400 bits per second. Datel provides service between the United States and most other major countries. It is used when the caller wants to connect directly into the public telephone network of the other country. Other variations of this service are available that allow transmission at 9600 bits per second.

International Switched Data Services are packet switched data services that provide speed and code conversion between similar data terminals operating at speeds up to 9600 bits per second. Users may access these services via private dedicated circuits, public dial up, or domestic packet switched networks. Several international carriers offer this service (see Figure 9–1). This is similar to the Datel service, except it is packet.

Picturephone is an offering that combines telephone voice transmission along with slow-scan television. It requires a special telephone that has a video tube. The Bell System offers picturephone.

Foreign Exchange (FX) service allows a user to call another central office via the dial-up telephone network without incurring any charges other than that of a local call.

If an organization is located in the suburbs of a major city but most of its customers are in the downtown metropolitan area, then it might lease an FX circuit. The telephones at the company in the suburbs are connected directly to the telephone company's central office in the downtown metropolitan area. In other words, an FX circuit is a circuit that runs from your telephone instrument to the telephone company's end office in another area. These circuits allow the suburban subscriber to have the same "free" dialing privileges as telephone subscribers in the downtown area. When the telephone instrument is picked up, the dial tone is directly from the distant city's end office. If you list your telephone number in that city's telephone book, people may call you at a local number. In reality, they reach you over the FX circuit.

The cost of FX service is the same as the cost of a voice grade private leased line plus the cost of a single telephone at the distant central office. While this may appear to be similar to intrastate Wide Area Telecommunications Services (WATS), it is not because the user does not have to dial the 800 area code and it will accept both incoming and outgoing calls.

CANADIAN COMMUNICATION SERVICES

Canadian Telecommunication is dominated by the geographical and political environment of Canada itself. In terms of population distribution, Canada has the bulk of its population spread across a thin ribbon 3500 miles wide and 250 miles long, stretching from the Atlantic to the Pacific across six time zones. Major cities, except in heavily populated central Canada (from Windsor to Quebec City, a distance of about 800 miles) are widely separated by rural areas. To the north of the population ribbon, the bulk of the Canadian land mass is frozen much of the year. Until recently, communication with the north (i.e., the Yukon, the Northwest Territories, northern Ontario, and northern Quebec) was by radiotelephone. In this respect, the larger part of Canada resembles Alaska. Only the advent of satellite communication has brought the North of Canada into the communication mainstream.

As in the United States, the Canadian federal government shares powers with the provinces. In many areas, including communication, the nature of this sharing still is being defined. Thus, it is not surprising to find a complex legal and regulatory climate with regard to telecommunications. The government of Canada regulates the telecommunications industry through the Canadian Radio-Telecommunications Commission (CRTC), a body created by the Parliament to oversee, among other pursuits, broadcasting and communications. The government also actively participates in the marketplace through Crown Corporations, direct financing, and research and development support. In these last two areas, the Federal Department of Communications provides policy research, grant monies, and expertise to private industry, in addition to operating several research facilities.

At provincial and municipal levels, government also is a participant as well as a regulator. In the Prairie Provinces (Alberta, Manitoba, and Saskatchewan) the provincial governments operate province-wide telephone companies. Even in these provinces, however, private telephone companies and rural cooperatives coexist. For example, the Alberta government operates Alberta Government Telephones (AGT) from its headquarters in Edmonton which is served by a municipal system (Edmonton Telephones). While most telephone systems in Canada are owned and operated by Canadian concerns (public or private), BCTel (British Columbia Telephone), the second largest provider of service, is owned by General Telephone of New York City. BCTel is *federally* chartered. So is giant Bell Canada, which operates more than half the telephone and data facilities in Canada. Bell Canada is centered in Ontario and Quebec but also operates the telephones in the eastern Northwest Territories and in several Maritime Provinces, through subsidiaries.

Complex cooperative arrangements exist between government and private concerns and among various providers of services. TeleSat Canada (satellite) is a semi-public body operated jointly by the government of Canada and the common carriers (joined informally in an industry association, the Canadian Telecommunications Carriers Association, or CTCA). The Trans-Canada Telephone System (TCTS) is an association of most of the larger telephone companies and TeleSat Canada. It operates all long-distance analog and digital services between provinces, is a member of CTCA, and is thus an owner of TeleSat Canada. Finally, CN Telecommunications, a subsidiary of Canadian National (a Crown Corporation operated at arm's length from the Parliament) has merged corporately with CP Telecommunications (a subsidiary of the investor-owned Canadian Pacific Railroad), to form CNCP Telecommunications. This is the only provider of Telex service in Canada and the only true competitor for private-line analog and digital services with TCTS. It, too, is a member of CTCA and thus is also an owner of TeleSat Canada. Teleglobe Canada (a Federal Crown Corporation) is an international telephone and data communication carrier.

Adding to this complexity of regulation and ownership is a fluid research and development situation. While there is little mainframe manufacturing in Canada, a silicon valley of sorts has grown up in the western suburbs of Ottawa, where advanced microcircuitry and telecommunications device manufacturing exists on a modest scale. Ottawa is the hometown of Bell-Northern Research, Ltd., a subsidiary of Bell Canada, similar in its corporate mandate to Bell Labs in the United States.

Recently, BCTel has opened a research facility in Vancouver and Alberta Government Telephone (AGT) has initiated the formation of a research-based subsidiary, possibly located in Calgary, which is served by AGT. A third source of R&D effort is coming from Toronto, home of the Toronto *Globe and Mail* (operators of InfoGlobe information retrieval service) and InfoMart (a Telidon service provider).

In summary, the telecommunication environment in Canada is complex, fluid, and vigorous. In spite of this, Canada was 75 percent wired with cable television before U.S. regulation was in place, and Canada's Telidon Videotext system has been successfully marketed overseas, again before any standards could be developed in the United States. Datapac was the first X.25 packet switching service to be brought up on a national basis in the world (1975) and Dataroute (1973) was the first digital network.

Because of the complex regulatory and economic climate, there are three aspects to the Canadian telecommunications industry, which are mutually interwoven. These are analog services (telephone, broadcast, cable), digital service (computer communications, electronic mail, and information supply), and satellites.

In addition to the complex ownership and regulatory situation, there is the troublesome issue of transborder data flow. While most data passing through Canadian networks flow in an east-west or west-east direction, a significant amount of raw data pass from Canada to American data centers for processing. In particular, centers such as Calgary (where a considerable number of American oil companies are located) are sources of a large amount of raw data. While Teleglobe Canada, a Crown Corporation handling most international data and voice communication, serves overseas locations, United States–Canada data flow is a matter of private concern. CNCP, for instance, operates three microwave relays between Canadian and American gateways for data transportation. Two of these are interfaced with Western Union facilities. The Canadian government wants to maintain strategic data in Canada. Where transborder data flow cannot be eliminated, duplicate data are maintained in Canada, for example that of the Western oil companies. This issue is far from resolution.

Related issues of an international nature include concern over the maintenance of massive American data banks on the economic status of Canadian citizens and the marketing of Telidon technology in the United States. Most of these issues are political rather than technical, but they affect manufacturers, service providers, and regulators.

Telephone service in Canada resembles that in the United States, especially from the user's viewpoint. In addition, the two networks are completely integrated. A salesman in Calgary can direct-dial a customer in San Francisco. While the charging method differs slightly, costs are comparable, although somewhat higher in Canada.

For the business data communication user, each of the dozen major and several hundred minor telephone companies offer different services and different pricing structures.

In terms of telephone penetration and usage, Canada enjoys one of the highest per capita installation and usage rates in the world. Nationwide, there are over 65 telephones per 100 inhabitants, placing Canada fourth behind the United States, Sweden, and Switzerland. In some cities, such as Calgary, the telephone system struggles to keep up with population growth. In general, however, the quality and quantity of telephones, especially for business use, are more than adequate and match or exceed that of the United States. All telephone companies are modernizing at a rapid rate, spurred on by the development of computer-based switching equipment.

Pricing structures vary from jurisdiction to jurisdiction and are subject to review either by provincial cabinet (in Saskatchewan), a public utilities board (in Alberta), or the CRTC (for Bell Canada, CNCP, BCTel, Northwest Tel, and Terre Nova Tel).

Cable penetration in Canada is quite high. In urban areas up to 90 percent of all households subscribe to cable; overall, the penetration is 65 percent. In general, cable companies are in the redistribution business. The popularity of cable can be attributed to its ability to deliver U.S. networks' programming into Canadian homes. Reacting strongly to this penetration, the CRTC has issued a number of regulations allowing for deletion of commercials on imported signals and a channel assignment algorithm which discriminates against American signals in favor of almost any Canadian signal. In most locations, too, cable enables local programming and teletext services such as news, weather, shopping tips, and so forth.

Computer communication services are offered to the public through TCTS and CNCP in roughly parallel and competitive services. TCTS offers a circuit switched data service called *DataLink*. CNCP offers a similar service known as *Infoexchange*. TCTS provides a packet switching facility known as *Datapac*; CNCP's corresponding service is called *Infoswitch*. TCTS has *Dataroute* and CNCP offers *Infodat* on a dedicated, leased basis. A wide range of data rates, connection possibilities, and service interconnections are available.

A major supplier of software and information services is InfoMart in Toronto. Another firm providing information services is the *Toronto Globe and Mail*, which distributes information on *Globe and Mail* articles, through InfoGlobe, in a fashion similar to the *New York Times* and the *Wall Street Journal*.

Bell-Northern Research, Ltd., is providing VIPS (Videotext Information Provider System) for creating and managing videotex graphics compatible with Telidon. Thus production, distribution, and management of software and services are all available through a single label: Telidon.

Several kinds of electronic mail are offered by the common carriers in Canada.

Telepost is offered by CNCP and is equivalent functionally to Western Union's Mailgram service. It is a combination of telephony and the telegram. Electronic transmission is used between remote centers, and local mail delivery completes the process.

Intelpost is an international fascimile-based service offered by CNCP, Canada Post, and Teleglobe Canada. For about $5 per page, any sort of printed material, including drawings, handwriting, signed correspondence, and so forth, can be transmitted. Satellite links and facsimile technology are used between service centers in eight major Canadian cities and London, Berne (Switzerland), Amsterdam, New York, and Washington. Transmission takes a matter of minutes.

Telex service is the teletyped telegram. There are over 400,000 Canadian and 1 million worldwide businesses subscribing to Telex.

Telepost and *Telex* are available through a CNCP agency or as a companywide service bureau. A true desk-originated electronic mail service is available through the CNCP *Infotex* and the TCTS *Envoy 100* offerings. Both are available in a growing Canada-wide network based upon common carrier data transmission facilities of a store and forward packet switching nature. Applications range from real-time message communication via dumb terminals to sophisticated word processing, personal filing, and control operations. Each is aimed at the automated office market of managers, secretaries, and executives. In addition, each provides a valuable interface to other services such as Telex (via Infotex) and the U.S. Tymnet and Telenet (via Envoy 100).

Telenet is a message-switching service offered by CNCP to complement its real-time Telex offering. It provides the common carrier equivalent of a private telex network arrangement.

Another CNCP offering related to electronic mail is *Broadband Exchange Service*. It is a circuit switched, distance and volume sensitive analog service intended for point-to-point communication on a dial-up basis. Data transmission may be via facsimile and receipt may be unattended. Speeds up to 9600 bps and full duplex operation are available.

Envoy 100 provides distance-insensitive packet switching. Charges are levied for each access, for creation of messages (by the 100 characters), and for each addressee. Envoy 100 adheres to X.25 (as do most Canadian products involving packet switching). In addition, Envoy 100 offers its message package in both English and French. Using Envoy 100, one can read messages, check bulletin boards, create individual or group directed messages, add to mailing lists, or manipulate personal files. Messages can be read and answered, filed or purged. Copies of messages may be forwarded to others. A number of connection-management facilities allow for timed delivery, return receipt, and registered delivery.

A 500-character message (about one-third of a typed page) costs 15 cents to create, 15 cents for each delivery, and 5 cents for each addressee. This 35-cent cost compares quite favorably with 30 cents for postage, and the speed of transmission, of course, cannot be compared to that of the postal service. A multi-addressee message offers significant savings: a three-addressee message of 500 characters costs 75 cents against a 90-cent postage fee (not to mention the cost of stationery, typing, and company handling). There is, however, a monthly charge of $20 for an account and $3 for each user.

Datapac is used to carry Envoy 100 messages. It is an X.25-compatible packet switching digital network available in most Canadian cities. With planned expansion, most data communication customers in southern Canada should have access to a local Datapac facility. In addition, Datapac may be accessed from Telenet or Tymnet in the United States. Datapac has been in operation since November 1976. The Datapac protocol is called SNAP (Standard Network Access Protocol). The DTE (data terminal equipment) connected to Datapac must be capable of creating SNAP packets or have access to a Datapac-compatible NIM (network interface machine). In general, packets may carry up to 2048 bits (256 8-bit bytes).

Since Datapac is a packet switched service with internal connections (rather than circuit switched), it accepts many different DTEs and data speed differences are unimportant.

WORLDWIDE COMMUNICATION SERVICES

There are many similarities in the way data communications activities have evolved in the United States and in Europe. There also are many differences. One of the primary differences is that the data communications industry in the United States operates as a series of private companies that are regulated by the government, whereas in Europe and most other countries of the world the PTTs (Postal and Telegraph Services) are a government monopoly that controls and sells all voice and data communications services.

If the industry is regulated, as in the United States, it may be more innovative, cost effective, and able to develop new services faster. On the other hand, if the industry is a monopoly as it is in most countries, it may be overburdened because it supports other government agencies (such as postal services) that drain its resources. Such a situation generally fosters an agency that is uneconomical and not very innovative or progressive with regard to developing new services. These statements are generalizations, however, and have notable exceptions. For example, the French data communications government agency is probably the most progressive in the world for switched packet service networks throughout an entire country.

The British PTT currently is up for sale, and the Australians have separated the Postal and Telegraph Services. In Mexico, a national policy to move companies out of Mexico City has been hampered because of inadequate communication facilities in the areas to which these companies are being relocated.

The United States has no single agency responsible for "message communications." In Europe this would be the role of the National Telecommunications Administrations (the PTTs). France, with its centrally planned communications, probably was better able to get a single packet switching network throughout the country than could the multiple companies of the United States. This is so because the United States has numerous packet switched networks with different protocols/software. These networks may be incompatible for many years, unless new hardware/software developments overcome the barriers.

The variety of services and developments in Europe are similar to those in the United States. Even though the European PTTs are not competing directly within a single geographical area, they are competitive among themselves. This is because of transnational transmission and national interests.

Other differences might stem from the fact that governments create pressure over unemployment, protection of industrial and technical markets, political considerations, national defense considerations, and cultural and social traits within a specific country.

The European environment is not a single sovereign market. It is plagued with many geographical/political/economic differences. For example, the situation could be compared with trying to interconnect and achieve total compatibility in the data communications facilities between Mexico, Canada, and the United States. It is for this reason that the European PTTs are more sensitive to the need for international standards than is the U.S. data communications industry.

There is a conference of European PTTs, called CEPT, which operates by mutual agreement and understanding for the cooperative exchange of different views. CEPT is gaining momentum because a number of its working group personnel have international interests.

The European Economic Community (EEC) is promoting *Euronet*, a database information network for member countries. This provides an opportunity to design and install a truly international service. Euronet is developing a videotext gateway so the average homeowner may have access to it in the future. Unfortunately, the European PTTs have not followed this example and are promoting international services bilaterally, thereby maintaining independent control. Euronet is a multinational network capable of interconnection with national networks. It uses packet switching protocols compatible with Transpac and other national networks. *Transpac* is the public packet switching network of France. England has the PSS network, and the Canadian packet switching system is called *Datapac*. In the United States the two basic public packet switching networks are *Tymnet* and *Telenet*. The Bell Packet Switched Services (BPSS) is another one. Perhaps some day the European Euronet will be interconnected with all of these other packet networks and we will have a truly international packet switched network.

Europe is ahead of the United States in videotext service to private homes because the goal of the national PTTs is to develop this service. The British are developing *System X*, which is a digital integrated network that allows digital switching, voice digital transmission, and data digital transmission. The Bell System in the United States is planning to convert to an all-digital system.

One perplexing problem is that of the different tariff levels for identical services between adjacent European countries. For multinational users this can be a deciding factor in situating their communication centers. The cost for leased private circuits varies by country. The European Economic Community (EEC) has expressed its desire to see these tariffs become more equitable, but the PTT's grip on communication costs will be very hard to loosen. In addition, the grade or quality of service of similar offerings between countries must be maintained and standardized.

One of the most complex networks that has been developed on a worldwide basis is the *Society for Worldwide Interbank Financial Telecommunications* (SWIFT) network. This network interconnects over 1,200 banks in 50 different countries. When the network started in 1977, it had an average volume of 60,000 messages per day. By 1984 the average volume will soar to 400,000 messages per day.

Figure 9–4 shows the basic diagram of the SWIFT communications network. There are three operating switching centers. From the switching centers, communication circuits go to a regional concentrator located in each country. The banks in the individual country interconnect with their regional concentrator. A regional concentrator in any country has a connection to its normal switching center and a backup connection to an alternate switching center.

If the Royal Bank of Canada had a funds transfer for either Midland Bank (England) or Security Pacific National Bank (United States), it would take the most direct path in Figure 9–4 (compare this with Figure 5–21).

Other international developments are:

- **Italy** A public packet switched data network became operational in 1983 with nodes in Rome, Milan, and Naples. Initially, there were 5000 users; that number is expected to double after three more nodes are added shortly after startup. Network costs are distance-independent. No decision has been made on how or whether to link the network to Euronet.

- **Australia** British Telecom's Prestel service will be sold in Australia. Prestel now has about 800 information providers, and 16,000 users accessing 226,000 pages of information. Australian clients will access Prestel via the Overseas Telecommunications Commission Midas service, which costs $0.20/min and $0.60/1000 characters of information transmitted.

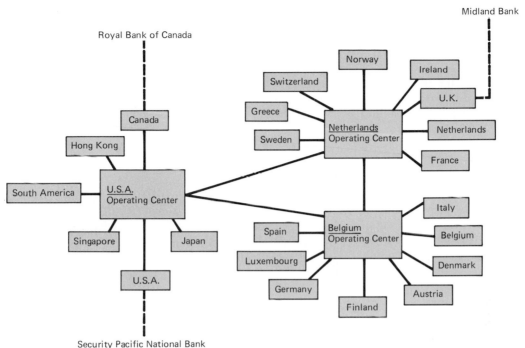

FIGURE 9–4 SWIFT network.

- **Hungary** Hungary is starting trials of France's Antiope and Britain's Prestel viewdata systems, becoming the first member of the Communist Block to begin the new TV-based service. It will provide TV viewers, by telephone, with 100 to 200 pages of information. Much is at stake because Hungary's choice will become the standard for the entire ten-nation Communist Bloc.

- **West Germany** Viewdata has become a true mass medium. In West Germany the video media are flourishing, with revenue from video sales passing $1.6 billion in 1981, when 20,000 different types of programs were available. Cable and pay TV are still in their infancy but are expected to find nationwide application by 1985.

- **Western Europe** This area is bypassing the United States in the establishment of public electronic mail services complying with the Teletex standard recommendations of CCITT, the international consultative committee of telecommunications common carriers. The state-owned common carriers in West Germany, Sweden, and the United Kingdom have Teletex services and all of Western Europe should be covered by 1985. Deutsche Bundespost expects about 40,000 Teletex terminals to be linked to its national network as early as 1985, and by 1990 about 130,000 Teletex units will be operating.

- **American Express** American Express has launched a viewdata system that enables people to check arrival and departure times at major U.K. airports on their TV sets via British Telecom's Prestel system. The service costs less than 25 cents per call. Skyguide is linked to London's Heathrow Airport and American Express expects to link another eleven airports in the U.K. and Ireland to the system in 1983. Also, American Express is in discussions with the major airports in the United States and in continental Europe about linking them to Skyguide.

FUTURE COMMUNICATION SERVICES

The Bell System is developing the Bell Packet Switched Service (BPSS), which will probably be the underlying network for the company's new "smart network" offerings. When AT&T Information Systems split off, Net 1000 was unveiled. The Bell Packet Switched Service probably will be offered by the Bell Operating Companies, whereas Net 1000 will be offered by AT&T Information Systems, the unregulated subsidiary that can engage in both data communications and computer services. Net 1000 will provide increased compatibility among many different types of terminals and host computers; it will permit integration and connection of diverse networks as well as access to multiple applications/databases from a single terminal. In addition, it will bring added flexibility and control to the data networks and reduce startup costs considerably. Net 1000 will be an intelligent network into which users might be able to connect any type of hardware using their own choice of software protocols (UNIX will be available).

Western Union Telegraph Company is developing a system for the first U.S. teletex service. They will join it with the international teletex service. Teletex, a new form of advanced Telex, is a 2400 bps service that will provide advanced functions such as buffered CRTs, compatability with data transmission protocols, and the ability to interface with existing business communication networks. Standards have been in development for several years by the CCITT. The basic service is designed to provide word processing type terminals with local screen editing and printer capability. So far there are about 6,000 Teletex terminals operating in Germany and 700 in Sweden, according to Philips Telecommunications. By 1985, over 30,000 are expected to be operating in Europe. In the United States, Western Union will provide 2,400-bps word processing communications to commercial and personal users. The basic goal here will be to deliver documents on-line at a cost that is competitive with air mail postage rates. If this can be achieved, it would mean about 35 cents per page for domestic teletex service and about $1 per page for service to Europe.

Homeowners' rooftop antennas should become more prevalent in the United

States because on June 16, 1982, the Federal Communication Commission approved in principle direct broadcasting from satellites to homeowners' rooftop antennas. Comsat (Communications Satellite Corporation) already has filed proposals in the area of direct broadcast systems (DBS). Comsat is after more than just the right to provide satellite facilities; it also wants to market its own programming directly to more than 20 million customers that the new television satellite services are expected to attract. This means that, besides being in competition with such giants as Western Union and RCA, Comsat also might be in competition with ABC, CBS, and NBC, the television networks.

Finally, as we stated earlier, other new service offerings will involve cable television, cellular radio, Digital Termination Systems, Videotext, optical fiber transmissions, and others. Between the 1960s and 1980s our concern was directed toward getting more "computer power." Between 1980 and 2000 our concern will be to get more "data communication transmission capacity." Just as computers run out of "memory space," data communication networks run out of "bandwidth."

QUESTIONS

1. What is a common carrier?
2. What is a tariff?
3. Identify and describe some of the private circuit (lease) services.
4. Identify various measured use services.
5. Identify some of the special services that are available.

10

NETWORK DESIGN FUNDAMENTALS

This chapter presents the fundamentals of designing data communication networks. A step-by-step systems approach is used and 13 individual steps are described. The systems approach starts with a feasibility study and goes through planning, current system review, designing, geographical considerations, message analysis, circuit loading, security controls, configurations, software/hardware, cost analysis, and implementation. Other features in this chapter are a listing of evaluation criteria, various forms/charts, and seven types of cost analysis.

THE SYSTEMS APPROACH TO DESIGN

When planning for a completely new data communication network, enhancement of a current network, or the use of publicly available networks, you should use the systems approach. Whether the network achieves success or just marginal utilization may be determined before a single piece of software or hardware is ordered. The key ingredient for success lies in planning based on the "system's interface with the users." Far too often, data processing-oriented personnel take an equipment-oriented approach or a technical software-oriented approach. In today's world of data communications the designer must take a user systems application approach.

For example, there are two major classes of users for a data communication network/system. These are the organization's "management" and "user" personnel.

Managers must accept the system and believe in it, or they will not trust the data/information/reports they receive from the system. Recall the word CATER that was discussed in Chapter 7. If the information received by management is not Consistent, Accurate, Timely, Economically feasible, and Relevant, then the system will not be accepted by management.

The "user" personnel who work with the system on a day-to-day basis must be able to accept the system or their productivity will fall drastically. When productivity falls, the cost of carrying out basic office functions may increase the cost of the final product or service by 10 to 50 percent. Office productivity recently has taken on added importance because we have been moving from a society that has most of its people engaged in manufacturing to one in which the majority of the people are engaged in information-type processes (service oriented). In other words, proportionately more people are involved in information-related work than in manufacturing/assembly work. We now need the industrial engineers from the factory environment to move into the automated business office.

These changes are the reason this book promulgates the systems approach to designing data communication networks. The discussions and recommended steps encompass the process needed to design a new network. If a current network is being enhanced, perhaps some of the steps can be omitted. If the decision has been made to use one of the publicly available networks, then it may also be possible to omit some of the steps. However, serious consideration should be given to all of the following 13 steps. Each step has a detailed explanation on how to carry it out. The 13 steps are:

1. Make a feasibility study
2. Prepare a plan
3. Understand the current systems
4. Design the data communication network

5. Identify the geographical scope
6. Analyze the messages
7. Determine traffic/circuit loading
8. Develop a control matrix
9. Develop network configurations
10. Consider software
11. Consider hardware
12. Do a cost analysis
13. Sell and implement the network

MAKE A FEASIBILITY STUDY

The first point that must be made with regard to a feasibility study is that it may not be necessary to conduct one. It may have already been performed by management in order to identify the problem or the purpose/objectives of the proposed system. Perhaps the scope of the proposed system already has been defined. Furthermore, it is entirely possible that either management or the realities of the economic/business environment have dictated that an on-line data communication network will be developed.

For example, can you imagine any major airline deciding that a network costs too much or does not meet its objectives? If one decided against a network, it would cease to be competitive with the other airlines; therefore, the feasibility of a "go/no go" network decision is decided even before the airline starts to think about a feasibility study.

Of course, in this case we are talking about a feasibility study that helps determine whether or not to proceed with a network, rather than a more elaborate feasibility study to identify which specific network should be set up. The decision as to which network is covered in the next 12 steps.

In proceeding with a feasibility study, a primary responsibility is to define the problem clearly and put it in writing. Problem definition involves identifying all the problems that have led to the need for a data communication network. Any of the following factors may be analyzed to determine if they contribute to the need for this new network.

- Increased volumes of inputs/outputs
- Inadequate data processing
- Obsolete hardware/software

- Inadequate file structures (database)
- Unsatisfactory movement of data/information throughout the organization
- Inadequate interfacing between application systems and staff within the organization
- Documentation not being available in a timely manner
- Current systems being unreliable
- Inability to maintain current systems
- Inadequate security/privacy
- Decreasing productivity
- Inadequate training
- Future growth requiring new methods
- Competition forcing the change
- Negative effect of old system on employee morale
- A new system being viewed as having a positive effect on investments, cash flow, etc.
- Inadequate floor space for personnel/files
- Future cost avoidance
- Need for more timely access to information for improved decision making
- Increasing flow of information/paperwork
- Need for expanded capacity for the business functions/manufacturing
- Necessity for increasing level of service quality/performance
- International operations requiring new methods and better exchange of information
- Reduction of inventories
- Need for a paperless office
- Desire to take advantage of the technology of the 1990s.

Once the problem has been defined in this way, the purpose and objectives of the new system are identified, the scope or boundaries that the system will encompass are established, and perhaps some preliminary "magnitudes" of cost are identified for the proposed data communication network.

The feasibility study might include some preliminary work on the geographical scope of the network, or the physical areas of the organization that will be interconnected by it. It may be appropriate to develop a rough draft geographical map of the intended network.

At the completion of this data gathering, a short feasibility study written report should be generated. This report is the medium by which you tell management what the problem is, what you have found its causes to be, and what you have to offer in the way of a solution. The feasibility study results might be presented verbally as well. This type of presentation provides management with an opportunity to ask questions or discuss issues that may have a bearing on whether to proceed.

Your solution is probably a "go/no go" decision as to whether a full program should be started for the design and development of a new data communication system. The feasibility study written report should contain:

- A statement of the problem that clearly demonstrates your understanding of it.

- A concise description of the purpose and objectives of the network. It is important to remember that the purpose is to improve inventory control, effect an improved cash flow, process orders faster, and the like, not to install modems or communication circuits.

- A clear statement of the scope or boundaries. Which application systems will use the network? At this point it may be appropriate to add the preliminary review of the geographical scope or the physical locations that will be interconnected.

- A clear statement of the various "magnitudes" of cost for such areas as software, hardware, communication circuits, restructuring of the internal business organization, and the redesign of current application systems. Any political problems or organizational costs should also be considered. For example, combining voice and data communications into one department has a political cost.

- Highlighting of special attention areas, unusual situations, or the interrelatedness of problems that were not seen before.

- Description of the entire system in generalities so management can visualize the overall data communication system.

- Recommendations as to whether a new system should be designed. In fact, at this point you should be able to recommend whether to enhance the current communication system, design a totally new communication system, or subscribe to a public packet data communication network.

- A suggested timetable, including some general milestones. Try to estimate the cost of reaching each milestone.

- An appendix with any pertinent geographical location charts, graphs, pictures, flow charts, floor plans, or other layouts that would add to management's understanding of this report.

PREPARE A PLAN

At this point the feasibility study has been completed and management has given its approval to proceed with the design and development of a data communication network. Be sure to note that in this chapter we are designing a totally new network. Some of the procedures that are discussed in the remainder of this chapter may be eliminated if you are merely enhancing a current network.

In developing the plan, remember that a successful plan always takes into account the following three factors:

- *Technical feasibility* of the network.
- *Operational feasibility* by the "users" who conduct their daily business using the network, and by "management" who has to rely on its reports.
- *Economic feasibility* to keep it within budgetary limitations.

The first step is to take the statement on the purpose/objectives of the network and write it down into three distinct goal areas. The *major* goal is the reason that the data communication network is being built. The objective is to ensure that the network meets these requirements. Next, *intermediate* goals are other gains the system can make while serving its major purpose, hopefully with little or no extra expense. Finally, *minor* goals are the functions that a communication network, along with data processing applications, can perform for the organization but for which it is not quite ready (future requirements). The major goals are mandatory. The intermediate goals are desirable. The minor goals are "wish list" items.

There is no way to outline the exact steps the plan should follow because it must be custom tailored to the organization and application systems that the network must service. The goal the network is to achieve should provide the framework for the plan. For example, referring to the three goals above, the major goal might be to speed up order entry and achieve improved cash flow through better collections. The intermediate goal might be to interface all of the accounting applications with the order entry operations. A minor goal might be to set up a voice mail/electronic mail system for the future. All too often network designers forget their priorities and concentrate on minor goals because of personal interest. Committing the goals to writing serves as a constant reminder to avoid this trap.

The first step in developing a "custom" plan might be to identify the various sources of information, types of information to be collected, and a schedule for performing various activities. It is likely that the designer will emulate the 13 key steps already listed in this chapter.

Finally, as the design begins, develop some evaluation criteria. If evaluation criteria are developed at the beginning, then there is a yardstick at completion so the design, development, and implementation of the data communication network/ system can be measured. The following evaluation criteria should be considered.

- **Time**   This could be elapsed time, transaction time, overall processing time, response time, or other operational times.
- **Cost**   This may be the annual cost of the system, per unit cost, maintenance cost, or other cost items such as operational, investment, and implementation.
- **Quality**   Is a better product or service being produced? Is there less rework because of the system? Has the quality of data/information improved?
- **Capacity**   This involves the capacity of the system to handle workloads, peak loads, and average loads, as well as long-term future capacity to meet the organization's needs in the next decade.
- **Efficiency**   Is the system more efficient than the previous one?
- **Productivity**   Has productivity of the user (information provider) and management (information user) improved? Is decision making faster and more accurate because of the information provided by this system?
- **Accuracy**   Are there fewer errors? Can management rely on this system more than the old one?
- **Flexibility**   Can the new system perform diverse operations that were not possible before?
- **Reliability**   Are there fewer breakdowns of this system compared with the previous one? Is uptime very high with this system? The reliability/uptime of an on-line network is probably the number one criterion by which to judge its design and development.
- **Acceptance**   Evaluate whether both the information providers and the information users have accepted the system.
- **Controls**   Are adequate security and control mechanisms in place in order to prevent threats to the system such as errors and omissions, fraud and defalcation, lost data, breaches of privacy, disastrous events, and the like?
- **Documentation**   Does the system have adequate written/pictorial descriptions documenting all of its hardware, protocols, software, circuits, and user manuals?
- **Training**   Are training courses adequate and are they offered on a continuous basis, especially for terminal operators? Are training manuals adequate and updated on a regular basis?

- **System Life**   Is the life of the system adequate? When two to five years are spent designing and implementing a system, the system life should be of adequate duration to take advantage of the economies of scale.

The above evaluation criteria can be used to evaluate the new data communication network after it has been developed. Also, it may be advisable to evaluate your own performance during the design and development of this new network. In that case, examine such items as whether development time schedules were on target. Were development costs within budget or was there a large cost overrun? Were any deviations from the original purpose/objectives and scope documented? Consider interactions with those affected by the system: Do they feel they were treated fairly and are they satisfied with you and your design? Was there a lot of turnover on the project team during the design and development?

In summary, as the plan is prepared (step-by-step approach), also develop evaluation criteria for the data communication network as well as the evaluation criteria by which to judge your own design efforts. If you ignore this step, someone else may do it and you may be judged by a set of criteria that do not relate well to your effort.

UNDERSTAND THE CURRENT SYSTEMS

The objective of this step of the design effort is to gain a complete understanding of the current operations (application system/messages) and any network that is functioning. This step provides a benchmark against which future design requirements can be gauged. It should provide a clear picture of the present sequence of operations, processing times, work volumes, current communication systems, existing costs, and user/management needs.

In order to be successful at this stage, begin by gathering general information or unique characteristics of the environment in which the system must operate. Next, identify the specific applications that will use the data communication network and any proposed future applications.

Learn something about the background of the industry in which the network will function (what competitors are doing in this regard), as well as about your individual company and about the departments that are responsible for the applications.

Determine if there are any legal requirements, such as local, state, federal, or international laws, that might affect the network.

With regard to the people in different departments who will be affected by the system, do not overlook the fact that there are formal organizations as shown on the organization chart, and there are also informal organizations within a specific department.

It is important to be aware that company politics might affect the design effort; people may tell you what they want for their personal interests rather than what is in the best interests of the organization.

Develop an input, processing, output model for *each* system that will utilize the data communication network. Figure 10–1 shows a typical input, processing, output model. Your task is to identify each generic input to the application sys-

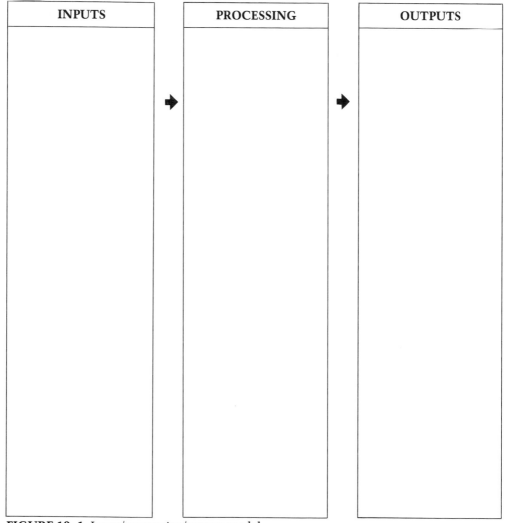

FIGURE 10–1 Input/processing/output model.

tem, the typical processing steps that are performed, and each generic output. Describe and list each input/process/output on the model in Figure 10–1.

Also, identify the "file formats" so database planners can start to design the database and database access methodologies. Transmission volumes increase dramatically when the network is used for database retrieval transactions.

Techniques used to complete this step might include interviewing user personnel, searching a variety of current records for message format and volumes, estimating and sampling for timings and volumes, and possibly comparing current systems with others that have been put on a data communication network.

The documentation gathered during each of the above tasks can serve as a future summary of the existing system. A written summary also should be developed. This summary should include everything of importance learned during this step of the design. It is your written understanding of the existing systems. It should include any design ideas, notes on whether currently used forms or transmittal documents are adequate or inadequate, who was helpful, who hindered progress, and any other overall impressions gained from interviews, meetings, records, flow charts, sampling, and the like. In general, the written summary should contain information that can be referred to during the detailed design steps for development of the data communication network. It is the benchmark to be used for later comparisons.

DESIGN THE DATA COMMUNICATION NETWORK

By the time the network design begins, certain items already should be established, such as the problem definition, purpose/objectives, scope of the network, general background information about the application systems that will use the network, and a thorough written understanding of the current systems. With these items in hand, a list of general system requirements can be developed.

The object of defining the general system requirements is to assemble an overall picture of the functions to be performed by the proposed network. At this point the input, processing, output models for each of the application systems might be of great value.

During the early stages of defining the general system requirements, a review of the organization's long-range and short-range plans is advised. This review helps provide the proper perspective in which to design a system that will not be obsolete in a couple of years and that will meet the future requirements of the organization. These long/short-range plans indicate such information as changes in company goals, strategic plans, development plans for new products or services, projections of changing sales, research and development projects, major capital expenditures, possible changes in product mix, emphasis on security, and future commitments to technology.

Once the system requirements have been identified, they should be prioritized. That is, they should be divided into mandatory system requirements, desirable system requirements, and wish list requirements. This information enables you to develop a minimum level of mandatory requirements and a negotiable list of desirable requirements that are dependent upon cost and availability. Match these against your major, intermediate, and minor goals mentioned earlier.

System requirements should be as precise as possible. For example, rather than stating "a large quantity of characters," state requirements in more precise figures such as "50 character per minute plus or minus 10 percent."

At this point, avoid presenting solutions; only requirements are needed. For example, a requirement might state that circuit capacity should be great enough to handle 5000 characters per minute which will triple by 1986. It would be a mistake to state this as a solution by saying that a 9600 bit per second, coaxial cable, voice grade circuit is required. Solutions should be left for later, during development of network configurations when software/hardware considerations must be interrelated with those configurations.

By definition, to design means to map out, plan, or arrange the parts into a whole which satisfies the objectives involved. The final 9 steps of designing a data communication network are described in the remainder of this chapter.

IDENTIFY THE GEOGRAPHICAL SCOPE

The scope of the applications systems that are to be included on the network have now been identified. The "very rough draft" geographical map that was developed during the feasibility study should be examined at this point, and a more detailed and accurate version should be prepared.

A data communication network has four basic levels of geographical scope:

- International (worldwide network)
- Country (within the boundaries and laws of a single country)
- City (within the boundaries of a specific city, state/province, or local governmental jurisdiction)
- Local facility (within a specific building or confined to a series of buildings located upon the same contiguous property)

Usually it is easiest to start with the highest level, international. Begin by making a network map drawing of all the international locations that must be connected. At this level it is necessary only to interconnect the major countries and/or cities around the world. It is sufficient to have a map that shows lines going between the countries/cities. Details such as the type of circuit, multiplex, multidrop, concentrators, and the like have not been decided yet. If the network does not cross international boundaries, then obviously this step can be omitted.

The next map you *might* prepare is the country map for each country. Interconnections should be drawn between all cities within the country or countries that require interconnection. Again, a single line drawn between the cities is quite adequate because the type of configuration has not yet been decided upon. Figure 10–2 is a typical example of a country map intermixed with an international map because of the closeness of the two countries.

The next map to prepare is one of the city or state/province. This can be divided into two levels. The first level uses a state map. It has lines drawn showing the

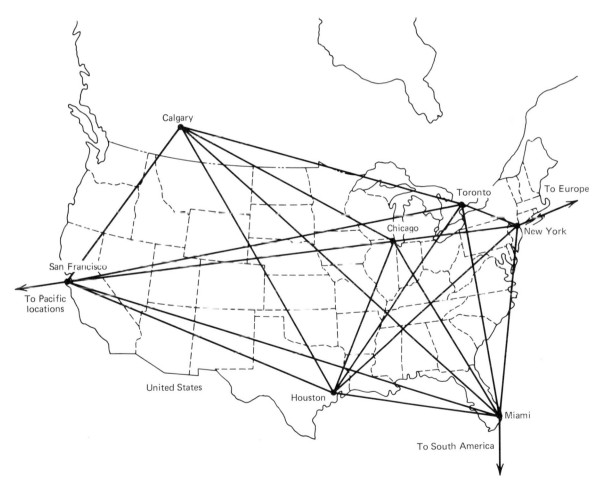

FIGURE 10–2 Country map (point-to-point).

interconnection among various cities within the state. City maps are used at the second level. They show interconnection of various "local facility" locations within the city. When two maps are used, it does add another level. The advantage, however, is that it also decreases the complexity of simultaneously trying to design both intrastate circuits and intracity circuits. If either of these levels is omitted, the state level maps may be the less vital. The city level maps are needed to identify concentrator sites and/or multidrop locations, as well as individual terminal locations. At this point, only lines are drawn between the various interconnect points because configurations have not been decided upon.

The local facility "maps" are really pictorial diagrams because usually blueprints or drawings of the building floor layouts are used. Specific terminal locations can be identified on these pictorial diagrams. It is too early to identify concentrator/multiplexer sites, so this should be left until a later time. It is appropriate at this point to identify the location of current telephone equipment rooms for incoming communication circuits (voice and data).

By the end of this task there are tentative locations for individual terminals and circuit paths for the local facility, intracity, intrastate, country and international needs. To date, little is known about the volume of data that must be transmitted; nor is anything known about the type of hardware/software that might be utilized by this system.

The next step is to analyze the specific messages, although this can be done simultaneously with the development of the geographical maps and pictorial diagrams.

ANALYZE THE MESSAGES

This step may be combined with the previous step on identifying the geographical scope, but it is more often combined with the following step, "determining traffic/circuit loading." The reason it is identified here as a separate step is so that you will understand clearly the level of detail that must be obtained during this very important step.

In this step each message that will be transmitted or received from each application system at each terminal location is identified. Also, each message field (data item/attribute) is identified, along with the average number of characters for each field. Further, it is necessary to identify message length and the volumes of messages transmitted per day or per hour. It probably will be necessary to visit each location where there is a system that will utilize the data communication network. These site visits are required in order to identify clearly each and every message type that will be transmitted or received.

If the system is a manual one, these messages might be forms in the current system, although they already might be electronically generated messages or video screen formats on terminals. Each message should be described by a short title, and a sample of the message should be attached if there is a current equivalent. If there is not an equivalent, *all* of the fields that will make up the message must be identified. Message analysis sometimes reveals that the system will have to handle a greater volume of data than was previously thought.

Before proceeding to the next step, we should discuss further the various fields/data items of the message. After each message is described and samples collected of messages that are in the current system, this data must be recorded. A simple form should be used to record this data, such as the one shown in Figure 10–3. The name of each individual message is listed along with the name of each field/data item that makes up that message. For each field in the message list the average number of characters in each field/data item. Most data items have only an average number of characters per message; few have a peak number. It is always worth the effort to determine if some of the individual data items have a peak number of characters per message. Peaks may occur during certain days of the year or hours of the day or any other time that is unique to the business situation.

TELLER INQUIRY SYSTEM			
Message Name	Message Fields (data item)	Average Characters/ Field	Peak Characters Field
Passbook savings inquiry	Password	4	4
	Customer account number	9	9
	Dollar amount	6	12
	Transaction code	3	3
	Total	22	28
Loan balance inquiry	Password	4	4
	Customer loan number	16	16

FIGURE 10–3 Message contents.

Once the messages have been described and recorded, the average/peak number of characters for each message can be calculated for each application system. The most important figure is the average number of characters per message, although sometimes peak numbers of characters per message must be taken into account.

It should be noted here that most systems are built using the average number of characters for their basis, because few organizations can afford the cost of a system built on the basis of the peak number of characters. The use of averages is even more prevalent when the choice is between average number of messages per day and peak number of messages per day.

The system designer should note that a pure character count may be misleading with regard to the number of characters contained in the transmitted message. Header characters (identifying overhead-type characters within the message) and the data communication network control characters must be taken into account. The control characters can be items such as a consecutive message number, the synchronization characters, carriage returns or tabulation characters when appropriate, and line control characters for the protocol utilized (although the protocol may be unknown at this point). As a rule of thumb, 20 or 30 line control characters might be added to each message transmitted, although when messages are transmitted in contiguous groups, this figure might be closer to 20 or less. There is no rule of thumb for message header characters. The best way to identify message content control information is probably to interview the people who run the current manual or computerized application system.

Determining the volumes of messages is critical. Now that the average number of characters for each message has been determined, the next step is to learn how many messages will be transmitted per day or per hour. To accumulate this information accurately, utilize Figure 10–4, which is a network link traffic table.

The first item in the upper left-hand corner is the identifier of a network link. This is nothing more than a first cut at determining where messages will go when users transmit from their local facility work area. The second column shows the name of the individual message type. The third and fourth columns show the average characters per message and the peak characters per message (if appropriate). The fifth and sixth columns show the average number of messages per day and the peak number of messages per day. The seventh and eighth columns show the average number of characters transmitted per day and the peak number of characters transmitted per day; these numbers are obtained by multiplying the characters per message by the number of messages per day.

Finally, when possible, these traffic statistics (characters transmitted per day) should be broken down into the hourly number of characters transmitted throughout the work day. This information can be used to help spot any problems with hourly peak volumes as the design progresses. For example, if a column total of the hourly number of characters transmitted between 9 to 10 A.M. has a volume that is 50 times the capacity of a single circuit network link, then a problem ex-

Network Link	Message Type	Characters/Message		Messages/Day		Characters/Day		Hourly Number of Characters Transmitted								
		Average	Peak	Average	Peak	Average	Peak	8-9	9-10	10-11	11-12	12-1	1-2	2-3	3-4	4-5
Calgary to San Francisco	Passbook Savings Inquiry	22	23	1500	1650	33000	46200		12500	4000	4000	2000	1500	9000		
	Loan Balance Inquiry															
Down Totals						330000	405000									

SUM OF THIS COLUMN

FIGURE 10–4 Network link traffic table.

ists. The problem can solve itself if you have some messages transmitted later, such as during the next several hours. Other solutions are to have some people work overtime or to design a network link that has the capacity to meet that very high one-hour volume, although cost may prohibit the latter solution.

Even though the most important figure is the average number of characters transmitted per day, there may be important factors that cause peak volumes at various times during the day, various days during the week, or various times during the month. There also may be seasonal times of peak volumes because of holidays or legal requirements. Recall that legal requirements were identified earlier during the "understand the current system" phase.

The designer should plan for varying volumes at different hours of the day. For example, in an on-line banking network, traffic volume peaks usually are in the midmorning (bank opening) and just prior to closing. Airline and rental car reservation system designers look for peak volumes of messages during holiday periods or during other vacation periods. A military system designer might look for extreme peaks in volume during crisis situations.

You can calculate message volumes by counting current messages in a current system or by estimating future messages. Whenever possible, take a random sample for several weeks of traffic and actually count the number of messages handled each day at each location.

If an on-line system is operating currently, network monitors/analyzers may be able to provide an actual circuit character count of the volume transmitted per hour or per day. Take care when selecting the sample of working days to ensure that it is not an "out of normal" situation. When estimating message volumes for a system that does not currently exist, you can use conglomerate estimating, comparison estimating, detailed estimating, or modeling.

- With *conglomerate estimating*, representatives from each application system that will use the network confer to develop estimates based on past experience.

- With *comparison estimating*, the network designer meets with people inside or outside the organization who have a similar system so they can supply estimates from their network.

- With *detailed estimating*, the network designer makes a detailed study of the overall application system and its future needs in order to develop subestimates, which are then totaled into the total volume of messages just as we have described above.

- *Modeling networks/response time* is described in Chapter 3.

When making estimates of volumes, be sure to take future growth into account so the system will meet needs of the next decade. Do not worry about the accuracy of estimates at this point, although you should make them as accurate as possible. Accuracy may not be a major concern, because of the stairstep nature of communication circuits. For example, assume a situation in which a voice grade circuit is used. It can be used to transmit at 16,000 bits per second, but, to meet data volumes, you need to transmit at 20,000 bits per second. This would require the lease of two voice grade circuits. The combined two voice grade circuits now have a maximum capacity of 32,000 bits per second, greatly exceeding the needed 20,000 bits per second. This example demonstrates that if actual message volumes are higher than estimated, there is plenty of spare capacity. On the other hand, the opposite problem may occur if estimates are too optimistic; the organization may be forced to lease two voice grade circuits when only one is needed.

Now that individual message contents and the network link traffic table have been developed, there should be some feeling for the total volume of characters per day transmitted on each link of the proposed network. These are the volumes of characters transmitted from and to each local facility (nodes) where terminals will be located. The next step, which usually is carried out simultaneously with this step, is to determine traffic/circuit loading.

DETERMINE TRAFFIC/CIRCUIT LOADING

Now that average/peak characters transmitted per day per link have been identified, work can begin on determining the circuit capacities that will be required to carry that traffic. They are based on the number of characters per message and the number of messages transmitted per hour or per day. They can also be augmented through the use of modeling, which we discussed earlier (Chapter 3).

At this point return to the geographical maps and pictorial diagrams (local facilities). Does this map or pictorial diagram still seem reasonable in light of the vast amount of further information that has been gathered during the message analysis? At this time some of the maps or pictorial diagrams might be reconfigured slightly in order to further solidify the geographic configuration of the network. Remember to evaluate all the geographical maps: international, country, city/ state, and local facilities.

The next step is to review all of the network links over which data will travel. This may have been done when the network link traffic table was completed. If so, double-check at this time to verify that each message type was cross-referenced to the proper network link (columns 1 and 2 in Figure 10–4). If the hour-to-hour variation is significant, it may be necessary to take hourly peaks into account or adjust working schedules and work flows. Match the characters per day for each network link in Figure 10–4 with each network link that was shown on the country map (Figure 10–2). It will be helpful, when examining alternate configurations, if you list the characters per day for each link shown on Figure 10–2.

It is the column totals (Figure 10–4) that really count. If the total number of characters transmitted in a single day on a single link is 330,000 or 405,000 then the network link has to operate at a speed that permits transmission of the 330,000 or 405,000 characters during the normal working hours. If it cannot meet this limit, certain adjustments have to be made. Now look at Figure 10–5. It shows: San Francisco/Miami, 890,000; San Francisco/Houston, 1,250,000; and Houston/Miami, 770,000 characters per day. Later in the design, if the San Francisco/Miami traffic is multidropped through Houston, the total traffic on the Houston/Miami link will be 1,660,000 characters per day (890,000 + 770,000).

To establish the circuit loading (the amount of data transmitted), the designer

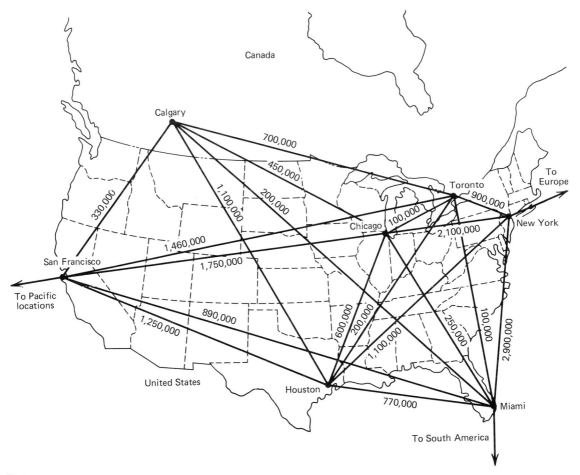

FIGURE 10–5 Link loading in characters per day.

usually starts with the total characters transmitted per day on each link, or if possible, the number of characters transmitted per hour if peaks must be met.

Starting with the total characters transmitted per day, the system designer first determines if there are any time zone differences between the various stations. This might be an international or a national system that has time zone differences that must be taken into account. For example, there is a three-hour time difference between Toronto and San Francisco. This means that if a host computer in Toronto operates from 7 A.M. until 4 P.M. (Toronto time), under normal circumstances there is only a five-hour working day in San Francisco, even assuming that someone is working through the lunch hour. By the time the people arrive at work in San Francisco at 8 A.M. it is already 11 A.M. in Toronto. Then Toronto shuts down its computer at 4 P.M. and it is still only 1 P.M. in San Francisco. This leaves the San Francisco facility with a workday that extends only from 8 A.M. until 1 P.M. The practical effect of this time difference is that the 1,460,000 characters (SF/TOR link of figure 10-5) of data must be transmitted during a five-hour period rather than the eight-hour day you might expect. These effects have to be taken into account or work schedules must be changed. Obviously, the Toronto host computer operating hours can be extended or the San Francisco staff can start work earlier. There is no perfect solution to time zone differences, but the system designer must account for them.

Other major factors that affect circuit loading include the basic efficiency of the code utilized and TRIBs (discussed in Chapter 3). Synchronous transmission is more efficient than asynchronous transmission. The number of line control characters involved in the basic protocol affects line loading. The application systems/business future growth factor must be considered so the system will have a reasonably useful lifetime. Forecasts should be made of expected message volumes three to seven years in the future. This growth factor may vary from 5 to 50 percent and, in some cases, exceed 100 percent for high-growth organizations.

Some extra time should be allowed for transmission line errors (error detection and retransmission) which may result in the retransmission of 1 to 2 percent of the messages. Retransmissions may be even higher where small common carriers are used or if transmission is into or out of developing countries. The network designer also should consider a 10 to 20 percent contingency factor for the "turnpike effect." The turnpike effect results when the system is utilized to a greater extent than was anticipated because the system is found to be available, is very efficient, and has electronic mail features. In other words, the system is now handling message types for which it was not originally designed.

Other factors to consider when evaluating line loading might be whether to include a message priority system. High-priority messages may require special identification and therefore may increase the number of characters per message. If the message mix changes and, over a period of time, most messages become high priority, then more characters will be transmitted during a working day. Also, a greater throughput may have to be planned to ensure that lower-priority messages get through in a reasonable period of time. The learning curve of new terminal operators may also affect line loading. Operator errors and retransmissions are greater when a new system is being learned.

Another factor that might affect circuit loading is an inaccurate traffic analysis (confidence intervals). Try to account for any business operating procedures that might affect the system and the volumes of data transmitted.

Other factors that must be taken into account include extra characters transmitted with regard to the system operation (line control characters) such as polling characters, turnaround time/synchronization characters, control characters in message frames and/or packets, modem turnaround time on half duplex circuits, message propagation time subtracts from the total useful hours for transmission of data, any printer time for carriage return/tabulation/form feeding, lost time when statistical time division multiplexers are overloaded, and periods of high error rates caused by atmospheric disturbances.

At this point the system designer should review and establish some of the response time criteria. These are required to meet the basic needs of the application systems that will utilize the network (response time was covered in Chapter 3).

Finally, begin recording, on the network maps and/or pictorial diagrams, some of the bit per second transmission rates that will be required for each circuit link. Sometimes it is useful to show the transmission capacity required for each link. In Figure 10–5 we show the characters per day per link. Now add the bit per second transmission rate necessary for each circuit link. This will help when alternative network configurations and software/hardware considerations are being developed and evaluated.

DEVELOP A CONTROL MATRIX

Because the network probably will be the "lifeline" of the entire information flow within the organization, security and control are mandatory. All of the security and control mechanisms that will be included in this data communication network must be taken into consideration during the design phase. As was stated earlier, we are well into an era in which information is the single most valuable resource within an organization. For this reason, it must be protected from all types of threats such as errors and omissions, message loss or change, disasters and disruptions, breaches of privacy, security/theft, unreliability, incorrect recovery/restart, poor error handling, and lack of data validation.

At this point the network control matrix is developed. How to develop this matrix was presented in Chapter 8 in the section "Matrix of Controls." Review that section now to be sure you understand how to set up and continue development of this matrix through the remainder of the design project. At this point, develop the basic blank matrix, naming only the threats and components. As the design effort continues, identify controls and relate them to their threats and components by placing them in the appropriate cell of the matrix. To assist in this area, also review Appendixes 1, 2, and 3, which relate to security and control for data communication networks.

DEVELOP NETWORK CONFIGURATIONS

During this step of the system approach to designing a data communication network, the designer utilizes all of the information collected to date. Of special value are the network maps and the traffic/circuit loading data. These are used to configure the network in such a way as to achieve the required throughput at a minimum circuit cost. Begin this step by reviewing the maps and pictorial diagrams that show the links between the station/node locations.

The object of this step is to configure the circuit paths between users and the host computer. The decision involves moving the stations/nodes about, and making judgments with regard to software and hardware. In reality, this step is performed simultaneously with the next two steps, "consider software" and "consider hardware." Some goals that the network designer tries to achieve with regard to an efficient and cost-effective network include:

- Minimum circuit mileage between the various stations/nodes. Computer programs/modeling can help here.
- Adequate circuit capacity to meet today's data transfer needs as well as those required three to seven years in the future.
- Reasonable response times at individual terminals and the response time needs must be met for each application.
- Reliable hardware that offers minimum cost, adequate speed and control features, a high mean time between failures (MTBF), and good diagnostic/serviceability features.
- Efficient software/protocols that can be used on a variety of circuit configurations including satellite circuits. One of the newer bit-oriented protocols that can interface with various international standards (X.25) might be used. This permits the network to interface with national/international networks as well as with electronic mail postal systems, to utilize multivendor hardware, and to connect to public packet switched networks.
- A "very high" level of reliability (network uptime) must be met. This may be the most important factor. The network designer should always remember that, when business operations move into an on-line, real-time data commu-

nications network, it is as if the company has closed its doors for business when the network is down.

- Reasonable costs (not necessarily the absolute lowest).
- Acceptance of the network by both day-to-day "user" personnel and "management" who utilize data/information from the system.

When one is developing different network configurations, there are a variety of choices. In other words, there is a "choice set" which is a set of all available alternatives. Each alternative is a different system or a slightly modified version of another alternative. During the deliberations, the following decisions must be considered:

- Determine the choice set, that is, all possible network configurations.
- Divide the choice set into attainable and unattainable sets. The attainable set(s) contain only those alternatives that have a reasonable chance of acceptance by management. Acceptance might be predicated on costs, software, hardware, circuit availability, or political factors within the organization.
- Review the attainable set of alternatives and place them in a ranked sequence from the most favored to the least favored, taking into account your evaluation criteria for choosing the most favored. Evaluation criteria were identified during the plan preparation phase.
- Present the most highly favored alternatives to management for review and, it is hoped, approval.

There is one other consideration in selecting the different network alternatives. The network designer must know whether the proposed alternative is going to maximize something, optimize something, or satisfice something, or if it will be a combination of the three. To *maximize* is to get the highest possible degree of use out of the system without regard to other systems. To *optimize* is to get the most favored degree of use out of the system taking into account all other systems; an optimal system does just the right amount of whatever it is supposed to do, which is not necessarily the maximum. To *satisfice* is to choose a particular level of performance for which to strive and for which management is willing to settle.

Finally, the network designer also must be aware that individual job tasks within the network may have three levels of dependence upon each other.

- *Random dependence*: a job task is required because of some other job tasks.
- *Sequential dependence*: one particular job task must precede or follow another job task.

- *Time dependence*: a job task is required at a set time with regard to another job task.

The network designer should assess various job tasks during the development and design of network configurations. Job task interrelationships must be studied with regard to future needs and growth. Job tasks that are dependent today may not be after a new application is completed next year. Job tasks require an open-ended approach.

Now that the network maps/pictorial diagrams and traffic/circuit loading have been reviewed, line controls and modes of operation can be considered. This probably involves software and such factors as full duplex versus half duplex, whether a satellite link is used, statistical multiplexers versus pure (transparent) multiplexers, modem speeds, intelligent terminal controllers, and how different configurations operate such as central control versus interrupt, multidrop, or point-to-point.

Various alternative configurations are shown in Figures 10–2, 10–6, 10–7, 10–8, and 10–9. Figure 10–2 shows a point-to-point configuration where each terminal node has its own communication circuit between all other nodes. Figure 10–6 shows the same configuration using a multidrop circuit with New York as a switching center. Figure 10–7 shows the same configuration again, using a multiplexed arrangement. The Houston site multiplexes San Francisco/Miami/Houston data to Chicago. Then Chicago multiplexes that data with the Chicago/Calgary/Toronto data and on to New York. Notice how the different configurations change overall circuit mileage. Circuits are paid for on the basis of dollars per mile per month; therefore, a minimum mileage configuration is also a minimum circuit cost configuration. Also notice that different numbers of modems are required in different configurations. For example, the point-to-point configuration requires many more modems than the multidrop configuration. Fewer modems save on modem costs.

Figure 10–8 shows a packet switching satellite network. This can be a public packet switcher or a private packet network. Figure 10–9 shows a combination of a local network, a packet satellite network, point-to-point, multiplex, and multidrop configurations.

This step involves choosing among various network alternatives. The main constraints are the availability of software packages, available hardware, and available circuit links. These three factors are all interconnected and must be considered along with the "performance and reliability" that must be obtained. All of these factors are also interrelated with regard to cost. Therefore, when alternative network configurations are developed, you must consider the software, hardware, circuits, performance, reliability, and interrelate these five factors through your cost/benefit analysis.

FIGURE 10–6 Multidrop configuration.

CONSIDER SOFTWARE

With regard to software, the type of host computer may be a major constraint. The protocols that the host can handle may limit the types of terminals or other hardware that can be utilized. This limitation may be overcome through the use of protocol converters and/or the purchase of a new front end communication processor that can interface with the host and a variety of software and hardware.

At this point the software will determine the line control methodology/mode of operation. Decisions must be made as to whether operations will be in full duplex or half duplex, asynchronous or synchronous, and at what speeds. For a new system one of the newer bit-oriented protocols should be selected, such as X.25, SDLC, HDLC, or the like. The older byte-oriented protocols (such as Binary Syn-

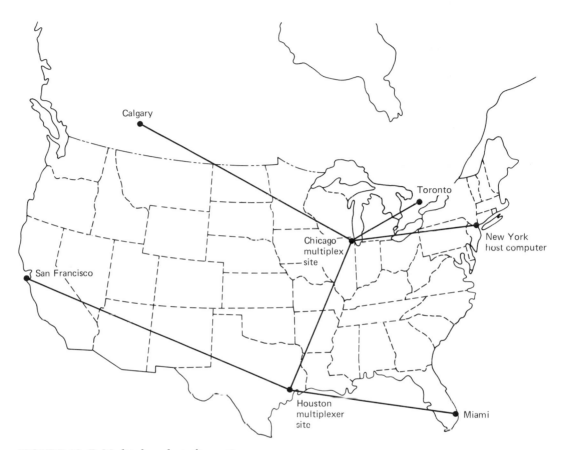

FIGURE 10–7 Multiplexed configuration.

chronous Communications—BSC) probably are not a good choice because of their limitations on satellite links, slow half duplex operation, and inability to meet international standards. It is desirable to select a protocol that is compatible with the International Organization for Standardization (ISO) seven layers, although reality might dictate that in order to be compatible with existing hardware another protocol must be utilized.

In addition to protocol/software, other network architecture/software that resides in the host computer and in the front end communication processors have to be considered. For example, telecommunication access programs and the teleprocessing monitors (evaluating teleprocessing monitors is covered in Appendix 6) may affect network operations. Security software packages in the host computer also can be a constraint. Finally, the host operating system itself may be a

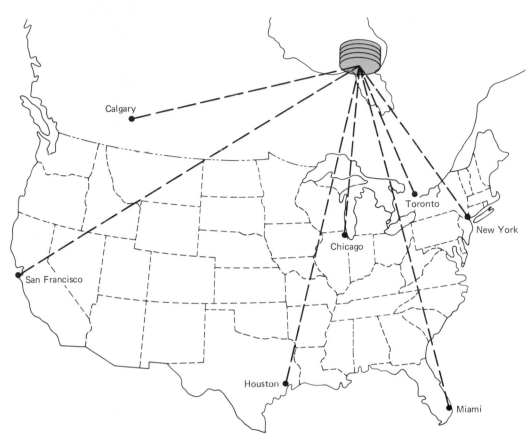

FIGURE 10–8 Public packet switching satellite configuration.

constraint to network control and operation, as might be the database management system software.

Any software programs that are located out in the network should be reviewed. These may be at remote concentrators, remote intelligent terminal control devices, statistical multiplexers, and terminals (microprocessor terminals). Microprocessor terminals also raise the question of distributed data processing/remote application programs.

The network designer can make a major contribution to the future by selecting a protocol that can grow, that is compatible with an internationally recognized standard, and that will not have to be changed for at least five to ten years. The protocol is crucial because the host computer network architecture must be able

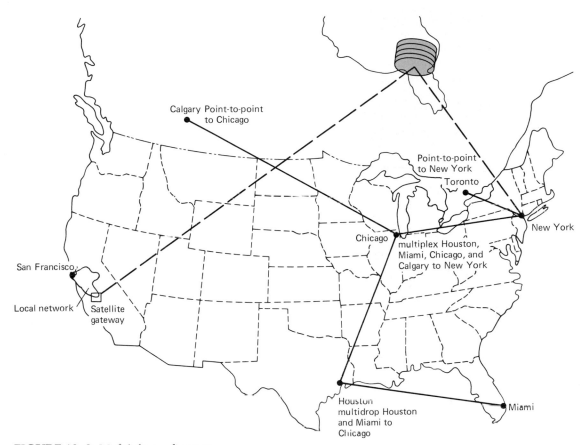

FIGURE 10–9 Multiple configurations.

to interface with it. For example, the telecommunication access programs and teleprocessing monitor should be compatible with international standards. This means that the International Organization for Standardization seven-layer model should be used as the basic skeleton when protocols are interfaced to host computer/front end software packages. Also, you might want to interface with a local area network sometime in the future.

Finally, software diagnostics and maintenance must not be overlooked. Determine how quickly either in-house people or the vendor can diagnose software problems and also how quickly they can fix these problems. Recall Mean Time To Diagnose (MTTD), Mean Time To Respond (MTTR), and Mean Time To Fix (MTTF) from Chapter 7; they also apply to protocols and other software packages.

CONSIDER HARDWARE

Hardware that interacts with the alternative network configurations is easier to handle than software because hardware is a tangible item. Some of the pieces of hardware that need to be considered are:

- Terminals/microprocessors
- Intelligent terminal controllers
- Modems (analog/digital)
- Multiplexers
- Intelligent multiplexers (STDM)/concentrators
- Line-sharing devices
- Protocol converters
- Hardware encryption boxes
- Automated switching devices
- PBX/CBX switchboards
- Data protectors
- Various communication circuit types
- Port-sharing devices
- Front end communication processors
- Host computers
- Testing equipment

With this in mind, the designer uses representations of the pieces of hardware and moves them about on the various network maps and pictorial diagrams. This experimentation with configurations should take into account the protocol/software considerations. The result should be a minimum-cost network that meets the organization's data communications (throughput) requirements. This is no trivial task. Many organizations use computer simulation and modeling to carry out this task successfully.

Before ordering hardware, the design team should decide how to handle diagnostics, troubleshooting, and repair. It should be remembered that MTTD, MTTR, and MTTF always apply to hardware because hardware usually fails more often than software. Vendor estimates of MTBF (Mean Time Between Failures) should be obtained by the design team. Issues that should be addressed include the types of test equipment that are necessary and the organizational structure of the network management group (see Chapter 7). Some hardware may have built-in diagnostic capabilities for its own internal electronic circuits, as well as the ability to identify problems on the communications circuit.

Diagnostics go hand in hand with network service. The vendor's MTBF and ability to respond to service calls are essential factors that affect downtime of the network.

In summary, a network configuration must be developed that takes both hardware and software into account. Costs are also analyzed during this effort.

DO A COST ANALYSIS

Usually it is a hindrance to propose cost limitations during the initial development of design alternatives. Of course, there always should be an effort to keep costs down; however, costs should not interfere with the *preliminary* design configuration alternatives (choice sets). The point is that the various alternatives should be identified first; then costs should be related to the *attainable* design configurations. The first task is to identify the attainable/workable configurations and the second is to identify the costs of those alternatives.

Estimating the cost of a network is much more complex than estimating the cost of a new piece of hardware. Many variables and intangibles are involved. Nevertheless, estimating the cost of a system is a necessary prerequisite to deciding whether implementation is justifiable. Some of the questions that must be considered are:

- What are the major cost categories of the overall system? These may include:
    Circuit costs
    Hardware costs
    Software costs
    Maintenance/network management costs
    Personnel costs
- What methods of estimating are available and what accuracy can be achieved?
- Can all costs be identified and accurately estimated?
- Can benefits be identified? Which benefits cannot be estimated in dollar terms?
- What criteria will management use when evaluating these estimates? (Refer to the set of criteria developed earlier during the "prepare a plan" step.)

Assembling the various costs for a network usually requires a very detailed analysis, although the estimation methodologies that were discussed during "analyze the messages" may be applicable (conglomerate estimating, comparison estimating, and detailed estimating). There are three major categories into which costs must be structured:

- *Implementation costs* are the one-time outlay to create or install new software, hardware, circuits, network control facilities, and the like.
- *Investment costs* are nonrecurring (one-time) outlays to acquire new software, hardware, and the like.
- *Operating costs* are recurring outlays required to operate the system, including the lease of software, hardware, circuits, and the like.

Figure 10–10 shows a formula for combining these three costs into a single annual cost.

$$\text{\$ Average annual cost of network} = \frac{\text{\$ Cost of implementation} + \text{\$ Cost of investment}}{\text{Estimated system life in years}} + \text{\$ annual operating cost}$$

FIGURE 10–10 Average annual cost of network.

Because the specific method used to gather and compile costs for presentation has to be tailored to meet the needs of the organization, several alternatives for cost analysis are presented here.

The methods/cost categories that we will discuss are called Network Cost Analysis, Payback Period, Network Cost Analyzer, a list of Cost/Benefits, Cost Chargeback Method, Circuit Mileage Costs, and Hardware Costs.

Network Cost Analysis   This network cost analysis[1] was developed as a guide for network personnel in estimating total private network costs. In a survey conducted during 1981 (by the CYLIX™ Communications Network people), it was found that a large majority of network managers were not aware of the total cost of their network. In response to this, CYLIX™ Market Research contacted over 50 companies nationwide for the purpose of analyzing all costs associated with private networks. The following analysis methodology shows how to segment costs

[1]This material is reprinted with permission of CYLIX™ Communications Network (an affiliate of Data Communications Corporation), 855 Ridge Lake Blvd., Suite 101, Memphis, TN 38119. CYLIX is a registered trademark.

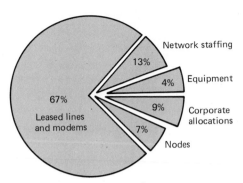

FIGURE 10–11 Network cost breakdown. In a survey conducted in January 1981, CYLIX™ Market Research contacted over 50 companies nationwide for the purpose of analyzing all costs associated with private networks. This analysis shows the average costs of the 50 + sample segmented into five major areas. (*Source:* Printed with permission of CYLIX Communications Network, 800 Ridge Lake Blvd./Suite 300, Memphis, TN 38119.)

into five major areas. Also, the sample costs collected in the survey of 50 companies are presented here. The five major cost categories are staffing (salaries and benefits), corporate allocations (personnel and space/utilities), network nodes, leased circuits and modem costs (leased lines, modems, and master modems), and equipment costs (test and backup). Figure 10–11 shows the percentage breakdown of these five major categories from the sample of 50 companies.

Figures 10–12 through 10–17 are forms that can be utilized to document your network costs in order to compare them with the sample survey. This network cost comparison consists of:

- Staffing costs (Figure 10–12)
    Salaries
    Benefits

- Corporate allocations (Figure 10–13)
    Personnel
    Space/utilities

- Nodes (Figure 10–14)

STAFFING COSTS	
Salaries	Benefits
The average salary of network personnel in the sample was $21,338 per year. Salaries ranged from a low of $10,000 to a high of $36,000. The number of people required to maintain a network was directly proportional to the number of remote locations as shown below.	Included in this category are FICA, Health Insurance, unemployment, etc. The cost of benefits ranged from 20–30% of salary. The sample cost uses 20% of salalry as shown below:

Number of Remote Locations	Network Staff Required
1– 20	0.5–1
21– 50	1–2
51–100	2–3
101 +	3–5

For the purpose of this sample of a 30 remote location network, only one person is used to maintain the network.

	Sample Cost	Your Cost		Sample Cost	Your Cost
Total yearly salary	$21,338	_____	Total yearly cost	$ 4,320	_____
Number of months	12	_____	Number of months	12	_____
Cost per month	$ 1,778	_____	Cost per month	$ 360	_____
Number of remotes	30	_____	Number of remotes	30	_____
Per month per remote	$ 59	_____	Per month per remote	$ 12	_____

	Sample Cost	Your Cost
Total Staffing Costs	**$71**	

FIGURE 10–12 Network cost comparison.

CORPORATE ALLOCATIONS	
Personnel	Space/Utilities
People in this category are not assigned to network communications but assist in the operation of the department. In the sample, one-third of one person's time (with an average salary and benefits of $25,000) was spent in network communications.	Included in this category are furniture, floor space, telephone, etc. The average cost per month per remote in the sample was $30.

	Sample Cost	Your Cost		Sample Cost	Your Cost
Total yearly cost	$ 8,333	___	Total yearly cost	$10,800	___
Number of months	12	___	Number of months	12	___
Cost per month	$ 694	___	Cost per month	$ 900	___
Number of remotes	30	___	Number of remotes	30	___
Per month per remote	$ 23	___	Per month per remote	$ 30	___

		Sample Cost	Your Cost
	Total Corp. Allocations	**$53**	☐

FIGURE 10–13 Network cost comparison.

- Leased line and modem costs (Figure 10–15)
    - Leased lincs
    - Modem
    - Master Modem

- Equipment costs (Figure 10–16)
    - Test
    - Backup

- Summary (Figure 10–17)

		NODES		
			Sample Cost	Your Cost
	Total yearly cost		$ 4,680	
	Number of months		12	
	Cost per month		$ 390	
	Number of remotes		30	
	Per month per remote		$ 13	
			Sample Cost	Your Cost
	Total Node Cost		**$39**	

These switching computers were utilized mainly by relatively larger networks in the sample. The average cost per month per remote location was $13.

The average cost per node was $13 as shown. In the sample network a control node* is used and two remote nodes. Therefore the total cost was $3 \times \$13 = \$39$.

A control node maintains tables for routing data. It also keeps status of network components and accounting data.

FIGURE 10–14 Network cost comparison.

Payback Period   A criterion that frequently is used to judge the profitability of a system is the payback. For example, if a new system costs $700,000 and is expected to yield $100,000 per year in savings, its payback period is seven years (before taxes). The payback period is defined as the number of years required to accumulate earnings sufficient to cover its costs.

The payback period criterion ranks projects in terms of the number of years to pay back. Two factors that must be examined closely are the rate at which the funds flow in and the income taxes (federal, state, and city).

The top half of Figure 10–18 shows two alternative configurations (A and B), each having a four-year payback period. Clearly, project B is preferred because a larger amount of investment is recovered during the first year; project B returns the investment money more promptly.

LEASED LINE AND MODEM COSTS

Leased Lines	Modem	Master Modem
The cost of leased telephone company lines ranged from $150 per month per remote location to $572 per month per remote location. The average was $220 in the sample.	The cost of modems ranged from $60 per month per remote location to $200 per month per remote location. The average cost was $140 in the sample. This example is for multipoint lines where one modem per remote location is required. In this case, a master modem is also required at the termination point of the multipoint line (usually at the host site).	In the case of multipoint lines a master modem is needed at the termination point. The cost of master modems ranged from $17 per month per remote location to $29 per month per remote location. The average was $20. The average number of remote locations per master modem was *seven*. **NOTE:** *This sample assumes multipoint lines. Most participants in the study used a combination of multipoint and point-to-point lines. On a point-to-point line, two modems are required.* *Since there is only one remote on a point-to-point line, the modem cost per remote is therefore doubled, making the monthly cost per remote $280.*

	Sample Cost	Your Cost		Sample Cost	Your Cost		Sample Cost	Your Cost
Total yearly cost	$79,200	___	Total yearly cost	$50,400	___	Total yearly cost	$ 7,200	___
Number of months	12	___	Number of months	12	___	Number of months	12	___
Cost per month	$ 6,600		Cost per month	$ 4,200		Cost per month	$ 600	
Number of remotes	30	___	Number of remotes	30	___	Number of remotes	30	___
Per month per remote	$ 220	___	Per month per remote	$ 140	___	Per month per remote	$ 20	___

	Sample Cost	Your Cost
Total Leased Line and Modem Cost	**$380**	☐

FIGURE 10–15 Network cost comparison.

EQUIPMENT COSTS							
Test				Backup			
Diagnostics, patch panels, tone generators, data scopes, etc. are included in this category. The average cost per month per remote in the sample was $14.				Backup consisting of the switch used to access backup nodes, the spare loop for modems and backup power collectively had an average cost of $10 per month per remote location.			
		Sample Cost	Your Cost			Sample Cost	Your Cost
Total yearly cost		$ 5,040		Total yearly cost		$ 3,600	
Number of months		12	____	Number of months		12	____
Cost per month	$	420	____	Cost per month	$	300	____
Number of remotes		30	____	Number of remotes		30	____
Per month per remote	$	14	____	Per month per remote	$	10	____
						Sample Cost	Your Cost
				Total Equipment Costs		$24	

FIGURE 10–16 Network cost comparison.

The second factor concerns income taxes. Income taxes lengthen the payback period. The bottom half of Figure 10–18 shows two different formulas used for calculating payback period, before taxes and after taxes. Notice that the project has a 12.96-year payback period after taxes and a 7.0-year payback before taxes.

Network Cost Analyzer This is more a worksheet than a methodology. It is an excellent way to document specific network costs after a *detailed* cost analysis. These are the costs that are considered hard costs to the network such as software/hardware/circuit costs.

Figure 10–19 shows that if each individual cost category is recorded horizontally by network link, a total link cost can be obtained for each network link. Some costs, like front ends or software, may have to be allocated among all of the

SUMMARY
To compare your total network cost with our survey average, just fill in the blanks below. Use the figures you've already put in the boxes. The costs associated with implementing and maintaining a private network are being compounded by line tariff increases, staffing shortages, and high start-up costs due to inflation.

	Sample Cost	Your Cost
Network staffing	$ 71	
Corporate allocations	$ 53	
Nodes	$ 39	
Equipment	$ 24	
Leased lines and modems	$380	
Total per month per remote	$567	

TOTAL YEARLY NETWORK COSTS

Cost Per Month Per Remote Location	×	Number Of Months	×	Number Of Remotes	=	Total Yearly Network Cost
_____	×	_____	×	_____	=	

FIGURE 10–17 Network cost comparison.

Project	Annual Return First Year	Annual Return Second Year	Annual Return Third Year	Annual Return Fourth Year	Total Return
A	$100,000	200,000	200.000	200,000	700,000
B	$400,000	100,000	100,000	100,000	700,000

Before taxes

$$P = \frac{I}{R}$$

After taxes

$$P = \frac{I}{(1-T)R}$$

where P = payback period, I = investment, R = average annual return on investment, and T = corporate tax rate in percent. For example, if the investment required for a network equals $700,000 ($I = 700,000$) and the average annual return is $100,000 ($R = 100,000$) and the corporate tax rate is 46% ($T = 0.46$), then the after taxes payback is:

$$P = \frac{700,000}{(1-0.46)\,100,000} = 12.96 \text{ years}$$

FIGURE 10–18 Payback period.

links on a fair basis. For example, a $300,000 front end serving 20 network links would be allocated at $15,000 per link.

Also, by totaling down a column, the total cost of a specific piece of hardware or software can be obtained. Finally, the grand total in the lower right corner shows the total cost of the entire network software, hardware, and circuits.

Cost/Benefit Categories Figure 10–20 shows various cost categories associated with data communication networks, as well as the various benefit categories. The most helpful items in this figure are the direct costs, the indirect costs, and the intangible benefits. Intangibles are sometimes very difficult to identify. The other benefits, such as direct and indirect cost reductions and revenue increases, must be identified in a manner that is unique to the organization for which the network is being designed. This figure can be used to ensure that no critical cost or important benefit has been overlooked. Figures 10–19 and 10–20 would be used together.

Cost Chargeback Method Some organizations may want to develop a charge-back methodology in order to charge users for their use of the data communication network.

Figure 10–21 depicts a user chargeback method for interactive systems that is

Network Link	Circuit Cost	Intelligent Controllers	Modems	Multi-plexer/ Concen-trators	Terminals	Front Ends	Software	Personnel and Facilities	Link Total Cost
Down Totals									Grand Total

FIGURE 10–19 Network cost analyzer.

Costs	Benefits
Direct costs	**Direct and indirect cost reductions**
• Computer equipment	• Elimination of clerical personnel and/or manual operations
• Communications equipment	• Reduction of inventories, manufacturing, sales, operations, and management costs
• Common carrier line charges	
• Software	
• Operations personnel costs	• Effective cost reduction, for example, less spoilage or waste, elimination of obsolete materials, and less pilferage
• File conversion costs	
• Facilities costs (space, power, air-conditioning, storage space, offices, etc.)	
	• Distribution of resources across demand for service
• Spare parts costs	
• Hardware maintenance costs	**Revenue increases**
• Software maintenance costs	• Increased sales due to better responsiveness
• Interaction with vendor and/or development group	• Improved services
• Development and performance of acceptance test procedures and parallel operation	• Faster processing of operations
	Intangible benefits
• Development of documentation	• Smoothing of operational flows
• Costs for backup of system in case of failure	• Reduced volume of paper produced and handled
• Costs of manually performing tests during a system outage	• Rise in level of service quality and performance
• Security and control	• Expansion capability
	• Improved decision process by provision of faster access to information
Indirect costs	
• Personnel training	• Ability to meet the competition
• Transformation of operational procedures	• Future cost avoidance
• Development of support software	• Positive effect on other classes of investments or resources such as better utilization of money, more efficient use of floor space or personnel, and so forth
• Disruption of normal activities	
• Increased system outage rate during initial operation period	
• Increase in the number of vendors (impacts fault detection and correction due to "finger pointing")	• Improved employee morale

FIGURE 10–20 Cost/benefit categories.

Terminals in California connect with headquarters computer in New York City. During peak day hours, 9 A.M. to 5 P.M., 100 input transactions of 75 characters each and 110 output transactions of 350 characters each are processed.
   Other elements of the algorithm include:

- Terminal charge of $243.00
- Modem charge of $15.00
- Administrative overhead charge of 5% on owned equipment
- Access charge of $400.00
- Character overhead (message address, special instructions, etc.) of 100 characters per transaction
- Daytime rate of $0.11/1,000 characters
- Month equals 22 days

Monthly communications charge would be:

1. Equipment		
Terminal	$243.00	
Modem	15.00	
Equipment costs	258.00	
Adminstrative overhead (5%)	12.90	
Equipment total		$270.90
2. Access charge (based on distance)		$400.00
3. Daytime character rate		
Input Transactions (100 × 75 char.)	7,500	
Overhead (100 × 100 char.)	10,000	
Total input characters	17,500	
Output Transactions (110 × 350 char.)	38,500	
Overhead (110 × 100 char.)	11,000	
Total output characters	49,500	
Total characters per day	67,000	
Daily charge ($0.11/1,000 char.)	$7.37	
Monthly char. charge (22 days)		$162.14
Total monthly charge		$833.04

FIGURE 10–21 Chargeback for interactive systems.

transaction, character, and distance sensitive. Monthly charges consist of three main elements: equipment, access charge, and number of characters transmitted. Character charges may be assessed at daytime, off-hours, and nighttime rates, thus providing an incentive for users to use the network during nonprime time hours.

An RJE user utilized 53 hours, or 3,180 minutes, on the computer for the month. Other elements of the algorithm include:

- Terminal charge of $640.00
- Administrative overhead charge of 5% on owned equipment
- Modems (two units) at a monthly lease cost of $135.00 each
- 44 long distance, 16 minute calls for a total of $460.00
- Connect time (session charge) to the host computer of $0.34 per minute

Monthly communications charge would be:

1. Equipment
Terminal	$640.00	
Administrative overhead (5%)	32.00	
Modems	270.00	
Equipment total		$942.00
2. Long-distance calls   460.00
3. Session costs (3,180 min.)
   Session charge ($0.34/min.)   1,081.20
4. Total monthly charge $2,483.20

FIGURE 10–22 Chargeback for remote job entry systems.

Figure 10–22 shows a remote job entry (RJE) dedicated chargeback scheme. This method is time sensitive in that users are charged for how long they are connected to the host computer and long-distance calls rather than for the number of characters transmitted.

Circuit Mileage Costs In this method the dollar cost per month is developed for each leased circuit. The major factors that have to be taken into account are the total mileage of each link, the cost per mile for the circuit type chosen, and line termination charges at each end of the circuit link.

For design purposes, you can determine the total air mileage by referring to a map or atlas (see Figure 10–23). In actual practice the common carriers use a system of vertical and horizontal coordinates to determine the air mileage between central offices. Figure 10–24 shows the method of using the vertical and horizontal coordinates to determine the air mileage. Federal Communications Commission Tariff Number 264 gives all of the vertical and horizontal coordinates for the cities in the United States. Typical vertical and horizontal coordinates are shown below:

Cities	Vertical	Horizontal
Phoenix	9135	6748
Los Angeles	9213	7878
San Diego	9468	7629
Oakland	8486	8695
San Francisco	8492	8719
Denver	7501	5899
Washington, D.C.	5622	1583
Kansas City, Kansas	7028	4212
Kansas City, Mo.	7027	4203
Dallas	8436	4034
Little Rock	7721	3451
Tulsa	7707	4173
Oklahoma City	7947	4373
Amarillo	8266	5076
Omaha	6687	4595
El Paso	9231	5665
Salt Lake City	7576	7065
Portland	6799	8914
Spokane	6247	8180
Seattle	6336	8896

The next step is to calculate the cost per mile for the type of circuit chosen. To carry out this calculation, identify the tariff that lists the proper cost for the circuit type and the methodology for calculating this cost. In Appendix 5 we have listed both the tariff and the methodology for calculating the cost of a voice grade (series 2000/3000) circuit, a digital circuit, a satellite circuit, a dial-up circuit, a wideband circuit (series 8000), and a packet switching network (public).

To calculate the circuit cost, refer to the final configurations (maps) and calculate the cost of each circuit link. The calculation must include various hardware costs such as modems, terminals, and the like. Software costs also should be added in at this point.

The Network Cost Analyzer (Figure 10–19) is invaluable during this step because all of the costs for each link can be accumulated onto a single document. Also, it is a good idea at this point to review the Cost/Benefit Categories (Figure 10–20) to ensure that nothing was overlooked.

Hardware Costs   Again, using the Network Cost Analyzer, identify and record the specific costs for each piece of equipment. Some equipment may be purchased and therefore is a one-time first-year only, cost. On the other hand, leased

	Albuquerque, N. Mex	Amarillo, Tex.	Atlanta, Ga.	Billings, Mont.	Birmingham, Ala.	Boston, Mass.	Buffalo, N.Y.	Burlington, Vt.	Charleston, S.C.	Charlotte, N.C.	Cheyenne, Wyo.	Chicago, Ill.	Cincinnati, Ohio	Cleveland, Ohio	Dallas, Tex.	Denver, Colo.	Des Moines, Iowa	Detroit, Mich.	El Paso, Tex.	Fargo, N. Dak.	Houston, Tex.	Indianapolis, Ind.	Jacksonville, Fla.
Albuquerque, N. Mex		273	1272	744	1138	1972	1580	1878	1539	1457	429	1129	1251	1421	588	334	837	1364	229	961	754	1169	1488
Amarillo, Tex.	273		999	809	866	1722	1338	1640	1266	1185	440	894	992	1173	334	358	626	1124	358	847	533	915	1219
Atlanta, Ga.	1272	999		1519	140	937	697	951	267	227	1229	587	369	554	721	1212	739	597	1291	1114	701	426	285
Billings, Mont.	744	809	1519		1425	1861	1473	1713	1761	1617	370	1073	1304	1369	1092	453	798	1283	973	565	1315	1204	1796
Birmingham, Ala.	1138	866	140	1425		1052	776	1049	402	361	1119	578	406	618	581	1095	670	641	1152	1060	567	433	374
Boston, Mass.	1972	1722	937	1861	1052		400	182	820	721	1735	851	740	551	1551	1769	1159	613	2072	1300	1605	807	1017
Buffalo, N.Y.	1580	1338	697	1473	776	400		304	699	538	1335	454	393	173	1198	1370	760	216	1692	919	1286	435	879
Burlington, Vt.	1878	1640	951	1713	1049	182	304		884	755	1612	749	690	476	1501	1654	1049	516	1995	1149	1580	739	1079
Charleston, S.C.	1539	1266	267	1761	402	820	699	884		177	1486	757	506	609	981	1474	967	681	1552	1317	936	594	177
Charlotte, N.C.	1457	1185	227	1617	361	721	538	755	177		1362	587	335	435	930	1358	819	504	1496	1153	927	428	341
Cheyenne, Wyo.	429	440	1229	370	1119	1735	1335	1612	1486	1362		891	1082	1199	726	96	583	1125	653	563	947	986	1493
Chicago, Ill.	1129	894	587	1073	578	851	454	749	757	587	891		252	308	803	920	309	238	1252	569	940	165	863
Cincinnati, Ohio	1251	992	369	1304	406	740	393	690	506	335	1082	252		222	814	1094	510	235	1335	820	892	100	626
Cleveland, Ohio	1421	1173	554	1369	618	551	173	476	609	435	1199	308	222		1025	1227	617	90	1521	835	1114	263	770
Dallas, Tex.	588	334	721	1092	581	1551	1198	1501	981	930	726	803	814	1025		663	632	999	572	972	225	763	908
Denver, Colo.	334	358	1212	453	1095	1769	1370	1654	1474	1358	96	920	1094	1227	663		610	1156	557	642	879	1000	1467
Des Moines, Iowa	837	626	739	798	670	1159	760	1049	967	819	583	309	510	617	632	610		546	983	397	821	411	1023
Detroit, Mich.	1364	1124	596	1283	641	613	216	516	681	504	1125	238	235	90	999	1156	546		1479	745	1105	240	831
El Paso, Tex.	229	358	1291	973	1152	2072	1692	1995	1552	1496	653	1252	1335	1525	572	557	983	1479		1163	676	1264	1473
Fargo, N. Dak.	961	847	1114	565	1060	1300	919	1149	1317	1153	563	569	820	835	972	642	397	745	1163		1183	725	1399
Houston, Tex.	754	533	701	1315	567	1605	1286	1580	936	927	947	940	892	1114	225	879	821	1105	676	1183		865	821
Indianapolis, Ind.	1169	915	426	1204	433	807	435	739	594	428	986	165	100	263	763	1000	411	240	1264	725	865		699
Jacksonville, Fla.	1488	1219	285	1796	374	1017	879	1079	177	341	1493	863	626	770	908	1467	1023	831	1473	1399	821	699	
Kansas City, Mo.	720	481	676	846	579	1251	861	1161	928	803	560	414	541	700	451	558	180	645	839	549	644	453	950
Knoxville, Tenn.	1280	1009	155	1447	235	818	548	815	316	180	1183	454	219	400	767	1178	651	442	1326	1004	790	290	410
Little Rock, Ark.	816	543	456	1143	325	1259	913	1214	723	649	813	552	524	740	293	780	478	723	847	869	388	483	690
Los Angeles, Calif.	664	937	1936	959	1802	2596	2198	2485	2203	2119	882	1745	1897	2049	1240	831	1438	1983	701	1427	1374	1809	2147
Louisville, Ky.	1178	915	319	1275	331	826	483	780	500	343	1033	269	90	311	726	1038	476	316	1254	818	803	107	594
Memphis, Tenn.	939	867	337	1213	217	1137	803	1100	604	521	902	482	410	630	420	879	485	623	976	882	484	384	590
Miami, Fla.	1598	1441	604	2085	665	1255	1181	1347	482	652	1763	1188	952	1087	1111	1726	1333	1152	1643	1716	968	1024	326
Minneapolis, Minn.	983	812	907	742	862	1123	731	985	1104	939	642	355	605	630	862	700	235	543	1151	214	1056	511	1191
Nashville, Tenn.	1119	848	214	1309	182	943	627	916	455	340	1032	397	238	459	617	1023	525	470	1169	902	665	251	499
New Orleans, La.	1029	776	424	1479	312	1359	1086	1361	630	649	1131	833	706	924	443	1082	827	939	983	1222	318	712	504
New York, N.Y.	1815	1560	748	1760	864	188	292	260	641	533	1604	713	570	405	1374	1631	1022	482	1905	1210	1420	646	838
Omaha, Nebr.	721	526	817	703	732	1282	883	1171	1058	918	463	432	622	739	586	488	123	669	878	390	794	525	1098
Philadelphia, Pa.	1753	1494	666	1727	783	271	279	328	562	451	1556	666	503	360	1299	1579	973	443	1836	1184	1341	585	758
Phoenix, Ariz.	330	598	1592	872	1456	2300	1906	2202	1857	1783	663	1453	1581	1749	887	586	1155	1690	346	1225	1017	1499	1794
Pittsburgh, Pa.	1499	1244	521	1479	608	483	178	445	528	362	1298	410	257	115	1070	1320	715	205	1590	949	1137	330	703
Portland, Oreg.	1107	1304	2172	686	2066	2540	2156	2385	2290	2290	947	1758	1985	2055	1633	982	1475	1969	1286	1239	1836	1885	2439
Raleigh, N.C.	1576	1306	356	1698	491	609	490	665	220	130	1461	642	396	428	1057	1463	902	510	1621	1210	1056	495	414
St. Louis, Mo.	942	685	467	1057	400	1038	662	966	704	568	795	262	309	492	547	796	273	455	1034	660	679	231	751
Salt Lake City, Utah	484	668	1583	387	1466	2099	1699	1969	1845	1727	371	1260	1453	1568	999	371	953	1492	689	863	1200	1356	1837
San Antonio, Tex.	617	444	882	1252	744	1766	1430	1729	1122	1105	882	1051	1039	1256	252	802	882	1238	503	1207	189	999	1011
San Francisco, Calif.	896	1157	2139	904	2013	2699	2300	2568	2405	2301	967	1858	2043	2166	1483	949	1550	2091	995	1446	1645	1949	2374
Seattle, Wash.	1184	1359	2182	668	2082	2493	2117	2333	2428	2285	973	1737	1972	2026	1681	1021	1467	1938	1376	1197	1891	1872	2455
Spokane, Wash.	1030	1176	1961	443	1865	2266	1888	2108	2204	2059	768	1508	1744	1796	1489	826	1240	1709	1239	969	1704	1644	2237
Syracuse, N.Y.	1718	1475	781	1600	875	264	138	177	738	595	1472	592	514	303	1326	1508	898	354	1828	1042	1403	567	928
Tulsa, Okla.	604	335	678	930	552	1398	1023	1327	945	853	588	598	661	853	236	550	396	813	674	741	432	591	921
Washington, D.C.	1653	1391	543	1669	661	393	292	432	453	330	1477	597	404	306	1185	1494	896	396	1728	1140	1220	494	647
Wichita, Kans.	549	304	776	801	658	1424	1036	1337	1039	933	465	591	702	873	340	437	334	821	661	634	559	620	1031

Cities	Vertical	Horizontal
Phoenix	9135	6748
Los Angeles	9213	7878
San Diego	9468	7629
Oakland	8486	8695
San Francisco	8492	8719
Denver	7501	5899
Washington, D.C.	5622	1583
Kansas City, Kansas	7028	4212
Kansas City, Mo.	7027	4203

FIGURE 10–23 U.S. air distance table.

Cities of the United States in Statute Miles

Kansas City, Mo.	Knoxville, Tenn.	Little Rock, Ark.	Los Angeles, Calif.	Louisville, Ky.	Memphis, Tenn.	Miami, Fla.	Minneapolis, Minn.	Nashville, Tenn.	New Orleans, La.	New York, N.Y.	Omaha, Nebr.	Philadelphia, Pa.	Phoenix, Ariz.	Pittsburgh, Pa.	Portland, Oreg.	Raleigh, N.C.	St. Louis, Mo.	Salt Lake City, Utah	San Antonio, Tex.	San Francisco, Calif.	Seattle, Wash.	Spokane, Wash.	Syracuse, N.Y.	Tulsa, Okla.	Washington, D.C.	Wichita, Kans.
720	1280	816	664	1178	939	1698	983	1119	1029	1815	721	1753	330	1499	1107	1576	942	484	617	896	1184	1030	1718	604	1653	549
481	1009	543	937	915	667	1441	812	848	776	1560	526	1494	598	1244	1304	1306	685	668	444	1157	1359	1176	1475	335	1391	304
676	155	456	1936	319	337	604	907	214	424	748	817	666	1592	521	2172	356	467	1583	882	2139	2182	1961	781	678	543	776
846	1447	1143	959	1275	1213	2085	742	1309	1479	1760	703	1727	872	1479	686	1698	1057	387	1252	904	668	443	1600	930	1669	801
579	235	325	1802	331	217	665	862	182	312	864	732	783	1456	608	2066	491	400	1466	744	2013	2082	1865	875	552	661	658
1251	818	1259	2596	826	1137	1255	1123	943	1359	188	1282	271	2300	483	2540	609	1038	2099	1766	2699	2493	2266	264	1398	393	1424
881	548	913	2198	483	803	1181	731	627	1086	292	883	279	1906	178	2156	490	662	1699	1430	2300	2117	1888	138	1023	292	1036
1161	815	1214	2485	780	1100	1347	985	916	1361	260	1171	328	2202	445	2385	665	966	1969	1729	2568	2333	2108	177	1327	432	1337
928	316	723	2203	500	604	482	1104	455	630	641	1058	562	1857	528	2425	220	704	1845	1122	2405	2428	2204	738	945	453	1039
803	180	649	2119	343	521	652	939	340	649	533	918	451	1783	362	2290	130	568	1727	1105	2301	2285	2059	595	853	330	933
560	1183	813	882	1033	902	1763	642	1032	1131	1604	463	1556	663	1298	947	1461	795	371	882	967	973	768	1472	588	1477	465
414	454	552	1745	269	482	1188	355	397	833	713	432	666	1453	410	1758	642	262	1216	1051	1858	1737	1508	592	598	597	591
541	219	524	1897	90	410	952	605	238	706	570	622	503	1581	257	1985	396	309	1453	1039	2043	1972	1744	514	661	404	702
700	400	740	2049	311	630	1087	630	459	924	405	739	360	1749	115	2055	428	492	1568	1256	2166	2026	1796	303	853	306	873
451	767	293	1240	726	420	1111	862	617	443	1374	586	1299	887	1070	1633	1057	547	999	252	1483	1681	1489	1326	236	1185	340
558	1178	780	831	1038	879	1726	700	1023	1082	1631	488	1579	586	1320	982	1463	796	371	802	949	1021	826	1508	550	1494	437
180	651	478	1438	476	485	1333	235	525	827	1022	123	973	1155	715	1475	902	273	953	882	1550	1467	1240	898	396	896	334
645	442	723	1983	316	623	1152	543	470	939	482	669	443	1690	205	1969	510	455	1492	1238	2091	1938	1709	354	813	396	821
839	1323	847	701	1254	976	1643	1157	1169	983	1905	878	1836	346	1500	1286	1621	1034	689	503	995	1376	1239	1828	674	1728	661
549	1004	869	1427	818	882	1716	214	902	1222	1210	390	1184	1225	949	1239	1210	660	863	1207	1446	1197	969	1042	741	1140	634
644	790	388	1374	803	484	968	1056	665	318	1420	794	1341	1017	1137	1836	1056	679	1200	189	1645	1891	1704	1403	442	1220	559
453	290	483	1809	107	384	1024	511	251	712	646	525	585	1499	330	1885	495	231	1356	999	1949	1872	1644	567	591	494	620
950	410	690	2147	594	590	326	1191	499	504	838	1098	758	1794	703	2439	414	751	1837	1011	2374	2455	2237	928	921	647	1031
	624	325	1356	480	369	1241	413	473	680	1097	166	1038	1049	781	1497	905	238	925	702	1506	1506	1287	998	216	945	177
624		479	1941	188	350	736	792	161	547	632	745	552	1607	375	2115	296	392	1547	959	2121	2114	1890	641	676	430	753
325	479		1480	435	129	949	708	325	355	1081	492	1007	1137	779	1759	774	291	1148	516	1688	1785	1573	1038	231	892	348
1356	1941	1480		1829	1603	2339	1524	1780	1673	2451	1315	2394	357	2136	825	2237	1589	579	1204	347	959	940	2336	1266	2300	1197
480	188	435	1829		320	919	605	154	623	652	580	582	1508	344	1950	429	242	1402	949	1986	1943	1717	603	582	476	633
369	350	129	1603	320		872	699	197	358	957	529	881	1263	660	1849	645	240	1250	631	1802	1867	1650	923	341	765	442
1241	736	949	2339	919	872		1511	815	669	1092	1397	1019	1982	1010	2708	695	1061	2089	1148	2594	2734	2520	1212	1176	923	1297
413	792	708	1524	605	699	1511		697	1051	1018	290	985	1280	743	1427	996	466	987	1110	1584	1395	1166	861	626	934	546
473	161	325	1780	154	197	815	697		469	761	607	685	1446	472	1969	457	254	1393	823	1963	1975	1752	739	515	569	594
680	547	355	1673	623	358	669	1051	469		1171	842	1089	1316	919	2063	776	598	1434	507	1926	2101	1898	1187	548	966	677
1097	632	1081	2451	652	957	1092	1018	761	1171		1144	83	2145	317	2445	426	875	1972	1584	2571	2408	2179	194	1231	205	1266
166	745	492	1315	580	529	1397	290	607	847	1144		1094	1036	836	1371	1008	354	833	828	1429	1369	1146	1021	352	1014	257
1038	552	1007	2394	582	881	1019	985	685	1089	83	1094		2083	259	2412	345	811	1925	1507	2523	2380	2151	220	1163	123	1204
1049	1607	1137	357	1508	1263	1982	1280	1446	1316	2145	1036	2083		1828	1005	1903	1272	504	849	653	1114	1019	2044	932	1983	879
781	375	779	2136	344	660	1010	743	472	919	317	836	259	1828		2165	330	559	1668	1291	2264	2138	1908	268	917	192	950
1497	2115	1759	825	1950	1849	2708	1427	1969	2063	2445	1371	2412	1005	2165		2377	1723	636	1720	534	145	290	2281	1531	2354	1411
905	296	774	2237	429	645	695	996	457	776	426	1008	345	1903	330	2377		667	1829	1235	2410	2367	2139	519	972	233	1044
238	392	291	1589	242	240	1061	466	254	598	875	354	811	1272	559	1720	667		1100	702	1744	1724	1600	796	261	712	394
925	1547	1148	579	1402	1250	2089	987	1393	1434	1972	833	1925	504	1668	636	1829	1162		1087	600	701	550	1835	917	1848	808
702	959	516	1204	949	631	1138	1110	823	507	1584	828	1507	829	1291	1720	1235	792	1087		1490	1787	1614	1553	486	1388	573
1506	2121	1688	347	1986	1802	2594	1584	1963	1926	2571	1429	2523	653	2264	534	2410	1744	600	1490		678	727	2435	1461	2442	1369
1506	2114	1785	959	1943	1867	2734	1395	1975	2101	2408	1369	2380	1114	2138	145	2367	1724	701	1787	070		229	2238	1560	2320	1437
1287	1890	1573	940	1717	1650	2520	1166	1752	1898	2179	1146	2151	1019	1908	290	2139	1500	550	1614	727	229		2010	1353	2100	1227
989	641	1038	2336	603	923	1212	861	739	1187	194	1021	220	2044	268	2281	519	796	1835	1553	2435	2238	2010		1157	290	1173
216	676	231	1266	582	341	1176	626	515	548	1231	352	1163	932	917	1531	972	361	917	486	1461	1560	1353	1157		1058	130
946	430	892	2300	476	765	923	934	569	966	205	1014	123	1983	192	2354	233	712	1848	1388	2442	2359	2100	290	1058		1006
177	753	348	1197	633	442	1297	546	594	677	1266	257	1204	879	950	1411	1044	394	808	573	1369	1437	1227	1173	130	1106	

Dallas	8436	4034
Little Rock	7721	3451
Tulsa	7707	4173
Oklahoma City	7947	4373
Amarillo	8266	5076
Omaha	6687	4595
El Paso	9231	5665
Salt Lake City	7576	7065
Portland	6799	8914
Spokane	6247	8180
Seattle	6336	8896

FIGURE 10–23 *(Continued)*

Location	Vertical	Horizontal
Tulsa	7707	4173
San Francisco	8492	8719

$$\text{Distance} = \sqrt{\frac{(V_1 - V_2)^2 + (H_1 - H_2)^2}{10}}$$

$$D - \sqrt{\frac{(7707 - 8492)^2 + (4173 - 8719)^2}{10}}$$

$$D = 1459 \text{ miles}$$

FIGURE 10–24 Mileage calculation.

equipment is an ongoing operating cost that will be faced each year. Appendix 5 lists some typical data communication equipment costs that may be used as a guideline or for working problems. Also, be sure to account for software costs.

The final step is to sell the organization's management and users on the new network and then to implement it.

SELL AND IMPLEMENT THE NETWORK

At this point there are two more subtasks:

- Sell the system both to management and to the users who will have to work with it. This is a verbal presentation.
- Implement the system. This probably is the most difficult task of all because the various pieces of hardware, protocol/software programs, network management/test facilities, and communication circuits must be assembled into a working network.

When presenting the system to gain management/user acceptance, the designer should be prepared for objections to the proposed system. Basic objections usually follow these lines:

- The cost is too high, or it appears too low for what the system is supposed to be able to do.
- The performance is not good enough, or it is more than is required at this time.
- The new network does not meet the goals, objectives, and policies of the organization/departments that will be using it.

- The response or processing time is either too slow or too fast with respect to other operations within the organization.
- The system is not flexible enough. If changes are made in other areas, the network may collapse and the investment will be wasted.
- The quality, capacity, efficiency, accuracy, or reliability of the new network does not meet management's criteria,
- Certain management personnel may dislike or distrust the network design team's motives, personalities, or presentation methods.
- Review the list of evaluation criteria that was prepared earlier. Be ready for questions on any that were not met.

The implementation process begins after management has accepted the new system. Implementation consists of the installation of the new system and the removal of the old system. It involves hardware, protocols/software, communication circuits, a network management/test facility, people, written procedures that specify how each task in the network is performed, training, and complete documentation of the working system.

The steps involved in implementing a new data communication network can be very complex and demanding. To enable implementation to proceed as smoothly as possible, a detailed implementation plan should be developed. The plan should specify who will do what and when they will do it. For this to be done properly, use should be made of Gantt charts, flowcharts, or the Program Evaluation Review Technique (PERT).[2]

The design/implementation team must take into account the earliest lead times that are required to order hardware, software, and circuits. In many cases, these items cannot be delivered immediately. Also, some lead time is needed for testing the protocols and software in order to ensure that they operate in conjunction with the hardware and circuits. Both hardware and circuits may have to be implemented in various parts of the city, a state, throughout the country, or even internationally. For this reason, it is imperative that a decision is made as to how the new system will be implemented. There are four basic approaches that can be used:

- All at once. All nodes and the host computer are started up at the same time (a one-for-one changeover).
- Chronologically, and in sequence, through the system. Start with the first application system, implementing those portions of the network that must be implemented with it, then move on to the second application system.

[2]Jerry FitzGerald, Ardra F. FitzGerald, and Warren D. Stallings, *Fundamentals of Systems Analysis* (New York: 1981), John Wiley & Sons. See Chapter 13 on charting for a complete description of these planning charts.

- In predetermined phases. Similar areas within the system are started up at one time, and other areas are started up later.
- Pilot operation. Set up a pilot or test facility (this later becomes one of the working nodes) to ensure that the operation is as expected before an all-at-once or chronological cutover is made.

Once the hardware is in place, the circuits have been installed, and the protocol/software is operating, training of the users can begin, although whenever possible it should be started earlier. It is advantageous to obtain test terminals so that terminal operators can use their particular application system in a training/test mode months before they do so in real life. Precise written procedures are required on how the terminal operators are to operate the system for data input and manipulation. Written descriptions on how to retrieve and interpret the information/data output should be provided to management.

The training should include individual operator training, extensive written training manuals, and a methodology for continual updating these manuals. At this point, the use of Computer Assisted Instruction (CAI) should be considered. With CAI all of the training techniques and procedures are stored in the computer system; there are no written manuals. Instead, terminal operators use their terminals for training as well as for standard business operations.

The network management/test center is a vital link in the network. This group must be in operation *before* the system is cut over to an operational status because reliability (uptime) is the single most important criterion for user acceptance.

Finally, after the system is operational, conduct follow-ups for the first six months or so to ensure that all parts of the new system actually are operating and that minor activities or operations have not been overlooked.

After the system is considered fully operational, a reevaluation should be performed. Reevaluation may come 6 or 12 months later. It is a critical review of operator/user complaints, management complaints, efficiency reports, network management trouble reports, an evaluation of statistics gathered on items such as errors during transmission and characters transmitted per link, and a review of peak load factors. Of course, it also should include a complete review of the original evaluation criteria so that you may determine the success of the design, development, and implementation of the new data communication network.

In summary, 13 steps have been completed for the design of a new data communication system. Although some steps may have been omitted because a current system was being enhanced, an orderly plan was followed. As the project is closed, pull all of the documentation together and set it up in a binder that contains 13 separate sections, one for each step that was carried out.

QUESTIONS

1. Describe the 13 steps that a designer performs when designing a new data communication network.

2. Identify two or three critical points that should appear in a feasibility study final report.

3. What are the three levels of goals that you should try to achieve when designing a network?

4. Identify and define five or six key evaluation criteria.

5. There are four levels of mapping. Can you identify and describe them?

6. When analyzing messages, which of the following peak volumes would be the most important, the character per message peak or the messages per day peak?

7. Why is minimum circuit mileage between various terminal locations important?

BIBLIOGRAPHY

SELECTED BOOKS

Aries, S. J. *Dictionary of Telecommunications.* Woburn, Mass.: Butterworth Publishers, Inc., 1981, 336 pp.

Arredondo, Larry A. *Telecommunications Management for Business and Government,* 2nd Ed. New York: The Telecom Library, Inc., 1981, 280 pp.

Becker, Hal B. *Functional Analysis of Information Networks: A Structured Approach to the Data Communications Environment.* New York: John Wiley & Sons, 1981 reprint of 1973 edition.

Black, Uyless D. *Data Communications, Networks and Distributed Systems.* Reston, Va.: Reston Publishing Company, Inc., 1983.

Bodson, Dennis, ed. *Fiber Optics and Lightwave Communications Vocabulary.* New York: McGraw-Hill Book Company, 1981, 149 pp.

Brock, Gerald W. *The Telecommunications Industry: The Dynamics of Market Structure.* Cambridge, Mass.: Harvard University Press, Harvard Economic Studies No. 151, 1981, 384 pp.

Brown, Martin B. *Compendium and Communication and Broadcast Satellites: Nineteen Fifty Eight to Nineteen Eighty One.* New York: John Wiley & Sons, 1981, 400 pp.

Chorafas, Dimitris. *Data Communication for Distributed Information Systems.* New York: McGraw-Hill Book Company, 1980, 235 pp.

Chu, Wesley W., ed. *Advances in Computer Communications and Networking.* Dedham, Mass.: Artech House, Inc., 1979, 653 pp.

Conard, James W. *Standards and Protocols for Communications Networks.* Madison, N.J.: Carnegie Press, Inc., 1982, 396 pp.

Data Communications: A Complete Systems Guide. Published by Datapro Research Corp., One Corporate Center, Moorestown, N.J. 08057.

Data Communications Primer. White Plains, N.Y.: IBM Corporation, Data Processing Division (IBM Manual No. GC20–1668).

Data Communications Standards Library. Westchester, Calif.: Remarkable Publications, 1981, 330 pp.

Davenport, William P. *Modern Data Communication: Concepts, Language, and Media.* New York: Hayden Book Company, 1971.

Davis, George R., ed. *The Local Network Handbook.* New York: McGraw-Hill Publications Company, 1982, 256 pp.

Deasington, R. J. *A Practical Guide to Computer Communications and Networking.* New York: John Wiley & Sons, 1982, 132 pp.

Dimension PBX and Alternatives, 3rd ed. Wellesley, Mass.: Q.E.D. Information Sciences, Inc., 1980.

Doll, Dixon R. *Data Communications: Facilities, Networks and Systems Design.* New York: John Wiley & Sons, 1978, 493 pp.

Easton, Anthony T. *The Home Satellite TV Book: How to Put the World in Your Backyard.* San Francisco: Wideview Books (The Easton Corporation), 1981.

FitzGerald, Jerry. *Designing Controls into Computerized Systems.* Redwood City, Calif.:Jerry FitzGerald and Associates, 1981, 157 pp.

———— *Internal Controls for Computerized Systems.* Redwood City, Calif.: Jerry FitzGerald and Associates, 1978, 93 pp.

FitzGerald, Jerry, Ardra F. FitzGerald, and Warren D. Stallings, Jr. *Fundamentals of Systems Analysis,* 2nd ed. New York: John Wiley & Sons, 1981, 590 pp.

Folts, Harold C., and Harry E. Karp, eds. *McGraw-Hill's Compilation of Data Communications Standards,* 2nd ed. New York: McGraw-Hill Book Company, 1982, 1923 pp.

Freeman, Roger L. *Telecommunication System Engineering: Analog and Digital Networks.* New York: John Wiley & Sons, 1980, 480 pp.

———— *Telecommunication Transmission Handbook,* 2nd ed. New York: John Wiley & Sons, 1981.

Galitz, Wilbert O. *Handbook of Screen Format Design.* Wellesley, Mass.: Q.E.D. Information Sciences, Inc., 1981.

Glass, Robert L. *Software Reliability Guidebook.* Englewood Cliffs, N.J.: Prentice-Hall, Inc., 1979.

Held, Gilbert. *Data Communications Procurement Manual.* New York: McGraw-Hill Book Company, 1979, 150pp.

Hollowell, Mary Louise, ed. *The Cable/Broadband Communications Book, Vol. 3, 1982–83; Vol. 2, 1980–81; Vol. 1, 1978–79.* White Plains, N.Y.: Knowledge Industry Publications, Inc., 1979–.

Intercity Services Handbook. Available from American Telephone and Telegraph Company, Long Lines Marketing Department, 110 Belmont Drive, Somerset, N.J. 08873. (Latest edition)

Karp, Harry R., ed. *Practical Applications of Data Communications: A User's Guide.* New York: McGraw-Hill Book Company, 1980, 424 pp.

Kuo, Franklin F., ed. *Protocols and Techniques for Data Communication Networks.* Englewood Cliffs, N.J.: Prentice-Hall, Inc., 1981, 468 pp.

Konheim, Alan G. *Cryptography: A Primer.* New York: John Wiley & Sons, 1981, 464 pp.

Lazer, Ellen A., Martin C. J. Elton, James W. Johnson, et al. *The Teleconferencing Handbook: A Guide to Cost-Effective Communication.* White Plains, N.Y.: Knowledge Industry Publications, Inc., 1983, 200 pp.

Lewin, Leonard, ed. *Telecommunications in the U.S.: Trends and Policies.* Dedham, Mass.: Artech House, Inc., 1981, 449 pp.

Martin, James T. *Computer Networks and Distributed Processing: Software, Techniques, and Architecture.* Englewood Cliffs, N.J.: Prentice-Hall, Inc., 1981, 544 pp.

_____. *Design and Strategy for Distributed Data Processing.* Englewood Cliffs, N.J.: Prentice-Hall, Inc., 1981, 672 pp.

_____. *Design of Man-Computer Dialogues.* Englewood Cliffs, N.J.: Prentice-Hall, Inc., 1973, 560 pp.

_____. *Design of Real-Time Computer Systems.* Englewood Cliffs, N.J.: Prentice-Hall, Inc., 1967, 629 pp.

_____. *Future Developments in Telecommunications,* 2nd ed. Englewood Cliffs, N.J.: Prentice-Hall, Inc., 1977, 624 pp.

_____. *Systems Analysis for Data Transmission.* Englewood Cliffs, N.J.: Prentice-Hall, Inc., 1972, 910 pp.

_____. *Telecommunications and the Computer,* 2nd ed. Englewood Cliffs, N.J: Prentice-Hall, Inc., 1976, 670 pp.

Matick, Richard E. *Transmission Lines for Digital and Communication Networks: An Introduction to Transmission Lines, High-Frequency and High-Speed Pulse Characteristics and Applications.* New York: McGraw-Hill Book Company, 1969.

McWhinney, Edward, ed. *International Law of Communications.* Dobbs Ferry, N.Y.: Oceana Publications, 1971.

Multi-Vendor DC Networks—Planning, Perspectives and Strategies. Wellesley, Mass.: Q.E.D. Information Sciences, Inc., 1982, 487 pp.

The 1982 Local Computer Network Vendor List. San Francisco, Calif.: Shotwell and Associates, 1982, 200 pp.

Nye, J. Michael. *Who, What and Where in Communications Security.* New York: The Telecom Library, Inc., 1981, 130 pp.

Phillips, Don T., and Alberto Garcia-Diaz. *Fundamentals of Network Analysis.* Englewood Cliffs, N.J.: Prentice-Hall, Inc., 1981, 496 pp.

Pless, Vera. *Introduction to the Theory of Error-Correcting Codes.* New York: John Wiley & Sons, 1981, 265 pp. (Wiley-Interscience Series in Discrete Mathematics)

Pooch, Udo W., William H. Greene, and Gary G. Moss. *Telecommunications and Networking.* Wellesley, Mass.: Q.E.D. Information Sciences, Inc., 1982, 450 pp.

Rosner, Roy. *Packet Switching: Tomorrow's Communications Today.* Belmont, Calif.: Lifetime Learning Publications, 1981, 352 pp.

Ruthberg, Zella G., ed. *Audit and Evaluation of Computer Security II: System Vulnerabil-*

ities and Controls. Proceedings of the NBS Invitational Workshop held at Miami Beach, Florida, November 28–30, 1978. Washington, D.C.: National Bureau of Standards Special Publication 500–57, April 1980, var. pag.

Self, Robert L. *Long Distance for Less: How to Choose Between Ma Bell and Those "Other" Carriers.* New York: The Telecom Library, Inc., 1982, 160 pp.

Sharma, Roshan Lal, Paulo J. T. de Sousa, and Ashok D. Inglé. *Network Systems: Modeling, Analysis and Design.* New York: Van Nostrand Reinhold Company, 1982, 321 pp.

Sherman, Kenneth. *Data Communications: A User's Guide.* Reston, Va.: Reston Publishing Company, Inc., 1980, 368 pp.

Sigel, Efrem, ed. *Videotext: The Coming Revolution in Home/Office Information Retrieval.* White Plains, N.Y.: Knowledge Industry Publications, Inc., 1980, 154 pp.

Sigel, Efrem, Mark Schubin, Paul F. Merrill, et al. *Video Discs: The Technology, the Applications, and the Future.* White Plains, N.Y.: Knowledge Industry Publications, Inc., 1980, 183 pp.

Sigel, Efrem, Peter Sommer, Jeffrey Silverstein, et al. *The Future of Videotext: Worldwide Prospects for Home/Office Electronic Information Services.* White Plains, N.Y.: Knowledge Industry Publications, Inc., 1983, 197 pp.

Sippl, Charles J. *Data Communications Dictionary.* New York: Van Nostrand Reinhold Company, 1980, 545 pp.

Skees, William D. *Computer Software for Data Communications: An Introduction for Programmers.* Belmont, Calif.: Lifetime Learning Publications, 1981, 192 pp.

Systems Network Architecture: Concepts and Products. White Plains, N.Y.: IBM Corporation, January 1981 (IBM Publication GC30–3072–0).

Systems Network Architecture: Technical Overview. White Plains, N.Y.: IBM Corporation, March 1982 (IBM Publication GC30–3073–0).

Talley, David. *Basic Carrier Telephony,* 3rd rev. ed. New York: Hayden Book Company, Inc., 1977.

———— *Basic Telephone Switching Systems,* 2nd ed. New York: Hayden Book Company, Inc., 1979.

Tanenbaum, Andrew S. *Computer Networks: Toward Distributed Processing Systems.* Englewood Cliffs, N.J.: Prentice-Hall, Inc., 1981, 517 pp.

Thomas, Rebecca and Jean Yates. *A User Guide to the UNIX System.* Berkeley, Calif.: Osborne/McGraw-Hill Publishing Company, 1982, 508 pp.

Transnational Data Regulation: The Realities. Wellesley, Mass.: Q.E.D. Information Sciences, Inc., 1979, 477 pp.

Tugal, Dogan, and Osman Tugal. *Data Transmission: Analysis, Design, and Applications.* New York: McGraw-Hill Book Company, 1982, 384 pp.

Tydeman, John, Hubert Lipinski, Richard P. Adler, Michael Nyhan, and Laurence Zwimpfer. *Teletext and Videotex in the United States: Market Potential, Technology, Public Policy Issues.* New York: McGraw-Hill Publications Company, 1982, 314 pp.

The Usage of International Data Networks in Europe. Washington, D.C.: Organization for Economic Cooperation and Development, Information Computer Communications Policy Series No. 2, 1979, 287 pp.

Whorwood, R. W. and P. K. Webb. *Transmission Planning of Telephone Networks.* New York: Institute of Electrical and Electronics Engineers, 1981, 256 pp.

SERIAL PUBLICATIONS

These publications are either devoted to data communications or have a regular monthly section on data communications.

ACM Computing Surveys, vol. 11, no. 4, December 1979. Special issue on cryptology.

Auerbach Data Communications Management. Published bimonthly by Auerbach Publishers, Inc., 6560 N. Park Drive, Pennsauken, N.J. 08109, 1975– .

Auerbach Data Communications Reports. Published monthly by Auerbach Publishers, Inc., 6560 N. Park Drive, Pennsauken, N.J. 08109, 1965– .

Bell Systems Technical Reference Catalog. Published annually by American Telephone and Telegraph Co., Murray Hill, N.J. 07974 (Pub 4000). This catalog lists the Bell System publications on data communications and voice communications.

Business Communications Review. Published monthly by BCR Enterprises, Inc., Hinsdale, Ill., 1971– .

Communications News. Published monthly by Harcourt Brace Jovanovich Publications, Inc., 124 S. First St., Geneva, Ill. 60134, 1964– .

Computer Decisions. Published monthly by Hayden Publishing Co., Inc., 50 Essex Street, Rochelle Park, N.J. 07662. 1968– .

Computer Fraud and Security Bulletin. Published monthly by Elsevier International Bulletins, Mayfield House, 256 Banbury Road, Oxford OX2 7DH England, 1978– .

Computer Networks: International Journal of Distribution Informatique. Published bimonthly by North-Holland Publishing Co., Box 211, 1100 AE Amsterdam, Netherlands, 1977– .

Computerworld. Published weekly by Computerworld, Inc., 797 Washington St., Newton, Mass. 02160, 1967– .

Data Channels. Published biweekly by Phillips Publishing, Inc., 7315 Wisconsin Ave., Suite 1200N, Bethesda, Md. 20014, 1973– .

Datacomm Advisor: IDC's Newsletter Covering Network Management—Products, Services, Applications. Published monthly by International Data Corp., 214 Third Ave., Waltham, Mass. 02254, 1977– .

Datacomm and Distributed Processing Report. Published monthly by Management Information Corp., 140 Barclay Center, Cherry Hill, N.J. 08034, 1976– .

Data/Comm Industry Report: A Monthly Digest of Significant Trends in Data Communications. Published monthly by EDP News Services, Inc., 7620 Little River Turnpike, Annadale, Va. 22003, 1978– .

Data Communications. Published monthly by McGraw-Hill Publications Co., 1221 Avenue of the Americas, New York, N.Y. 10020, 1972– .

DataComm User. Published monthly by Communications Trends, Inc., 214 Third Ave., Waltham, Mass. 02154, 1969– .

Datamation. Published monthly by Technical Publishing, 875 Third Ave., New York, N.Y. 10022, 1957– .

Datapro Reports on Data Communications. Published monthly by Datapro Research Corp., 1805 Underwood Blvd., Delran, N.J. 08075.

EDPACS: The EDP Audit, Control and Security Newsletter. Published monthly by Automation Training Center, Inc., 11250 Roger Bacon Drive, Suite 17, Reston, Va. 22090, 1973– .

Fiber/Laser News. Published biweekly by Phillips Publishing, Inc., 7315 Wisconsin Ave., Suite 1200N, Bethesda, Md. 20014, 1981– .

The Guide to Communication Services. A monthly updated guide to communication costs by the Center for Communications Management, Inc., 79 North Franklin Turnpike, Ramsey, N.J. 07446.

Handbook of Intercity Telecommunications Rates and Services. Published monthly by Economics and Technology, Inc., 101 Tremont St., Boston, Mass. 02108.

IBM Systems Journal. Published quarterly by the IBM Corp., Armonk, N.Y. 10504, 1962– .

International Data Networks News. Published monthly by Information Gatekeepers, Inc., 167 Corey Road, Brookline, Mass. 02146, 1978– .

Journal of Telecommunication Networks. Published quarterly by Computer Science Press, Inc., 11 Taft Court, Rockville, Md., 1982– .

Networks: An International Journal. Published quarterly by John Wiley & Sons, 605 Third Ave., New York, N.Y. 10158, 1970– .

PBX Systems Guide. Published quarterly by Marketing Programs and Services Group, Inc., Gaithersburg, Md. 20760, 1972– .

Satellite Communications. Published monthly by Cardiff Publishing Co., 3900 S. Wadsworth, Suite 560, Denver, Colo. 80235, 1977– .

Satellite News. Published weekly by Phillips Publishing, Inc., 7315 Wisconsin Ave., Suite 1200N, Bethesda, Md. 20014, 1978– .

Telecommunications. Published monthly by Horizon House, 610 Washington St., Dedham, Mass. 02026, 1968– .

Telephone Engineer and Management: The Telephone Industry Magazine. Published semimonthly by Harcourt Brace Jovanovich, Inc., 402 W. Liberty Drive, Wheaton, Ill. 60187, 1909– .

Telephone News. Published biweekly by Phillips Publishing, Inc., 7315 Wisconsin Ave., Suite 1200N, Bethesda, Md. 20014, 1980– .

Telephony: Journal of the Telephone Industry. Published weekly by Telephony Publishing Corp., 53 E. Jackson Blvd., Chicago, Ill. 60604, 1901– .

Trends in Communications Regulation. Published monthly by Economics and Technology, Inc., 101 Tremont St., Boston, Mass. 02108, 1975– .

Videonews. Published biweekly by Phillips Publishing, Inc., 7315 Wisconsin Ave., Suite 1200N, Bethesda, Md. 20014, 1970– .

Western Electric Engineer. Published quarterly by Western Electric Co., Inc., 222 Broadway, New York, N.Y. 10038, 1975– .

COMPUTERIZED LITERATURE RESOURCES

These bibliographic databases are available from several on-line database vendors such as DIALOG, ORBIT, and BRS. Contact your librarian to obtain further information on them.

Abstracted Business Information. This index, which is available only in computerized form, cites references to articles related to the business aspects of data communications such as the workplace environment and management in the data communications environment. References contain bibliographic information and informative abstracts. 1971– .

Books in Print. This publication lists books, symposia, and other monographic materials that are currently being sold by publishers in the United States. It is an important tool for learning about what books are available in a particular field. The paper version is available in indexes by author, title, and subject. The computerized version extends its usefulness because terms can be used for which adequate subject indexing is unavailable since the subject is either a narrow one or a new one.

COMPENDEX (COMPuterized ENgineering inDEX). As the title implies, emphasis of this index is on the engineering aspects of data communications. It is useful especially for determining how others have applied data communications in an industrial or factory-type situation. An example would be an industrial control application. Indexing includes a large number of international journals, technical symposia, reports, government documents, and other materials. References include abstracts. 1970– .

INPEC. The printed counterparts of this computerized index are *Physics Abstracts*, *Electrical and Electronics Abstracts*, and *Computer and Control Abstracts*. Because data communications rely so heavily on computers, and because computers enable the process of data communications, the *Computer and Control Abstracts* portion of this database is an excellent resource. It includes the technical aspects: applications, techniques, hardware, software, technological developments, and architectures. Recently, emphasis has been added in the areas of economics and the practical aspects of implementing such systems. References are international in scope and contain abstracts. 1969– .

PROMT and *Funk and Scott Index.* These complementary indexes are essentially the same database. *PROMT* abstracts marketing-oriented articles with a slant toward products, processes, and services for sale by business firms. Indexing is by product, country/state, and "event" (sales, new product/process, demand, profits, cost per unit, industry structure/members, legal actions, regulatory actions, etc.). Bibliographic information is brief, but abstracts are informative. The *Funk and Scott Index* is strictly an indexing tool; it indexes the same references found in the *PROMT* portion of the index, plus others that are too short for an abstract. The latter would include a one-line announcement indicating that one firm has contracted with another for so many dollars to obtain a certain product. Items that are indexed are international in scope and include journals, trade literature, government documents, and so on. Subjects may be in such areas as data communications equipment, networks, office automation, teletext, or new offerings. 1972– .

Ulrich's International Periodicals Directory. This publication is a guide to journals published in all countries. It provides the name of the journal, publisher information, frequency of publication, price, and whether it includes such items as advertisements or illustrations. Since it is arranged and indexed by subject, it is an excellent means of locating the leading journals in a field. Because it is updated frequently, it is an important tool for locating new journals that may not yet be indexed in other databases. The computer enhances retrieval since one can obtain the titles of all journals having a particular word in the title.

GLOSSARY

ACK An error code character indicating a positive acknowledgment, i.e., that a message has been received correctly.

Acoustic Coupler A type of modem that permits use of a telephone handset as a connection to the public telephone network for data transmission.

ACU See **Automatic Calling Unit**.

Address
1. A coded representation of the destination of data, or of their originating terminal. Multiple terminals on one communication line, for example, must each have a unique address.
2. (Sometimes referred to as *called number*). That group of digits which makes up a telephone number. For example, an address may consist of area code, central office, and line number.

ADU See **Automatic Dialing Unit**.

American Standard Code for Information Interchange See **ASCII**.

Amplifier A device used to boost the strength of a signal. Amplifiers are spaced at intervals throughout the length of a communication circuit. Also called *repeater/amplifier*.

Amplitude Modulation See **Modulation, Amplitude**.

Analog Pertaining to representation by means of continuously variable physical quantities, such as varying voltages or varying frequencies.

Analog Signal A signal in the form of a continuously varying physical quantity such as voltage, which reflects variations in the loudness of the human voice.

Analog Transmission Transmission of a continuous variable signal as opposed to a discretely variable signal. Physical quantities such as temperature are continuously variable and so are described as "analog." The normal way of transmitting a telephone, or voice, signal has been analog, but now digital encoding (using PCM) is coming into use over trunks.

ARQ Automatic repeat request. A system employing an error-detecting code so conceived that any false signal initiates a repetition of the transmission of the message incorrectly received.

ASCII American Standard Code for Information Interchange. Usually pronounced "ask'-ee." An eight-level code for data transfer adopted by the American Standards Association to achieve compatibility among data devices.

Asynchronous Transmission Transmission in which each information character, or sometimes each word or small block, is individually synchronized, usually by the use of start and stop elements. The gap between each character or word is not a fixed length. (Compare with **Synchronous transmission**.)

Attenuation The difference between the transmitted and received power due to loss through the equipment, communication circuits, or other devices. May be expressed in decibels.

Automatic Calling Unit (ACU) A device which permits a business machine to dial calls automatically.

Automatic Dialing Unit (ADU) A device capable of automatically dialing digits.

Automatic Equalization Equalization of a transmission channel which is adjusted while sending data signals.

Automatic Repeat Request See **ARQ**.

Bandwidth The difference between the highest and the lowest frequencies in a band, such as 3000 hertz bandwidth in a voice grade circuit.

Baseband Signaling Transmission of a signal at its original frequencies, i.e., a signal in its original form, not changed by modulation. It can be an analog or digital signal and is usually direct electrical voltages.

Baud Unit of signaling speed. The speed in bauds is the number of discrete conditions or signal elements per second. (This is applied only to the actual signals on a communication line.) If each signal event represents only one bit condition, *baud* is the same as *bits per second*. When each signal event represents other than one bit, *baud* does not equal *bits per second*.

Baudot Five-bit, 58-character alphanumeric code used in transmission of information.

BCD Binary coded decimal. Six-bit alphanumeric code.

Bell System The telephone operating companies controlled by American Telephone and Telegraph Corporation and American Bell.

BERT Bit Error Rate Testing. Testing a data line with a pattern of bits which are compared before and after the transmission in order to detect errors.

Binary Coded Decimal See **BCD**.

Bipolar Coding A method of transmitting a binary stream in which binary zero is sent as a negative pulse and binary one is sent as a positive pulse.

Bit
1. An abbreviation of the words *binary digit*.
2. A single element in a binary number.
3. A single pulse in a group of pulses.
4. A unit of information capacity of a storage device.
5. Zeros and ones.

Bit Error Rate Testing See **BERT**.

Bit Rate (See also bps) The rate at which bits (binary digits) are transmitted over a communication path. Normally expressed in bits per second (bps). The bit rate is not to be confused with the data signaling rate (*baud*), which measures the rate of signal changes being transmitted.

Bit Stream A continuous series of bits being transmitted on a transmission line.

BLERT Block Error Rate Testing. Testing a data link with groups of information arranged into transmission blocks for error checking,

Block Some sets of contiguous bits and/or bytes which make up a definable quantity of information such as a message.

Block Error Rate Testing See **BLERT**.

bps Bits per second. The basic unit of data communications rate measurement. Usually refers to rate of information bits transmitted.

Broadband Circuit A communication circuit that has a bandwidth of greater than 4000 hertz.

Buffer A temporary storage area for a block of data.

Burst Error A series of consecutive errors in data transmission. Refers to the phenomenon on communication lines in which errors are highly prone to occurring in groups or clusters.

Byte A small group of bits of data that are handled as a unit. In most cases it is an 8-bit byte and it is known as a *character*.

Cable Assembly of one or more conductors (usually wire) within an enveloping protective sheath.

Cable Television See **CATV**.

Carrier An analog signal at some fixed amplitude and frequency which is then combined with an information-bearing signal in the modulation process to produce an intelligent output signal suitable for transmission of meaningful information. Also called *carrier wave* or *carrier frequency*.

Carrier Frequency The basic frequency or pulse repetition rate of a signal bearing no intelligence until it is modulated by another signal which does impart intelligence.

Carrier Wave The basic frequency or pulse repetition rate of a signal bearing no intelligence until it is modulated by another signal which does impart intelligence.

CATV Originally *Community Antenna Television*. Now also *Cable Television*. It refers to the use of coaxial cable loops to deliver television or other signals to subscribers.

CCITT Consultive Committee for International Telephone and Telegraph. An international standards group.

C Conditioning A North American term for a type of conditioning that controls attenuation, distortion, and delay distortion so that they lie within specified limits.

CCSA See *Common Control Switching Arrangement*

Central Office The telephone company switching office for the interconnection of direct dial-up calls.

Centrex A telephone PABX equipment service that allows dialing within the system, direct inward dialing, and automatic identification of outward dialing, and that can be used to limit outward long-distance dialing.

Channel
1. A path for transmission of electromagnetic signals. Synonym for *line* or *link*. Compare with *circuit*.
2. A data communications path. Channels may be divided into subchannels.

Character A member of a set of elements upon which agreement has been reached and that is used for the organization, control, or representation of data. Characters may be letters, digits, punctuation marks, or other symbols. (Also called *byte*.)

Character Parity A technique of adding a redundant bit to a character code to provide error-checking capability.

Checking, Echo A method of checking the accuracy of transmission data in which the received data are returned to the sending end for comparison with the original data.

Checking, Parity A check that tests whether the number of ones (or zeros) in an array of binary digits is odd (or even).

Checking, Polynomial A checking method using polynomial functions of the data transmitted to test for changes in data in transmission. Also called *cyclical redundancy check* (CRC).

Circuit A means of two-way communication between two data terminal installations. Compare with *channel, line, link*.

Circuit Switching A method of communications whereby an electrical connection between calling and called stations is established on demand for exclusive use of the circuit until the connection is terminated.

Clear to Send See **CTS**.

Coaxial Cable Two-conductor wire whose longitudinal axes are coincident. Cable with a shield against noise around a signal-carrying conductor.

Code A transformation or representation of information in a different form according to some set of preestablished conventions.

Code, Constant Ratio   A code in which the ratio of ones and zeros in each character is maintained constant.

Code Conversion   A hardware box or software that converts from one code to another, such as from ASCII to EBCDIC.

Code, EBCDIC   An acronym for Extended Binary Coded Decimal Interchange Code. A standard code consisting of a character set of 8-bit characters used for information representation and interchange among data processing and communication systems. Very common in IBM equipment.

Common Carrier   An organization in the business of providing regulated telephone, telegraph, telex, and data communications services. This term is applied most often to U.S. and Canadian commercial organizations but sometimes it is used to refer to telecommunications entities (such as government-operated PTTs) in other countries. In the United States these organizations are regulated by the U.S. Federal Communications Commission or state public utility commissions.

Common Control Switching Arrangement (CCSA)   A dedicated switched network leased by a user to handle communication requirements among various locations.

Communication Processor, Front End   An auxiliary processor that is placed between a computer central processing unit and transmission facilities. This device normally handles housekeeping functions such as management of lines, translation of codes, etc. which would otherwise interfere with efficient operation of the central processing unit. Synonym for *front-end computer*.

Communication Services   The population or entire group of all the various transmission facilities that are available for lease or purchase.

Communications Satellite   An earth satellite designed to act as a telecommunications radio relay. Most communication satellites are in geosynchronous orbit approximately 22,400 miles above the equator so that they appear from earth to be stationary in space.

Compaction   In SNA, the transformation of data by packing two characters in a byte. The most frequently sent characters are compacted.

Compression   In SNA, the replacement of a string of up to 64 repeated characters by an encoded control byte to reduce the length of the data stream sent to the LU–LU session partner. The encoded control byte is followed by the character that was repeated (unless that character is the prime compression character, typically the space character).

Computer, Central   In data transmission, the computer that lies at the center of the network and generally does the basic centralized functions for which the network was designed. Synonym for *host computer*.

Comsat   Communications Satellite Corporation, a private U.S. company established by statute as the exclusive international satellite carrier and representing the United States in Intelsat.

Concentrator   A device that multiplexes several low-speed communications lines onto a single high-speed trunk. An RDC (Remote Data Concentrator) is similar in function to a multiplexer but differs in that host computer software usually must be rewritten to accommodate an RDC. An RDC differs from a statistical multiplexer in that the total capacity of the high-speed outgoing line, in characters per second, is generally less than or equal to the total capacity of the incoming low-speed lines. Output capacity of a statistical multiplexer (stat mux), on the other hand, may exceed the total capacity of the incoming lines.

Conditioning   A technique of applying electronic filtering elements to a communications line to improve the capability of that line to support higher transmission rates of data (See **Equalization**.)

Connector Cable   The cable that goes between the terminal and the modem. It is usually either the RS232C or RS449 standard.

Consultive Committee for International Telephone and Telegraph   See **CCITT**.

Control Character   A character whose occurrence in a particular context initiates, modifies, or stops a control operation—e.g., a character to control carriage return.

Control Matrix A two-dimensional matrix that shows the relationship between all of the controls in the data communication network and the specific threats that they mitigate.

cps Characters-per-second. A data rate unit used where circuits carry bits forming a data character.

CPU Central processing unit.

CRC Cyclical Redundancy Check. An error-checking control technique utilizing a specifically binary prime divisor which results in a unique remainder.

CTS Clear to Send. A control line between a modem and a controller used to operate over a communications line.

Cyclical Redundancy Check See **CRC**.

DAA Data Access Arrangement. A telephone-switching system protective device used to attach non-telephone-company-manufactured equipment to the carrier network.

Data
1. Specific individual facts or a list of such items.
2. Facts from which conclusions can be drawn.

Database A set of logically connected files that have a common access. They are the sum total of all the data items that exist for several related systems. In other words, a database might have several data items that can be assembled into many different record types.

Data Circuit-Terminating Equipment See **DCE**.

Data Communications
1. The movement of encoded information by means of electrical transmission systems.
2. The transmission of data from one point to another.

Data Compression The technique which provides for the transmission of fewer data bits without the loss of information. The receiving location expands the received data bits into the original bit sequence. Also see **Compression**.

Data Protectors Devices that protect the telephone company circuits from extraneous electrical signals. They limit the amount of power that can be transmitted to the telephone company central office.

Data Terminal Equipment See **DTE**.

db See **Decibel**.

dBm Power level measurement unit in the telephone industry based on 600 ohms impedance and 1000 hertz frequency. Zero dBm is 1 milliwatt at 1000 hertz terminated by 600 ohms impedance.

DCE Data Circuit-terminating Equipment. The equipment installed at the user's premises which provides all the functions required to establish, maintain, and terminate a connection, including the signal conversion and coding between the data terminal equipment (DTE) and the common carrier's line, e.g., data set, modem.

D Conditioning A U.S. term for a type of conditioning that controls harmonic distortion and signal-to-noise ratio so that they lie within specified limits.

Decibel (db) A tenth of a bel. A unit for measuring relative strength of a signal parameter such as power, voltage, etc. The number of decibels is ten times the logarithm (base 10) of the ratio of the power of two signals, or ratio of the power of one signal to a reference level. The reference level must always be indicated, such as 1 milliwatt for power ratio.

DDD See **Direct Distance Dialing**.

Dedicated Circuits A leased communication circuit that goes from your premises to some other location. It is a clear unbroken communication path that is yours to use 24 hours per day, seven days per week.

Delay Distortion A distortion on communication lines that is due to the different propagation speeds of signals at different frequencies. Some frequencies travel more slowly than others in a given transmission medium and therefore arrive at the destination at slightly different times. Delay distortion is measured in microseconds of delay relative to the delay at 1700 Hz. This type of distortion does not affect voice but can have a serious effect on data transmissions.

Delay Equalizer A corrective network which is

designed to make the phase delay or envelope delay of a circuit or system substantially constant over a desired frequency range. (See **Equalizer**.)

Delphi Group A small group of experts (3 to 5 people) who get together in order to develop a consensus in an area where it may be impossible or too expensive to collect more accurate data. For example, a Delphi Group of communication experts might assemble in order to reach a consensus on the various threats to a communication network, the potential dollar losses for each occurrence of each threat, and the estimated frequency of occurrence for each threat.

Dial Tone A 90 Hz signal (the difference between 350 Hz and 440 Hz) sent to an operator or subscriber indicating that the receiving end is ready to receive dial pulses.

Dial-Up Telephone Network See **Direct Distance Dialing**.

Dibit A group of two bits. In four-phase modulation each possible dibit is encoded as one of four unique carrier phase shifts. The four possible states for dibit are 00, 01, 10, and 11.

Digital Signal A discrete or discontinuous signal. A signal whose various states are discrete intervals apart such as $+15$ volts and -15 volts.

Digital Termination Systems See **DTS**.

Direct Distance Dialing (DDD) A telephone exchange service which enables the telephone user to call other subscribers outside the local area without operator assistance. In the United Kingdom and some other countries, this is called *subscriber trunk dialing* (STD).

Discrete Files A set of data items and record types for one specific application. A discrete file is a separate, individual file for one application.

Distortion The unwanted modification or change of signals from their true form by some characteristic of the communication line or equipment being used for transmission, e.g., delay distortion, amplitude distortion.

Distortion Types
1. *Bias*: a type of distortion resulting when the intervals of modulation do not all have exactly their normal durations.
2. *Characteristic*: distortion caused by transients (disturbances) which, as a result of modulation, are present in the transmission circuit.
3. *Delay*: distortion occurring when the envelope delay of a circuit or system is not consistent over the frequency range required for transmission.
4. *End*: distortion of start/stop signals. The shifting of the end of all marking pulses from their proper positions in relation to the beginning of the start pulse.
5. *Jitter*: a type of distortion which results in the intermittent shortening or lengthening of the signals. This distortion is entirely random in nature and can be caused by hits on the line.
6. *Harmonic*: the resultant process of harmonic frequencies (due to nonlinear characteristics of a transmission circuit) in the response when a sinusoidal stimulus is applied.

DTE Data Terminal Equipment. Equipment comprising the data source, the data sink, or both that provides for the communication control function (protocol). Data termination equipment is actually any piece of equipment at which a communications path begins or ends, such as a terminal. The data sink is the receiving device.

DTS Digital Termination Systems. A form of local loop. They interconnect private homes and/or business locations to the common carrier switching facility.

EBCDIC See **Code, EBCDIC**.

Echo Checking See **Checking, Echo**.

Electronic Switching System (ESS) A type of telephone switching system which uses a special-purpose stored program digital computer to direct and control the switching operation. ESS permits the provision of custom calling services such as speed dialing, call transfer, three-way calling, etc.

Emulation A method by which one computer processes the computer instructions of another computer. One machine duplicating another machine's actions.

Encryption The technique of modifying a known bit stream on a transmission line so that it appears to an unauthorized observer to be a random sequence of bits.

End Office The telephone company switching office for the interconnection of direct dial-up calls.

Equalization The process of reducing frequency and/or phase distortion of a circuit by the introduction of networks to compensate for the difference in attenuation and/or time delay at the various frequencies in the transmission band.

Equalizer Any combination (usually adjustable) of coils, capacitors, and/or resistors inserted in the transmission line or amplifier circuit to improve its frequency response.

Error Control An arrangement that will detect the presence of errors. In some systems, refinements are added that will correct the detected errors, either by operations on the received data or by retransmission from the source.

ESS See **Electronic Switching System**.

Exchange Office The telephone company switching office for the interconnection of direct dial-up calls.

Facsimile A system for the transmission of images. The image is scanned at the transmitter, reconstructed at the receiving station, and duplicated on some form of paper.

Facsimile Devices Devices that will transmit an exact copy of a handwritten or printed 8½ × 11 sheet of paper.

FCC See **Federal Communications Commission**.

FDM See **Multiplexer**.

FDX See **Full Duplex**.

FEC See **Forward Error Correction**.

Federal Communications Commission (FCC) A board of seven commissioners appointed by the U.S. President under the Communication Act of 1934, having the power to regulate all interstate and foreign electrical communication systems originating in the United States.

Fiber Optics Plastic or glass fibers which carry visible light containing information in cables.

Firmware A set of software instructions set permanently or semipermanently into a read-only memory.

Foreign Exchange Service (FX) A service that connects a customer's telephone to a remote exchange. This service provides the equivalent of local telephone service to and from the distant exchange.

Forward Error Correction (FEC) A technique which can identify errors at the received station and automatically correct those errors without retransmitting the message.

Four-Wire Circuit A circuit using two pairs of conductors, one pair for the "go" channel and the other pair for the "return" channel. A telephone circuit carries voice signals both ways. In the local loops this two-way transmission is achieved over only two wires because the waveforms traveling each direction can be distinguished. In the trunk network, where amplifiers are needed at intervals and multiplexing is common, it is easier to separate the two directions of transmission and effectively use a pair of wires for each direction. At this point it is a four-wire circuit.

Frequency The rate at which a current alternates, measured in hertz, kilohertz, megahertz, etc. Older units of measure are cycles, kilocycles, or megacycles, where hertz and cycles are synonymous

Frequency Division Multiplexing See **FDM**.

Frequency Modulation See **Modulation, Frequency**.

Frequency Shift Keying (FSK) A method of transmission whereby the carrier frequency is shifted up and down from a mean value in accordance with the binary signal; one frequency represents a binary one, and the other represents a binary zero.

Front End See **Communication Processor**.

FSK See **Frequency Shift Keying**.

Full Duplex (FDX) The capability of transmission in both directions at one time. Contrast with *half duplex*.

FX See **Foreign Exchange Service**.

Gaussian Noise See **Noise, Gaussian**.

Geosynchronous Orbit A satellite's orbit that is

over the equator and traveling in the same direction as the earth's surface, so that the satellite appears to be stationary over a point on the earth.

Guard Frequency   A single frequency carrier tone used to indicate that the analog line is prepared to send data. Also the frequencies between subchannels in FDM systems used to guard against subchannel interference.

Half Duplex (HDX)   A circuit that permits transmission of a signal in two directions but not at the same time. Contrast with *full duplex*.

Hamming Code   A forward error correction (FEC) technique named for its inventor.

Handshaking   Line termination interplay to establish data communication path.

HDLC   High Level Data Link Control. A CCITT data communication line protocol standard.

HDX   See **Half Duplex**.

Hertz (Hz)   Same as cycles per second; e.g., 3000 hertz is 3000 cycles per second.

High Level Data Link Control   See **HDLC**.

Hot Line   A service offered by Western Union that provides direct connection between customers in various cities using a dedicated line network.

Impedance   The total opposition offered by a circuit to the flow of an alternating or varying current; a combination of resistance and reactance.

Impulse Noise   See **Noise, Impulse**.

Information   A meaningful aggregation of data. Contrast with *data*.

Intelligent Terminal Controller   A microprocessor-based intelligent device that controls a group of terminals.

INTELSAT   The International Telecommunications Satellite Consortium was established in 1964 to establish a global communications satellite system.

Interexchange Channel (IXC)   A channel between exchanges (central offices).

Intermodulation Distortion   An analog line impairment where two frequencies create a third erroneous frequency which in turn distorts the data signal representation.

IXC   See **Interexchange Channel**.

Jitter   Type of analog communication line distortion caused by the variation of a signal from its reference timing positions which can cause data transmission errors, particularly at high speeds. This variation can be amplitude, time, frequency, or phase.

Jumbo Group   Six U.S. master groups frequency division multiplexed together in the Bell System. A jumbo group can carry 3600 telephone calls.

Kbps   Kilo bits per second. A data rate equal to 10^3 bps.

Key Management   The process of controlling the secret keys used in encryption.

Kilo Bits per Second   See **Kbps**.

LAN   See **Local Area Network**.

Large-Scale Integration   See **LSI**.

Laser   Light amplification by stimulated emission of radiation. A device which transmits an extremely narrow and coherent beam of electromagnetic energy in the visible light spectrum. (Coherent means that the separate waves are in phase with one another rather than jumbled as in normal light.)

Leased Circuits   Leased communication circuits that go from your premises to some other location. It is a clear, unbroken communication path that is yours to use 24 hours per day, seven days per week.

Line   A channel or link.

Line Control Codes/Characters   The 8-bit characters that control messages. They appear at the beginning of a message and end of the message to show the beginning and end. They might also appear within the message to show such things as the beginning and end of the text.

Line Loading   The total amount of transmission traffic carried by a line, usually expressed as a percentage of the total theoretical capacity of that line.

Line Protocol   A control program used to perform data communications functions over network lines. Consists of both handshaking and line-control functions which move the data between transmit and receive locations.

Link   A channel or a line, normally restricted in use to a point-to-point line.

Local Area Network (LAN)   A loop network that is usually contained within a single building. The data move around this network in one direction only and there might be 100 or more individual terminal stations connected to it.

Local Loop   That part of a communication circuit between the subscriber's equipment and the equipment in the local central office.

Log
1. A record of everything pertinent to a system function.
2. A collection of messages that provides a history of message traffic.

Logical Unit (LU)   In SNA, a port through which an end user accesses the SNA network in order to communicate with another end user and through which the end user accesses the functions provided by system services control points (SSCPs). An LU can support at least two sessions—one with an SSCP, and one with another logical unit—and may be capable of supporting many sessions with other logical units.

Logical Unit (LU) Services   In SNA, capabilities in a logical unit to:
1. Receive requests from an end user and, in turn, issue requests to the system services control point (SSCP) in order to perform the requested functions, typically for session initiation.
2. Receive requests from the SSCP, for example to activate LU-LU sessions via bind session requests.
3. Provide session presentation and other services for LU-LU sessions.

Longitudinal Redundancy Check (LRC)   A system of error control based on the formation of a block check following preset rules. The check formation rule is applied in the same manner to each character. In a simple case, the LRC is created by forming a parity check on each bit position of all the characters in the block (e.g., the first bit of the LRC character creates odd parity among the 1-bit positions of the characters in the block).

LRC   See **Longitudinal Redundancy Check**.

LSI   Large Scale Integration. A type of electronic device comprising many logic elements in one very small package (integrated circuit) to be used for data handling, storage, and processing.

LU   See **Logical Unit**.

mA   Milliampere. Electric current measurement unit.

Master Group   An assembly of ten supergroups occupying adjacent bands in the transmission spectrum for the purposes of simultaneous modulation and demodulation.

Matrix of Controls   A two-dimensional matrix that shows the relationship between all of the controls in the data communications network and the specific threats that they mitigate.

Mbps   A data rate equal to 10^6 bps. Sometimes called mega bits per second.

Message   A communication of information from a source to one or more destinations, usually in code. A message is usually composed of three parts:
1. A heading, containing a suitable indicator of the beginning of the message together with some of the following information: source, destination, date, time, routing.
2. A body containing information to be communicated.
3. An ending containing a suitable indicator of the end of the message.

Message Switching   In this operation the entire message that you are transmitting is switched to the other location without regard as to whether the circuits were actually interconnected at the time of your call. This usually involves a message store and forward facility.

Metered Service (WATS)   Wide Area Telecommunications Service by AT&T. A combination of offerings of bulk long-distance service under various terms involving flat-rate charges per hour of usage.

MHz   A unit of analog frequency equal to 10^6 hertz. Sometimes referred to as megahertz.

Microprocessor   A single or multiple chip set which makes up a microcomputer. Usually has an 8- or 16-bit word length.

Milliampere See **MA**.

Modem A contraction of the words modulator-demodulator. A modem is a device for performing necessary signal transformation between terminal devices and communication lines. They are used in pairs, one at either end of the communication line.

Modulation, Amplitude The form of modulation in which the amplitude of the carrier is varied in accordance with the instantaneous value of the modulating signal.

Modulation, Frequency A form of modulation in which the frequency of the carrier is varied in accordance with the instantaneous value of the modulating signal.

Modulation, Phase A form of modulation in which the phase of the carrier is varied in accordance with the instantaneous value of the modulating singal.

Modulation, Pulse See **Pulse**.

Multidrop (Multipoint) A line or circuit interconnecting several stations.

Multiplexer A device that combines data traffic from several low-speed communications lines onto a single high-speed line. The two popular types of multiplexing are FDM (frequency division multiplexing) and TDM (time division multiplexing). In FDM, the voice grade link is divided into subchannels, each covering a different frequency range in such a manner that each subchannel can be employed as if it were an individual line. In TDM, separate time segments are assigned to each terminal. During these times, data may be sent without conflicting with data sent from another terminal.

Multiplexing The subdivision of a transmission channel into two or more separate channels. This can be achieved by splitting the frequency range of the channel into narrower frequency bands (*frequency division multiplexing*) or by assigning a given channel successively to several different users at different times (*time division multiplexing*).

Multiprocessing Strictly, this term refers to the simultaneous application of more than one processor in a multi-CPU computer system to the execution of a single "user job," which is possible only if the job can be effectively defined in terms of a number of independently executable components. The term is more often used to denote multiprogramming operation of multi-CPU computer systems.

Multiprogramming A method of operation of a computer system whereby a number of independent jobs are processed together. Rather than allow each job to run to completion in turn, the computer switches between them so as to improve the utilization of the system hardware components.

Multithreading Concurrent processing of more than one message (or similar service requested) by an application program.

NAK See **Negative Acknowledgment**.

NAU See **SNA Network**.

NCP See **Network Control Program**.

Negative Acknowledge (NAK) In the method of error control which relies on repeating any message received with (detectable) errors, the return signal that reports an error is a NAK (the opposite of *ACK*, or acknowledge).

Network
1. A series of points connected by communications channels.
2. The switched telephone network is the network of telephone lines normally used for dialed telephone calls.
3. A private network is a network of communications channels confined to the use of one customer.

Network Control Program (NCP) The program within the software system for a data processing system which deals with the control of the network. Normally it manages the allocation, use, and diagnosis of performance of all lines in the network and of the availability of the terminals at the ends of the network. *NCP* is also used as a specific term referring to a component of SNA (Systems Network Architecture).

Node In a topological description of a network a node is a point of junction of the links. The word has also come to mean a switching center in the context of data networks, particularly in the context of packet switching.

Noise The unwanted change in waveform that occurs between two points in a transmission circuit.

Noise, Cross Talk Noise resulting from the interchange of signals on two adjacent channels.

Noise, Echo On voice grade lines with improper echo suppression, the "hollow" or echoing characteristic that results.

Noise, Gaussian Noise that is characterized statistically by a Gaussian, or random distribution.

Noise, Impulse Noise caused by individual impulses on the channel.

Noise, Intermodulation Noise resulting from the intermodulation products of two signals. This is a result of harmonic reinforcements and cancellation of frequencies.

Off Hook Activated (in regard to a telephone set). By extension, a data set automatically answering on a public switched system is said to go "off hook." The off-hook condition indicates a "busy" condition to incoming calls.

Office, Central (or End) The common carrier switching office closest to the subscriber.

Office, Tandem End A switching office that terminates a tandem trunk.

Office, Toll A switching office that terminates a toll trunk.

Ohm A unit of resistance, such that one ampere through it produces a potential difference of one volt. Ohm's law is applicable to electric components carrying direct current and it states that in metallic conductors at a constant temperature and zero magnetic field, the resistance is independent of the current.

On Hook Deactivated (in regard to a telephone set). A telephone not in use is "on hook."

On-Line

1. Pertaining to equipment or devices under the direct control of a central processing unit.
2. Pertaining to a user's ability to interact with a computer.
3. Pertaining to a user's access to a computer via a terminal.

On-Line System A system in which the input data enter the computer directly from the point of origin or in which output data are transmitted directly to where they are used.

Open-Wire Communications lines not insulated and formed into cables, but mounted on aerial cross arms on utility poles.

Optical Fibers Hair-thin strands of very pure glass (sometimes plastic) over which light waves travel. They are used as a medium over which information is transmitted.

Out-of-Band Signaling A method of signaling which uses a frequency that is within the passband of the transmission facility but outside of a carrier channel normally used for voice transmission.

Overhead Computer time used to keep track of or run the system as compared with computer time used to process data.

PABX See **Private Automatic Branch Exchange**.

Packet A group of binary digits, including data and control signals, which is switched as a composite whole. The data, control signals, and possibly error control information are arranged in a specific format.

Packet Assembler/Disassembler See **PAD**.

Packet Switching Process whereby your message is broken into finite size packets that are always accepted by the network. The message packets are forwarded to the other party over a multitude of different circuit paths. At the other end of the circuit, the packets are reassembled into the message, which is then passed on to the receiving terminal.

Packet Switching Network A network designed to carry data in the form of packets. The packet and its format is internal to that network. The external interfaces may handle data in different formats, and conversion is done by an interface computer.

PAD Packet Assembler/Disassembler. Equipment providing packet assembly and disassembly facilities.

PAM See **Pulse Amplitude Modulation**.

Parity Bit A binary bit appended to an array of

bits to make the sum of all the bits always be odd or even for an individual character.

Parity Check   Addition of noninformation bits to a message in order to detect any changes in the original bit structure from the time it left the sending device to the time it was received.

Path Control (PC) Network   In SNA, the part of the SNA network that includes the data link control and path control layers.

PBX   See **Private Branch Exchange**.

PBX/CBX   Private Branch Exchange/Computer Branch Exchange. One of the newer digital switchboards that interconnect both voice communications and digital data communications.

PC   See **Path Control (PC) Network**.

PCM   See **Pulse Code Modulation**.

PDM   See **Pulse Duration Modulation**.

Phase Modulation   See **Modulation, Phase**.

Physical Unit (PU)   In SNA, the component that manages and monitors the resources (such as attached links and adjacent link stations) of a node, as requested by an SSCP via an SSCP-PU session. Each node of an SNA network contains a physical unit.

Physical Unit Control Point (PUCP)   In SNA, a component that provides a subset of system services control point (SSCP) functions for activating the physical unit (PU) within its node and its local link resources. Each peripheral node and each subarea node without an SSCP contains a PUCP.

Point-to-Point   Denoting a channel or line that has only two terminals. A link.

Polling   Any procedure that sequentially contacts several terminals in a network.

Polling, Hub Go-Ahead   Sequential polling in which the polling device contacts a terminal, that terminal contacts the next terminal, and so on.

Polling, Roll Call   Polling accomplished from a prespecified list in a fixed sequence, with polling restarted when the list is completed.

Polynomial Checking   See **Checking, Polynomial**.

Port   One of the circuit connection points on a front end communication processor or local intelligent terminal controller.

PPM   See **Pulse Position Modulation**.

Private Automatic Branch Exchange (PABX)   A private automatic telephone exchange that provides for the transmission of calls internally and to and from the public telephone network (switchboard).

Private Branch Exchange (PBX)   A small manual telephone exchange installed on a customer's premises to allow internal dialing from station to station within the customer's premises and connection to outgoing and incoming lines.

Private Leased Circuit   A leased communication circuit that goes from your premises to some other location. It is a clear unbroken communication path that is yours to use 24 hours per day, seven days per week.

Propagation Delay   The time necessary for a signal to travel from one point on the circuit to another.

Protocol   A formal set of conventions governing the format and control of inputs and outputs between two communicating devices. This includes the rules by which these two devices communicate as well as the handshaking and line discipline. Some protocols are "bit-oriented," with a change of a single bit within a character conveying a new control message to the other end of the communication circuit. Some protocols are "byte-oriented," with the entire 8 bits of a character having to be changed in order to convey a control message to the other end of the communication network.

Protocol Converter   A hardware device that changes the protocol of one vendor to the protocol of another. For example, if you want to interconnect an IBM-oriented data communication network to a Honeywell data communication network, the protocol converter will convert the message format so they are compatible. It is similar to a person who translates between French and English for two people who do not speak each other's language.

PU   See **Physical Unit**.

PUCP   See **Physical Unit Control Point**.

Pulse   A brief change of current or voltage produced in a circuit to operate a switch or relay or which can be detected by a logic circuit.

Pulse Amplitude Modulation (PAM) Amplitude modulation of a pulse carrier.

Pulse Code Modulation (PCM) Representation of a speech signal (or other analog signal) by sampling of a regular rate and converting of each sample to a binary number.

Pulse Duration Modulation (PDM) Pulse width modulation. A form of pulse modulation in which the durations of pulses are varied.

Pulse Modulation The modulation of the characteristics of a series of pulses in one of several ways to represent the information-bearing signal. Typical methods involve modifying the amplitude (PAM), width or duration (PDM), or position (PPM). The most common pulse modulation technique in telephone work is a pulse code modulation (PCM). In PCM, the information signals are sampled at regular intervals and a series of pulses in coded form are transmitted, representing the amplitude of the information signal at that time.

Pulse Position Modulation (PPM) A form of pulse modulation in which the positions in time of pulses are varied, without their duration being modified.

Queue A line of items. In data communications there can be message input queues, output queues, and various other queues whenever the system cannot handle all of the transactions that are arriving.

RDC See **Concentrator**.

Real-Time The entry of information into a network from a terminal and immediate processing of the task.

Redundancy The portion of the total information contained in a message that can be eliminated without loss of essential information.

Reliability The characteristic of equipment, software, or systems that relates to the integrity of the system against failure. Reliability is usually measured in terms of mean-time-between-failures, the statistical measure of the interval between successive failures of the system under consideration.

Remote Job Entry (RJE) Submission of jobs (e.g., computer production tasks) through an input unit (terminal) that has access to a computer through data communication facilities.

Repeater A device used to boost the strength of a signal. Repeaters are spaced at intervals throughout the length of a communication circuit. Also called *repeater/amplifier*.

Request to Send See **RTS**.

Resistance The metallic resistance that is inherent in a circuit.

Response Time The time the system takes to react to a given input. If a message is keyed into a terminal by an operator and the reply from the computer, when it comes, is typed at the same terminal, response time may be defined as the time interval between the operator's pressing the last key and the terminal's typing the first letter of the reply. It is the interval between an event and the system's response to the event. Response time thus defined includes: (1) transmission time to the computer; (2) processing time at the computer, including access time to obtain any file records needed to answer the inquiry; and (3) transmission time back to the terminal.

Reverse Channel A feature of certain modems which allows simultaneous transmission (usually of control or parity information) from the receiver to the transmitter over a half duplex data link. Generally the reverse channel is a low-speed channel.

RJE See **Remote Job Entry**.

Rotary Switching System An automatic telephone switching system which is generally characterized by the following features:

1. The selecting mechanisms are rotary switches.
2. The switching pulses are received and stored by controlling mechanisms which govern the subsequent operations necessary in establishing a telephone connection.

RS232 Interface The interface cable between a modem and the associated data terminal, as defined by the Electronics Industries Association Standard RS232.

RTS Request to Send. An RS232 control line between a modem and user digital equipment which initiates the data transmission sequence in a communications line.

Satellite See **Geosynchronous Orbit**.

Satellite Microwave Radio Microwave or beam radio system using geosychronously orbiting communications satellites.

Satisfice To choose a particular level of performance for which to strive and for which management is willing to settle.

SC See **Session Control**.

SDLC See **Synchronous Data Link Control**.

Serial
1. Pertaining to the sequential performance of two or more activities in a single device.
2. Pertaining to the sequential or consecutive occurrence of two or more related activities in a single device or channel.
3. Pertaining to the sequential processing of the individual part of the whole, such as the bits of a character, or the characters of a word using the same facilities for successive parts.

Session A logical connection between two terminals. This is the part of the message transmission where the two parties are exchanging messages. It takes place after the communication circuit has been set up and is functioning.

Session Control (SC) In SNA, one of the components of transmission control. Session control is used to purge data flowing in a session after an unrecoverable error occurs, to resynchronize the data flow after such an error, and to perform cryptographic verification.

Signal A signal is something that is sent over a communication circuit. It might be a message that you are transmitting or it might be a control signal used by the system to control itself.

Signaling Supplying and interpreting the supervisory and address signals needed to perform the switching operation.

Signaling, In-Band Signaling that utilizes frequencies within the intelligence band of a channel, usually within the voice channel.

Signal-to-Noise Ratio The ratio, expressed in dB, of the usable signal to the noise signal present.

Simplex A circuit capable of transmission in one direction only. Contrast with *Half Duplex*, *Full Duplex*.

SN See **Switching Node**.

SNA See **System Network Architecture**.

SNA Network In SNA, the part of a user-application network that conforms to the formats and protocols of Systems Network Architecture. It enables transfer of data among end users and provides protocols for controlling the resources of various network configurations. The SNA network consists of network addressable units (NAUs), boundary-function components, and the path control network.

Software A generic, somewhat slang term for a computer program, sometimes taken to include also documentation and procedures associated with such programs.

Special Common Carrier An organization other than the public telephone companies, that is registered to sell or lease communication facilities.

SSCP System services control point.

SSCP-LU Session In SNA, a session between a system services control point (SSCP) and a logical unit (LU); the session enables the LU to request the SSCP to help initiate a LU-LU session.

SSCP-PU Session In SNA, a session between a system services control point (SSCP) and a physical unit (PU); SSCP-PU sessions allow SSCPs to send requests to and receive status information from individual nodes in order to control the network configuration.

SSCP Services In SNA, the components within a system services control point (SSCP) that provide configuration, maintenance, management, network, and session services for SSCP-LU, SSCP-PU, and SSCP-SSCP sessions.

SSCP-SSCP Session In SNA, a session between the system services control point (SSCP) in one domain and the SSCP in another domain. An SSCP-SSCP session is used to initiate and terminate cross-domain LU-LU sessions.

Start Bit A bit preceding the group of bits representing a character used to signal the arrival of the character in asynchronous transmission.

Start-Stop (Signaling) Signaling in which each group of code elements corresponding to an alphabetical signal is preceded by a start signal which serves to prepare the receiving mechanism for the reception and registration of a character, and is followed by a stop signal which serves to bring the receiving mechanism to rest in preparation for the reception of the next character (contrast with *Synchronous Transmission*). Start-stop transmission is also referred to as *asynchronous transmission*.

Station One of the input or output points on a network.

Station Terminal The plug supplied by the common carrier into which the modem plugs (Series 2000/3000 channels).

Statistical Multiplexer Or Stat Mux. A TDM that dynamically allocates communication-line time to each of the various attached terminals, according to whether a terminal is active or not at a particular moment. Buffering and queuing functions also are included.

Stop Bit A bit used following the group of bits representing a character, to signal the end of a character in asynchronous transmission.

Store and Forward A data communications technique which accepts messages or transactions, stores them until they are completely in the memory system, and then forwards them on to the next location as addressed in the message or transaction header.

Super Group An FDM carrier multiplexing level containing 60 voice frequency channels. It is the assembly of five 12-channel groups occupying adjacent bands in the spectrum for the purpose of simultaneous modulation and demodulation.

Switchboard Equipment on which switching operations are performed by operators.

Switched Network Any network in which switching is present and which is used to direct messages from the sender to the ultimate recipient.

Switched Network, Circuit-Switched A switched network in which switching is accomplished by disconnecting and reconnecting of lines in different configurations in order to set up a continuous pathway between the sender and the recipient.

Switched Network, Store-and-Forward A switched network in which the store-and-forward principle is used to handle transmissions between senders and recipients.

Switching Identifying and connecting independent transmission links to form a continuous path from one location to another.

Switching Node (SN) The intelligent interface point where your equipment is connected to a public packet switching network for your own private packet switching network. The switching node is a type of front end but its primary purpose is the packetizing, routing, and scheduling on the packet switching network.

Synchronous Data Link Control (SDLC) A discipline for managing synchronous, code-transparent, serial-by-bit information transfer over a link connection. Transmission exchanges may be full duplex or half duplex over switched or nonswitched links. The configurations of the link connection may be point-to-point, multipoint, or loop. SDLC conforms to subsets of the Advance Data Communication Control Procedures (ADCCP) of the American National Standards Institute and High-Level DataLink Control (HDLC) of the International Organization for Standardization.

Synchronous Transmission Form of transmission in which data are sent continuously against a time base that is shared by transmitting and receiving terminals. If no legitimate data are available to be sent at a given time, "synch" or "idle" characters are sent to keep the transmitter and receiver in time synchronization.

System Network Architecture (SNA) The term applied by IBM to the conceptual framework used in defining data communication interaction with computer systems.

System Services Control Point (SSCP) In SNA, a focal point within an SNA network for managing the configuration, coordinating network operator and problem determination requests, and providing directory support and other session services for end users of the network. Multiple SSCPs, cooperating as peers with one another, can divide the network into domains of control, with each SSCP having a hierarchical control relationship to the physical units and logical units within its own domain. Also see **SSCP** entries.

Tariff   The schedule of rates and regulations pertaining to the services of a communication common carrier.

TASI   Time Assisted Speech Interpolation. The process of sending two or more voice calls on the same telephone circuit simultaneously.

T-Carrier (Bell System)   A hierarchy of digital systems designed to carry speech and other signals in digital form, designated T1, T2 and T4. The T1 carrier has 24 PCM voice channels.

TDM   See **Multiplexer**.

Telecommunication Access Programs   The software programs located in the front end communication processor. They handle all of the tasks associated with the routing, scheduling, and movement of messages between the remote terminal sites and the central host computer.

Telecommunication Monitors   See **Teleprocessing Monitors**

Teleprocessing Monitors   A set of software programs (usually located in the host computer) that handle the various tasks required for incoming and outgoing messages. For example, the teleprocessing monitor would build the input/output queues of messages in the host computer and relieve the computer operating system of many of its tasks as related to the data communication network.

Telidon   A two-way dialogue through a television set to a central site that offers various services in the home. This is a videotext service offered in Canada.

Time Assisted Speech Interpolation   See **TASI**.

Trunk   A communication channel between switching devices or central offices.

Turnaround Time   The time required to reverse the direction of transmission from send to receive or vice versa on a half duplex circuit.

Two-Wire Circuit   A circuit formed by two conductors insulated from each other. It is possible to use the two conductors as either a one-way transmission path, a half duplex path, or a full duplex path.

Unipolar Coding   A method of transmitting a binary stream in which binary 0 is sent as no pulse and binary 1 is sent as a positive pulse.

UNIX   A very flexible operating system that can be used with microprocessors and is applicable to local area networks.

Value Added Common Carrier   A corporation which sells services of a value added network. Such a network is built using the communications offerings of traditional common carriers, connected to computers which permit new types of telecommunication tariffs to be offered. The network may be a packet-switching or message-switching network. Services offered include transmission of data charged for by the packet, and transmission of facsimile documents.

Vertical Redundancy Checking   See **VRC**.

Videotex   A two-way dialogue through a television set to a central site that offers various services in the home.

Virtual   Conceptual or appearing to be, rather than actually being.

Virtual Circuit   A proposed CCITT definition for a data transmission service. The user presents a data message for delivery, with a header of a specified format. The system delivers the message as though a circuit existed to the specified destination. One of many different routes and techniques could be used to deliver the message, but the user need not know which is employed. It appears to the user as though an actual circuit existed.

Virtual Storage   A computer user may employ a computer as though it has a much larger memory than its real memory. The difference is made up by software rapidly moving pages in and out, to and from a backing store. The apparent memory which the user can employ is called *virtual memory*.

Virtual Terminal A terminal that is defined as a standard on a network that can handle diverse terminals. Signals to and from each nonstandard terminal are converted to equivalent standard-terminal signals by an interface computer. The network protocols then operate as though all terminals were the standard "virtual" terminals.

Voice Grade A telecommunications link with a bandwidth (about 4 kHz) appropriate to an audio telephone line.

Voice Grade (Series 2000/3000) The term applied to channels suitable for transmission of speech and digital or analog data or facsimile, generally with a frequency range of about 300 to 3,000 hertz contained within a 4,000 hertz channel.

VRC Vertical Redundancy Checking A method of character parity checking.

WAK Positive acknowledgment but stop sending. Contrast with *ACK* and *NAK*.

WATS See **Metered Service**.

Wideband (SERIES 8000) The term applied to channels provided by common carriers capable of transferring data at speeds from 19,200 bps up to the 1 million bps region (19.2 kHz to 1000 kHz).

Wideband Circuit A communication circuit that has a bandwidth of greater than 4000 hertz.

Word

1. In communications, six characters (five plus a space).

2. In computers, the unit of information transmitted, stored, and operated upon at one time.

X.3 The International Organization for Standardization standard that describes the function of a packet assembler/disassembler (PAD).

X.25 The International Organization for Standardization standard protocol that defines the set of rules on how two machines communicate with each other. This standard relates specifically to the interconnection between terminals and public packet switching networks.

X.28 The International Organization for Standardization standard that defines the interface between a nonintelligent terminal and a packet assembly/disassembly (PAD).

X.29 The International Organization for Standardization standard that defines the procedure for controlling the exchange of data between a packet assembly/disassembly (PAD) and a packet mode terminal.

X.75 The International Organization for Standardization standard controlling the interface between a public packet switching network and the entry point to some other network. It is characteristically called a *gateway*. Local area networks will have gateways to the public packet switching networks.

APPENDIX 1

DATA COMMUNICATION CONTROL MATRIX

This section outlines the control review matrix to be used when reviewing the data communication network that interconnects remote terminals and the central computer system or the various portions of an on-line distributed network. The controls/safeguards listed in this matrix are specifically designed for review of the data communication network.

THE MATRIX APPROACH

The internal control area to be reviewed using this matrix covers the data communication links between the computer and the input/output terminals. These data communication-oriented controls may involve hardware controls, software controls, and personnel controls. When reviewing the data communication controls, match each resource/asset with its corresponding concern/exposure as

Source: Appendix 1 is taken from Chapter 4 of the book, *Internal Controls for Computerized Systems*, by Jerry FitzGerald. This book contains eight other matrices on subjects such as software, physical security, database, application program controls, etc. This book is available from Jerry FitzGerald & Associates, 506 Barkentine Lane, Redwood City, CA 94065.

listed in Figure A1–1, Data Communication Control Matrix. This matrix lists the resources in relation to the potential exposures and cross-relates these with the various controls/safeguards that should be considered when reviewing the data communication controls (see Chapter 8 for an explanation on how to use the control matrix approach).

Following is a definition of each of the concerns/exposures that are listed across the top of the matrix and each of the resources/assets that are listed down the left vertical column of the matrix. Following these definitions is a complete numerical listing and description of each of the controls/safeguards that are listed numerically in the cells of the matrix.

CONCERNS/EXPOSURES (THREATS)

The following concerns/exposures are those that are directly applicable to the data communication network review of an on-line system. The definition for each of these exposures, listed across the top of the matrix, is as follows:

- **Errors and Omissions** The accidental or intentional transmission of data that is in error, including the accidental or intentional omission of data that should have been entered or transmitted on the on-line system. This type of exposure includes, but is not limited to, inaccurate data, incomplete data, malfunctioning hardware, and the like.
- **Message Loss or Change** The loss of messages as they are transmitted throughout the data communication system, or the accidental/intentional changing of messages during transmission.
- **Disasters and Disruptions (natural and man-made)** The temporary or long-term disruption of normal data communication capabilities. This exposure renders the organization's normal data communication on-line system inoperative.
- **Privacy** The accidental or intentional release of data about an individual, assuming that the release of this personal information was improper to the normal conduct of the business at the organization.
- **Security/Theft** The security or theft of information that should have been kept confidential because of its proprietary nature. In a way, this is a form of privacy, but the information removed from the organization does not pertain to an individual. The information might be inadvertently (accidentally) released, or it might be the subject of an outright theft. This exposure also in-

THREATS/CONCERNS/EXPOSURES

	Errors and Omissions	Message Loss or Change	Disasters and Disruptions	Privacy	Security/ Theft	Reliability (Uptime)	Recovery and Restart	Error Handling	Data Validation and Checking
Central System	1-4, 7, 39, 41-43, 47, 48	1-5, 7, 37, 39, 48, 49, 89	1, 8, 11, 13, 16, 29, 43, 48, 50, 51, 54, 57, 58, 64, 65, 79, 85	6, 8, 24, 35, 53, 56, 60, 62, 68, 70, 72-74, 78-80	6, 8, 24, 35, 53, 56, 60, 62, 68, 70, 72-74, 77-80	1, 13, 16, 29, 38, 40, 50, 51, 63-65, 68, 81, 88	50, 51, 63-65, 68	48, 85, 89	6, 24, 39, 41, 47, 88
Software	1-4, 7, 39, 41-43, 46-49, 52	1-5, 7, 37, 39, 41, 42, 48, 49, 52, 54, 89	1, 8, 16, 40, 48, 50-54, 57-59, 63, 85	6, 8, 24, 35, 53, 56, 60, 62, 68, 70, 72-74, 78-80	6, 8, 24, 35, 39, 53, 56, 60, 62, 68, 70, 72-74, 78-80	1, 38, 40, 50, 51, 56-59, 61, 63, 68, 88	50-52, 61, 63, 64, 68	48, 61, 85, 89	6, 24, 39, 41, 47-49, 52, 53, 55, 60, 68
Front End Communication Processor	1-4, 7, 34, 39, 41-44, 46-48	1-5, 7, 34, 37, 39, 41, 42, 49, 89	1, 8, 13, 16, 29, 40, 44, 48, 50, 51, 54, 57, 58, 64, 65, 79, 85	6, 8, 24, 35, 37, 45, 60, 62, 68, 70, 72-74, 78-80	6, 8, 24, 29, 35, 37, 39, 45, 50, 62, 68, 70, 72-74, 78-80	1, 13, 16, 29, 30, 34, 36, 40, 43, 44, 50, 51, 63-65, 81, 88	37, 50, 51, 63-65	43, 48, 85, 89	6, 24, 39, 41, 45, 47, 48, 88
Multiplexer, Concentrator Switch	1-4, 7, 37, 39, 41, 44, 46, 47	1-5, 7, 37, 39, 41, 42, 49, 89	1, 8, 13, 16, 29, 30, 32, 33, 40, 44, 48, 50, 51, 54, 57, 58, 65, 79, 85	6, 8, 24, 35, 37, 45, 60, 62, 68, 70, 72-74, 78-80	6, 8, 24, 29, 35, 37, 39, 45, 60, 62, 68, 70, 72-74, 78-80	1, 13, 16, 29, 30, 32-34, 36, 40, 44, 50, 51, 63-65, 81, 88	37, 50, 51, 63, 64	48, 85, 89	6, 24, 39, 41, 45, 47, 48, 88
Communication Circuits (lines)	12, 26	28, 70, 91	10, 15, 16, 18, 26, 63, 64, 66, 75, 76, 79, 91	25, 28, 68, 70, 75, 76, 78-80, 91	25, 28, 68, 70, 75, 76, 78-80, 91	15, 16, 20, 21, 23, 26, 27, 63, 64, 66-68, 88	63, 64, 66, 68	85	
Local Loop	12	25	25, 75, 85	25, 76	25, 29, 75, 76	68, 38	63, 64, 68	85	
Modems	12, 18	18, 24	8-11, 13-16, 18	24	24, 29	9-11, 13-18, 20, 21, 23, 36, 88	9-11, 14, 15, 63, 64	18-20, 22, 23	
People	5, 39	5, 7, 31, 39, 70	79-87	6, 8, 24, 53, 69-71, 74, 77, 79, 80	6, 8, 24, 29, 53, 69-71, 74, 77, 79, 80	81, 82, 85-87	50, 51, 86, 87	49, 86, 87, 89, 90	6, 88
Terminals/ Distributed Intelligence		2		6, 8, 24, 45, 53, 56, 62, 70	6, 8, 24, 29, 45, 53, 56, 62, 70	1, 40, 88	63, 64		6, 24, 45

COMPONENTS/RESOURCES/ASSETS

FIGURE A1-1 Data Communication Control Matrix.

cludes the theft of assets such as might be experienced in embezzlement, fraud, or defalcation.

- **Reliability (Uptime)** The reliability of the data communication network and its "uptime." This includes the organization's ability to keep the data communication network operating and the mean time between failures (MTBF) as well as the time to repair equipment when it malfunctions. Reliability of hardware, reliability of software, and the maintenance of these two items are chief concerns here.

- **Recovery and Restart** The recovery and restart capabilities of the data communication network, should it fail. In other words, How does the software operate in a failure mode? How long does it take to recover from a failure? This recovery and restart concern also includes backup for key portions of the data communication network and the contingency planning for backup, should there be a failure at any point of the data communication network.

- **Error Handling** The methodologies and controls for handling errors at a remote distributed site or at the centralized computer site. This may also involve the error handling procedures of a distributed data processing system (at the distributed site). The object here is to ensure that when errors are discovered they are promptly corrected and reentered into the system for processing.

- **Data Validation and Checking** The validation of data either at the time of transmission or during transmission. The validation may take place at a remote site (intelligent terminal), at the central site (front end communication processor), or at a distributed intelligence site (concentrator or remote front end communication processor).

RESOURCES/ASSETS (COMPONENTS)

The following resources/assets are those that should be reviewed during the data communication control review. The definition for each of these assets, listed down the left vertical column of the matrix, is as follows:

- **Central System** Most prevalent in the form of a central computer to which the data communication network transmits and from which it receives information. In a distributed system, with equal processing at each distributed node, there might not be an identifiable central system (just some other equal-sized distributed computer).

- **Software**   The software programs that operate the data communication network. These programs may reside in the central computer, a distributed-system computer, the front end communication processor, a remote concentrator or statistical multiplexer, and/or a remote intelligent terminal. This software may include the telecommunications access methods, an overall teleprocessing monitor, programs that reside in the front end processors, and/or programs that reside in the intelligent terminals.

- **Front End Communication Processor**   A hardware device that interconnects all the data communication circuits (lines) to the central computer or distributed computers and performs a subset of the following functions: code and speed conversion, protocol, error detection and correction, format checking, authentication, data validation, statistical data gathering, polling/addressing, insertion/deletion of line control codes, and the like.

- **Multiplexer, Concentrator, Switch**   Hardware devices that enable the data communication network to operate in the most efficient manner. The *multiplexer* is a device that combines, in one data stream, several simultaneous data signals from independent stations. The *concentrator* performs the same functions as a multiplexer except that it is intelligent and therefore can perform some of the functions of a front end communication processor. A *switch* is a device that allows the interconnection between any two circuits (lines) connected to the switch. There might be two distinct types of switch: a switch that performs message switching between stations (terminals) might be located within the data communication network facilities that are owned and operated by the organization; a circuit or line switching switch that interconnects various circuits might be located at (and owned by) the telephone company central office. For example, organizations perform message switching and the telephone company performs circuit switching.

- **Communication Circuits (Lines)**   The common carrier facilities used as links (a *link* is the interconnection of any two stations/terminals) to interconnect the organization's stations/terminals. These communication circuits include, not to the exclusion of others, satellite facilities, public switched dial-up facilities, point-to-point private lines, multiplexed lines, multipoint or loop configured private lines, WATS services, and many others.

- **Local Loop**   The communication facility between the customer's premises and the telephone company's central office or the central office of any other special common carrier. The local loop is usually assumed to be metallic pairs of wires.

- **Modems**   A hardware device used for the conversion of data signals from terminals (digital signal) to an electrical form (analog signal) which is acceptable for transmission over the communication circuits that are owned and maintained by the telephone company or other special common carrier.

- **People** The individuals responsible for inputting data, operating and maintaining the data communication network equipment, writing the software programs for the data communications, managing the overall data communication network, and those involved at the remote stations/terminals.

- **Terminals/Distributed Intelligence** Any or all of the input or output devices used to interconnect with the on-line data communication network. This resource would specifically include, without excluding other devices, teleprinter terminals, video terminals, remote job entry terminals, transaction terminals, intelligent terminals, and any other devices used with distributed data communication networks. These may include microprocessors or minicomputers when they are input/output devices or if they are used to control portions of the data communication network.

CONTROLS/SAFEGUARDS

The following controls/safeguards should be considered when reviewing the data communication network review of an on-line system. This numerical listing describes each control.

It should be noted that implementation of various controls can be both costly and time-consuming. It is of great importance that a realistic and pragmatic evaluation be made with regard to the probability of a specific exposure affecting a specific asset. Only then can the control for safeguarding the asset be evaluated in a cost-effective manner.

The controls, as numerically listed in the cells of the matrix, are as follows:

1. Ensure that the system can switch messages destined for a down station/terminal to an alternate station/terminal.

2. Determine whether the system can perform message switching to transmit messages between stations/terminals.

3. In order to avoid lost messages in a message-switching system, provide a store and forward capability. This is where a message destined for a busy station is stored at the central switch and then forwarded at a later time when the station is no longer busy.

4. Review the message or transaction logging capabilities to reduce lost messages, provide for an audit trail, restrict messages, prohibit illegal mes-

sages, and the like. These messages might be logged at the remote station (intelligent terminal), they might be logged at a remote concentrator/remote front end processor, or they might be logged at the central front end communication processor/central computer.

5. Transmit messages promptly to reduce risk of loss.

6. Identify each message by the individual user's password, the terminal, and the individual message sequence number.

7. Acknowledge the successful or unsuccessful receipt of all messages.

8. Utilize physical security controls throughout the data communication network. This includes the use of locks, guards, badges, sensors, alarms, and administrative measures to protect the physical facilities, data communication networks, and related data communication equipment. These safeguards are required for access monitoring and control to protect data communication equipment and software from damage by accident, fire, and environmental hazard either intentional or unintentional.

9. Consider using modems that have either manual or remote actuated loopback switches for fault isolation to ensure the prompt identification of malfunctioning equipment. These are extremely important in order to increase the uptime and to identify faults.

10. Use front panel lights on modems to indicate if the circuit/line is functioning properly (carrier signal is up). This may not be a viable alternative with organizations that have hundreds of modems.

11. Consider a modem with alternative voice capabilities for quick troubleshooting between the central site and a major remote site.

12. When feasible, use digital data transmission, because it has a lower error rate than analog data transmission.

13. For data communication equipment, check the manufacturer's mean time between failures (MTBF) in order to ensure that the data communication equipment has the largest MTBF.

14. Consider placing unused backup modems in critical areas of the data communication network.

15. Consider using modems that have an automatic or semiautomatic dial backup capability in case the leased line fails.

16. Review the maintenance contract and mean time to fix (MTTF) for all data communication equipment. Maintenance should be both fast and available. Determine from where the maintenance is dispatched, and determine

if tests can be made from a remote site (for example, in many cases modems have remote loopback capabilities).

17. Increase data transmission efficiency. The faster the modem synchronization time, the lower will be the turnaround time and thus more throughput to the system.

18. Consider modems with automatic equalization (built-in microprocessors for circuit equalization and balancing) in order to compensate for amplitude and phase distortions on the line. This will reduce the number of errors in transmission and may decrease the need for conditioned lines.

19. With regard to the efficiency of modems, review to see if they have multiple-speed switches so the transmission rate can be lowered when the line error rates are high.

20. Utilize four-wire circuits in a pseudo-full duplex transmission mode. In other words, keep the carrier wave up in each direction on alternate pairs of wires in order to reduce turnaround time and gain efficiency during transmission.

21. If needed, use full duplex transmission on two-wire circuits with special modems that split the frequencies and thus achieve full duplex transmission.

22. Increase the speed of transmission. The faster the speed of transmission by the modem, the most cost-effective are the data communications, but error rates may increase with speed, and therefore you may need more error detection and correction facilities.

23. Utilize a reverse channel capability for control signals (supervisory) and to keep the carrier wave up in both directions.

24. Consider the following special controls on dial-up modems when the data communication network allows incoming dial-up connections: change the telephone numbers at regular intervals; keep the telephone numbers confidential; remove the telephone numbers from the modems in the computer operations area; require that each "dial-up terminal" have an electronic identification circuit chip to transmit its unique identification to the front end communication processor; do not allow automatic call receipt and connection (always have a person intercept the call and make a verbal identification); have the central site call the various terminals that will be allowed connection to the system; utilize dial-out only where an incoming dialed call triggers an automatic dial-back to the caller (in this way the central system controls those telephone numbers to which it will allow connection).

25. Physically trace out and, as best as possible, secure the local loop communication circuits/lines within the organization or facility. After these lines leave the facility and enter the public domain, they cannot be physically secured.

26. Consider conditioning the voice grade circuits in order to reduce the number of errors during transmission (this may be unnecessary with the newer microprocessor-based modems that perform automatic equalization and balancing).

27. Use four-wire circuits in such a fashion that there is little to no turnaround time. This can be done by using two wires in each direction and keeping the carrier signal up.

28. Within an organizational facility, fiber optics (laser) communication circuits can be used to totally preclude the possibility of wiretapping.

29. Ensure that there is adequate physical security at remote sites and especially for terminals, concentrators, multiplexers, and front end communication processors.

30. Determine whether the multiplexer/concentrator/remote front end hardware has redundant logic and backup power supplies with automatic fallback capabilities in case the hardware fails. This will increase the uptime of the many stations/terminals that might be connected to this equipment.

31. Consider logging inbound and outbound messages at the remote site.

32. Consider uninterruptible power supplies at large multiplexer/concentrator type remote sites.

33. Consider multiplexer/concentrator equipment that has diagnostic lights, diagnostic capabilities, and the like.

34. If a concentrator is being used, is it performing some of the controls that are usually performed by the front end communication processor and therefore increasing the efficiency and correctness of data transmissions?

35. See if the polling configuration list can be changed during the day in order to exclude or include specific terminals. This would allow the positive exclusion of a terminal as well as allowing various terminals to come on-line and off-line during the working day.

36. Can the front ends, concentrators, modems, and the like handle the automatic answering and automatic outward dialing of calls? This capability would increase the efficiency and accuracy when it is preprogrammed into the system.

37. Ensure that all inbound and outbound messages are logged by the central

processor, the front end, or remote concentrator in order to ensure against lost messages, to keep track of message sequence numbers (identify illegal messages), and to use for system restart should the entire system crash.

38. For efficiency, ensure that the central system can address either a group of terminals (group address), several terminals at a time (multiple address), one terminal at a time (single address), or send a broadcast message simultaneously to all stations/terminals in the system.

39. See that each inbound and outbound message is serial numbered as well as time and date stamped at the time of logging.

40. Ensure that there is a "time out" facility so the system does not get hung up trying to poll/address a station. Also, if a particular station "times out" four or five consecutive times, it should be removed from the network configuration polling list so time is not wasted on this station (improves communication efficiency).

41. Consider having concentrators and front ends perform two levels of editing. In the first level the front end may add items to a message, reroute the message, or rearrange the data for further transmission. It may also check a message address for accuracy and perform parity checks. In the second level of editing, the concentrator or front end is programmed to perform specific edits of the different transactions that enter the system. This editing is an application system type of editing and deals with message content rather than form and is specific to each application program being executed.

42. Have the concentrators, front ends, and central computers handle the message priority system, if one exists. A priority system is set up to permit a higher line utilization to certain areas of the network or to ensure that certain transactions are handled before other transactions of lesser importance.

43. See that the front end collects message traffic statistics and performs correlations of traffic density and circuit availability. These analyses are mandatory for the effective management of a large data communication network. Some of the items included in a traffic density report might be the number of messages handled per hour or per day on each link of the network, the number of errors encountered per hour or per day, the number of errors encountered per program or per program module, the terminals or stations that appear to have a higher than average error record, and the like.

44. Ensure that the front ends and concentrators can perform miscellaneous functions such as triggering remote alarms if certain parameters are ex-

ceeded, performing multiplexing operations internally, signaling abnormal occurrences to the central computer, slowing up input/output messages when the central computer is overburdened due to heavy traffic, and the like.

45. Ensure that the concentrators and front ends can validate electronic terminal identification.

46. Ensure that there is a message intercept function for inoperable terminals or invalid terminal addresses.

47. See that messages are checked for valid destination address.

48. Ensure adequate error detection and control capabilities. These might include echo-checking, where a message is transmitted to a remote site and the remote site echoes the message back for verification, or it might include forward error correction, where special hardware boxes can automatically correct some errors upon receipt of the message, or it might include detection with transmission. Detection with transmission is the most common and cost-effective form of error detection and correction. This may include identification of errors by reviewing the parity bit or utilizing a special code to identify errors in individual characters during transmission. A more prevalent form is to utilize a polynominal (mathematical algorithm) to detect errors in message blocks. Whichever way is used, when a message error is detected, it is retransmitted until it is received correctly.

49. When reviewing error detection in transmission, first determine whatever error rate can be tolerated, then determine the extent and pattern of errors on the communication links used by the organization, and then review the error detection and correction methodologies in use and determine if they are adequate for the application systems utilizing the data communication network. In other words, a purely administrative message network (no critical financial data) would not require error detection and correction capabilities equal to a network that transmits critical financial data.

50. Ensure that there are adequate restart and recovery software routines to recover from items such as a trapped machine check, where instead of bringing down the entire data communication system, a quick recovery can be made and only the one transaction need be retransmitted.

51. Ensure that there are adequate restart and recovery procedures to effect both a warm start and a cold start. In other words, a data communication system should never completely fail so the user has to perform a cold start (start up as if it is a new day, all message counters cleared). The system should go into a warm start procedure, where only parts of the system are

disabled and recovery can be made while the system is operating in a degraded mode.

52. Ensure that there is an audit trail logging facility to assist in the reconstruction of data files and the reconstruction of transactions from the various stations. There should be the capability to trace back to the terminal end user.

53. Provide some tables for checking for access by terminals, people, database, and programs. These tables should be in protected areas of memory.

54. Safe store all messages. All transactions/messages should be protected in case of a disastrous situation such as power failure.

55. Protect against concurrent file updates. If the data management software does not provide this protection, the data communication software should.

56. For convenience, flexibility, and security, ensure that terminals can be brought up or down dynamically while the system is running.

57. Make available a systems trace capability to assist in locating problems.

58. Ensure that the documentation of the system software is comprehensive.

59. Provide adequate maintenance for the software programs.

60. Ensure that the system supports password protection (multilevel password protection).

61. Identify all default options in the software and their impact if they do not operate properly.

62. For entering sensitive or critical systems commands, restrict these commands to one master input terminal and ensure strict physical custody over this terminal. In other words, restrict those personnel who can use this terminal.

63. Ensure that there are adequate recovery facilities and/or capabilities for a software failure, loss of key pieces of hardware, and loss of various communication circuit/lines.

64. Ensure that there are adequate backup facilities (local and remote) to back up key pieces of hardware and communication circuits/lines.

65. Consider backup power capabilities for large facilities such as the central site and various remote concentrators.

66. Consider installing the capabilities to fall back to the public dial network from a leased-line configuration.

67. When utilizing multidrop or loop circuits, review the uptime problems. These types of configurations are more cost-effective than point-to-point

configurations, but when there is a circuit failure close to the central site, all terminals/stations downline are disconnected.

68. Review the physical security (local and remote) for circuits/lines (especially the local loop), hardware/software, physical facilities, storage media, and the like.

69. For personnel that work in critical or sensitive areas, consider enforcing the following policies: insist that they take at least five consecutive days of vacation per year, check with their previous employers, perform an annual credit check, and have them sign hiring agreements stating that they will not sell programs, etc.

70. With regard to data security, consider encrypting all messages transmitted.

71. Develop an overall organizational security policy for the data communication network. This policy should specifically cover the security and privacy of information.

72. Ensure that all sensitive communication programs and data are stored in protected areas of memory or disk storage.

73. Ensure that all communication programs or data, when they are off-line, are stored in areas with adequate physical security.

74. Ensure that all communication programs and data are adequately controlled when they are transferred to microfiche.

75. Lock up telephone equipment rooms and install alarms on the doors of those telephone equipment rooms that contain the basic data communication circuits.

76. Do not put communication lines through the public switchboard unless it is a new electronic switchboard (ESS) and the intent is to gain verbal identification of incoming dial-up data communication calls.

77. Review the communication system's console log that shows "network supervisor terminal commands" such as: disable or enable a line or station for input or output, alternately route traffic from one station to another, change the order and/or frequency of line or terminal service (polling, calling, dialing), and the like.

78. Consider packet switching networks which use alternate routes for different packets of information from the same message; this would offer a form of security in case someone were intercepting messages.

79. Ensure that there is a policy for the use of test equipment. Modern test equipment may offer a new vulnerability to the organization. This test equipment is easily connected to communication lines, and all messages

can be read in clear "English language." Test equipment should not be used for monitoring lines "for fun"; it should be locked up (key lock or locked hood) when it is not in use and after normal working hours when it is not needed for testing and debugging; programs written for programmable test equipment should be kept locked up and out of the hands of those who do not need these programs.

80. Review the operational procedures, for example, the administrative regulations, policies, and day-to-day activities supporting the security/safeguards of the data communication network. These procedures may include:

    - Specifying the objectives of the EDP security for an organization, especially as they relate to data communications.
    - Planning for contingencies of security "events," including recording of all exception conditions and activities.
    - Assuring management that other safeguards are implemented, maintained, and audited, including background checks, security clearances and hiring of people with adequate security-oriented characteristics; separation of duties; mandatory vacations.
    - Developing effective safeguards for deterring, detecting, preventing, and correcting undesirable security events.
    - Reviewing the cost effectiveness of the system and the related benefits such as better efficiency, improved reliability, and economy.
    - Looking for the existence of current administrative regulations, security plans, contingency plans, risk analysis, personnel understanding of management objectives, and then reviewing the adequacy and timeliness of the specified procedures in satisfying these.

81. Review the preventive maintenance and scheduled diagnostic testing such as cleaning, replacement, and inspection of equipment to evaluate its accuracy, reliability, and integrity. This may include schedules for testing and repair, adequate testing of software program changes submitted by the vendor, inventories of replacement parts (circuit boards), past maintenance records, and the like.

82. Determine whether there is a central site for reporting all problems encountered in the data communication network. This usually results in faster repair time.

83. Review the financial protection afforded from insurance for various hardware, software, and data stored on magnetic media.

84. Review the legal contracts with regard to the agreements for performing a specific service and a specific costing basis for the data communication

network hardware and software. These might include bonding of employees, conflict of interest agreements, clearances, nondisclosure agreements, agreements establishing liability for specific security events by vendors, agreements by vendors not to perform certain acts that would incur a penalty, and the like.

85. Review the organization's fault isolation/diagnostics, including the techniques used to ascertain the integrity of the various hardware/software components comprising the total data communication entity. These techniques are used to audit, review, and control the total data communication environment and to isolate the offending elements either on a periodic basis or upon detection of a failure. These techniques may include diagnostic software routines, electrical loopback, test message generation, administrative and personnel procedures, and the like.

86. Review the training and education of employees with regard to the data communication network. Employees must be adequately trained in this area because of the high technical competence required for data communication networks.

87. Ensure that there is adequate documentation, including a precise description of programs, hardware, system configurations, and procedures intended to assist in the prevention of problems, identification of problems, and recovery from problems. The documentation should be sufficiently detailed to assist in reconstructing the system from its parts.

88. Review the techniques used for testing to validate the hardware and software operation to ensure integrity. Testing, including that of personnel, should uncover departures from the specified operation.

89. Review error recording to reduce lost messages. All errors in transmission of messages in the system should be logged and this log should include the type of error, the time and date, the terminal operator, and the number of times the message was retransmitted before it was correctly received.

90. Review the error correction procedures. A user's manual should specify a cross-reference of error messages to the appropriate error code generated by the system. These messages help the user interpret the error that has occurred and suggest the corrective action to be taken. Ensure that the errors are in fact corrected and the correct data reentered into the system.

91. Consider backing up key circuits/lines. This circuit backup may take the form of a second leased line, modems that have the ability to go to the public dial-up network when a leased line fails, or manual procedures where the remote stations can transmit verbal messages using the public dial-up network.

APPENDIX 2

CONTROL LISTS FOR DATA COMMUNICATION NETWORKS

SOFTWARE CONTROLS, DATA COMMUNICATION

The software programs that operate the data communication network portion of the system may reside in the central computer, a distributed-system computer, the front end communication processor, a remote concentrator or statistical multiplexer, and/or a remote intelligent terminal. This software is concerned specifically with the telecommunication access methods, the telecommunication monitors that may oversee the entire data communication function, any front end communication processor software, or programs that reside in the intelligent terminals. Front end software might be located remotely with regard to the central communication center. This software review also may involve data communication software located at remote concentrator sites or the data communication software located at remote intelligent terminal devices.

Source: The following 14 lists of controls were excerpted from the book, *Designing Controls into Computerized Systems*, by Jerry FitzGerald. This book contains 101 different lists enumerating 2500 controls for computerized systems. It is available from Jerry FitzGerald & Associates, 506 Barkentine Lane, Redwood City, CA 94065.

1. Ensure that the system can properly switch messages destined for a down station/terminal to an alternate station/terminal.

2. In order to avoid lost messages in a message-switching system, provide a store and forward capability, where a message destined for a busy station is stored at the central switch and then forwarded at a later time when the station is no longer busy.

3. Review the message or transaction logging capabilities to reduce lost messages, provide for an audit trail, restrict messages, prohibit illegal messages, and the like. These messages might be logged at the remote station (intelligent terminal), they might be logged at a remote concentrator/remote front end processor, or they might be logged at the central front end communication processor/central computer.

4. Identify each message by the individual user's password, the terminal, and the individual message sequence number.

5. Acknowledge the successful or unsuccessful receipt of all messages.

6. See if the polling configuration list can be changed during the day in order to exclude or include specific terminals. This would allow the positive exclusion of a terminal as well as allowing various terminals to come on-line and off-line during the working day.

7. Consider having concentrators and front ends perform two levels of editing. In the first level, the front end may add to a message, reroute the message, or rearrange the data for further transmission. It also may check a message address for accuracy and perform parity checks. In the second level of editing, the concentrator or front end is programmed to perform specific edits to the different transactions that enter the system. This editing is an application system type of editing and deals with message content rather than form. It is specific to each application program being executed.

8. See that messages are checked for valid destination address.

9. Ensure adequate error detection and control capabilities. These might include echo-checking, where a message is transmitted to a remote site and the remote site echoes the message back for verification, or it might include forward error correction, where special hardware boxes can correct automatically some errors upon receipt of the message, or it might include detection with retransmission. Detection with retransmission is the most common and cost-effective form of error detection and correction. This may include identification of errors by reviewing the parity bit or utilizing a special code to identify errors in individual characters during transmis-

sion. A more prevalent form is to utilize a polynomial (mathematical algorithm) to detect errors in message blocks. Whichever way is used, when a message error is detected, it is retransmitted until is received correctly.

10. Ensure that there are adequate restart and recovery software routines to recover from items such as a trapped machine check, where instead of bringing down the entire data communication system, a quick recovery can be made and only the one transaction need be retransmitted.

11. Ensure that there are adequate restart and recovery procedures to effect both a warm start and a cold start. In other words, a data communication system never should fail completely so the user has to perform a cold start (start up as if it is a new day, all message counters cleared). The system should go into a warm start procedure, where only parts of the system are disabled and recovery can be made while the system is operating in a degraded mode.

12. Ensure that there is an audit trail logging facility to assist in the reconstruction of data files and the reconstruction of transactions from the various stations. There should be the capability to trace back to the terminal end user.

13. Provide some tables for checking for access by terminals, people, database, and programs. These tables should be in protected areas of memory.

14. Provide adequate maintenance for the software programs.

15. Identify all default options in the software and their impact if they do not operate properly.

16. Ensure that all sensitive communication programs and data are stored in protected areas of memory or disk storage.

17. Review the techniques for testing used to validate the hardware and software operation to ensure integrity. Testing, including that of personnel, should reveal departures from the specified operation.

18. Review error recording to reduce lost messages. All errors in transmission of messages in the system should be logged. This log should include the type of error, the time and date, the terminal, the circuit, the terminal operator, and the number of times the message was retransmitted before it was received correctly.

19. Maintain a checksum count of the "bits" in the software packages. In this way, a quick check can be made to see if there are the same number of bits. If it tallies, the organization probably can rest assured that there have been no software program modifications.

20. When feasible, conduct either source code comparisons or object code comparisons (some organizations have conducted source-to-object code comparisons). This is to determine if there have been any changes since the last source or object code comparison. This control is very time-consuming and expensive because it involves the validation of a specific program on a line-by-line basis. That same program is compared, at some future time, to the validated version.

21. When sensitive software is utilized at distributed sites, consider downline loading that software from the central site. This would provide the assurance that no illegal program changes have been made at the remote site. Also, new programs could be downline loaded every time a vendor conducted maintenance.

22. Utilize generalized audit software to review various functions of the systems software packages. Distribute these generalized audit software packages to personnel at remote sites. At the central site the auditors or system designers would conduct this function.

23. Regularly review the logs of system restarts and accountings of rerun time caused by system malfunctions.

24. Ensure that there is a trouble log regarding software. It should contain the diagnosis of each problem and the person, software component, or device that caused the malfunction. Consider developing statistical reports from these logs and initiate appropriate actions if patterns emerge. Each malfunction should be isolated.

25. Ensure that all security features that were built into any of the system software packages have been considered. If they are not being used, determine the reason or reasons why.

26. Determine whether there are cleanly programmed and well-defined interfaces between any system software packages such as between operating systems, data communication software, distributed intelligence software, database management systems, and the like.

27. Determine whether the system software programmers have enumerated all the known loopholes in any of the software. They must ascertain the degree of exposure attributable to each loophole and make possible corrections.

28. If the system is running any type of queuing system, such as paging or data communication input/output transactions, review the queues, space management, and other dynamic allocation spaces in order to ensure that a user

cannot get out of his or her address space and violate another user space or the operating system.

29. Force the queuing system to fail to determine if it leaves sensitive information spread throughout the computer system.

30. Following a system catastrophe, ensure that a terminal not logged on before the catastrophe cannot get logged on following the catastrophe without the full authentication sequence.

31. Should a communication circuit fail, ensure that the communication software does not give that open port to the next terminal signing onto the system.

32. Ensure that there are adequate maintenance and vendor support for all system software.

33. Check remote access. Terminal access to a system introduces a new dimension to system software security problems. While these may not be related directly to software security, they do pose the problem that software has the potential of being attacked from a long distance away from the physical area where the software resides. A variety of possible penetration paths may exist so the software might check the various circuits to ensure that they are the proper circuits and that the proper input/output transactions are being entered on these circuits; consider encryption of the data; limit the use of dial-up modems when they can be connected to highly sensitive and secure software systems; ensure that, when a communication circuit drops, there is an absolute physical dropping of the computer software so the next person dialing into the system will not be connected to the software or to another user's program; utilize passwords and various log-in codes, etc.

34. Examine checkpoint/restart modifications. In this case, the penetrator takes a checkpoint of his or her program during its execution and subsequently operates upon the checkpoint file as if it were a regular data file. Since for restart purposes the system must include all status and system registers in the checkpoint file, the penetrator can change effectively the contents of status or system registers (which could not otherwise be modified) by appropriately manipulating the checkpoint file. Then by restarting the program from the modified checkpoint file, the penetrator can execute in a supervisory mode or whatever status desired. Log whenever a restart is executed or status/systems registers are dumped.

35. Check removal or addition of software code. However achieved, the removal or addition of code from the software could pose a security threat.

COPYRIGHT © 1981 JERRY FITZGERALD

The use of a checksum bit count with regard to sensitive software packages might be an appropriate measure to ensure against unauthorized program changes.

36. Look for exploitable logic errors. In any major software system there are, at any point in time, some "bugs" or errors. Some of these may be documented but not yet corrected. A logic error may be exploited by a penetrator in order to compromise the integrity of the software. Logic errors should be evaluated with regard to their potential security risk if they must remain uncorrected for any period of time.

37. Check software generation options. Vendor software packages usually contain many options that can be called upon when generating the software system. All the security options must be reviewed, evaluated, and a positive decision made whenever one of them is not to be used to its fullest extent. Also, any options that are left to the default state should be tested. In other words, the default mechanism should be tested in order to ensure that it is operating correctly and not branching to some protected or restricted area of memory or someone else's program space.

38. Check interrupt handling. During the handling of interrupt, various parameters are stored. Ensure that these parameters are stored in protected memory space, protected registers, or other protected areas.

DISASTERS AND DISRUPTIONS, DATA COMMUNICATION

Disasters or disruptions refer to either natural or human-created disruption of normal data communication capabilities. They could be either temporary or long-term. They would render the organization's normal data communication on-line system inoperable.

1. Ensure that the system can switch messages destined for a down station/terminal to an alternate station/terminal.

2. Utilize physical security controls throughout the data communication network. This includes the use of locks, guards, badges, sensors, alarms, and administrative measures to protect the physical facilities, data communication networks, and related data communication equipment. These safeguards are required for access monitoring and control to protect data communication equipment and software from damage by accident, fire, and environmental hazard either intentional or unintentional.

3. Consider using modems that have either manual or remote actuated loop-back switches for fault isolation to ensure the prompt identification of malfunctioning equipment. These are extremely important in order to increase the uptime and to identify faults.

4. Use front panel lights on modems to indicate if the circuit/line is functioning properly (carrier signal is up). This may not be a viable alternative with organizations that have hundreds of modems.

5. Consider a modem with alternate voice capabilities for quick troubleshooting between the central site and a major remote site.

6. For data communication equipment, check the manufacturer's mean time between failures (MTBF) in order to ensure that the data communication equipment has the largest MTBF.

7. Consider placing unused backup modems in critical areas of the data communication network.

8. Consider using modems that have an automatic or semi-automatic dial backup capability in case the leased line fails.

9. Review the maintenance contract and mean time to fix (MTTF) for all data communication equipment. Maintenance should be both fast and available. Determine from where the maintenance is dispatched, and determine if tests can be made from a remote site (for example, in many cases modems have remote loopback capabilities).

10. Consider modems with automatic equalization (built-in microprocessors for circuit equalization and balancing) in order to compensate for amplitude and phase distortions on the line. This will reduce the number of errors in transmission and may decrease the need for conditioned lines.

11. Physically trace out and, as best as possible, secure the local loop communication circuits/lines within the organization or facility. After these lines leave the facility and enter the public domain, they cannot be physically secured.

12. Consider conditioning the voice grade circuits in order to reduce the number of errors during transmission (this may be unnecessary with the newer microprocessor-based modems that perform automatic equalization and balancing).

13. Ensure that there is adequate physical security at remote sites and especially for terminals, concentrators, multiplexers, and front end communication processors.

14. Determine whether the multiplexer/concentrator/remote front end hardware has redundant logic and backup power supplies with automatic fallback capabilities in case the hardware fails. This will increase the uptime of the many stations/terminals that might be connected to this equipment.

15. Consider uninterruptible power supplies at large multiplexer/concentrator type remote sites.

16. Consider multiplexer/concentrator equipment that has diagnostic lights, diagnostic capabilities, and the like.

17. Ensure that the front ends and concentrators can perform miscellaneous functions such as triggering remote alarms if certain parameters are exceeded, performing multiplexing operations internally, signaling abnormal occurrences to the central computer, slowing up input/output messages when the central computer is overburdened because of heavy traffic, and the like.

18. Review the organization's fault isolation/diagnostics, including the techniques used to ascertain the integrity of the various hardware/software components comprising the total data communication entity. These techniques are used to audit, review, and control the total data communication environment and to isolate the offending elements either on a periodic basis or upon detection of a failure. These techniques may include diagnostic software routines, electrical loopback, test message generation, administrative and personnel procedures, and the like.

19. Review the training and education of employees with regard to the data communication network. Employees must be trained adequately in this area because of the high technical competence required for data communication networks.

20. Ensure that there is adequate documentation, including a precise description of programs, hardware, system configuration, and procedures intended to assist in the prevention of problems, identification of problems, and recovery of problems. The documentation should be sufficiently detailed to assist in reconstructing the system from its parts.

21. Ensure that there are adequate restart and recovery software routines to recover from items such as a trapped machine check, where instead of bringing down the entire data communication system, a quick recovery can be made and only the one transaction need be retransmitted.

22. Ensure that there are adequate restart and recovery procedures to effect both a warm start and a cold start. In other words, a data communication system should never completely fail so the user has to perform a cold start

(start up as if it is a new day, all message counters cleared). The system should go into a warm start procedure, where only parts of the system are disabled and recovery can be made while the system is operating in a degraded mode.

23. Ensure that there is an audit trail logging facility to assist in the reconstruction of data files and the reconstruction of transactions from the various stations. There should be the capability to trace back to the terminal end user.

24. Safe store all messages. All transactions/messages should be protected in case of a disastrous situation such as power failure.

25. Make available a systems trace capability to assist in locating problems.

26. Ensure that there are adequate recovery facilities and/or capabilities for a software failure, loss of key pieces of hardware, and loss of various communication circuit/lines.

27. Ensure that there are adequate backup facilities (local and remote) to back up key pieces of hardware and communication circuits/lines.

28. Consider backup power capabilities for large facilities such as the central site and various remote concentrators.

29. Consider installing the capabilities to fall back to the public dial network from a leased-line configuration.

30. Lock up telephone equipment rooms and install alarms on the doors of those telephone equipment rooms that contain the basic data communication circuits.

31. Do not put communication lines through the public switchboard unless it is a new electronic switchboard (EES) and the intent is to gain verbal identification of incoming dial-up data communication calls.

32. Protect all electrical circuits from malicious vandalism where someone might open the circuits and cut the power. This means providing locking circuit-control boxes and locating circuit-control boxes in locked rooms.

33. Consider backing up key circuits/lines. This circuit backup may take the form of a second leased line, modems that have the ability to go to the public dial-up network when a leased line fails, or manual procedures whereby the remote stations can transmit verbal messages using the public dial-up network.

34. Review the operational procedures, for example, the administrative regulations, policies, and day-to-day activities supporting the security/safeguards of the data communication network. These procedures may include:

- Specifying the objectives of the EDP security for an organization, especially as they relate to data communications.
- Planning for contingencies of security "events," including recording of all exception conditions and activities.
- Assuring management that other safeguards are implemented, maintained, and audited, including background checks, security clearances and hiring of people with adequate security-oriented characteristics; separation of duties; mandatory vacations.
- Developing effective safeguards for deterring, detecting, preventing, and correcting undesirable security events.
- Reviewing the cost effectiveness of the system and the related benefits such as better efficiency, improved reliability, and economy.
- Looking for the existence of current administrative regulations, security plans, contingency plans, risk analysis, and personnel understanding of management objectives, and then reviewing the adequacy and timeliness of the specified procedures in satisfying these.

35. Review the preventive maintenance and scheduled diagnostic testing such as cleaning, replacement, and inspection of equipment to evaluate its accuracy, reliability, and integrity. This may include schedules for testing and repair, adequate testing of software program changes submitted by the vendor, inventories of replacement parts (circuit boards), past maintenance records, and the like.

36. Determine whether there is a central site for reporting all problems encountered in the data communication network. This usually results in faster repair time.

MODEMS

A modem is a hardware device used for the conversion of data signals from terminals (digital signal) to a form that is acceptable for transmission over the communication circuits that are owned and maintained by the telephone company or other special common carrier.

1. Consider using modems that have either manual or remote actuated loopback switches for fault isolation to ensure the prompt identification of malfunctioning equipment. These are extremely important in order to increase the uptime and to identify faults.

2. Use front panel lights on modems to indicate if the circuit/line is functioning properly (carrier signal is up). This may not be a viable alternative with organizations that have hundreds of modems.

3. Consider a modem with alternate voice capabilities for quick troubleshooting between the central site and a major remote site.

4. When feasible, use digital data transmission, because it has a lower error rate than analog data transmission.

5. For data communication equipment, check the manufacturer's mean time between failures (MTBF) in order to ensure that the data communication equipment has the largest MTBF.

6. Consider placing unused backup modems in critical areas of the data communication network.

7. Consider using modems that have an automatic or semi-automatic dial backup capability in case the leased line fails.

8. Review the maintenance contract and mean time to fix (MTTF) for all data communication equipment. Maintenance should be both fast and available. Determine from where the maintenance is dispatched, and if tests can be made from a remote site (for example, in many cases modems have remote loopback capabilities).

9. Increase data transmission efficiently. The faster the modem synchronization time, the lower will be the turnaround time and, thus, more throughput to the system.

10. Consider modems with automatic equalization (built-in microprocessors for circuit equalization and balancing) in order to compensate for amplitude and phase distortion on the line. They will reduce the number of errors in transmission and may decrease the need for conditioned lines.

11. With regard to the efficiency of modems, review to see if they have multiple-speed switches so the transmission rate can be lowered when the line error rates are high.

12. Utilize four-wire circuits in a pseudo-full duplex transmission mode. In other words, keep the carrier wave up in each direction on alternate pairs of wires in order to reduce turnaround time and gain efficiency during transmission.

13. If needed, use full duplex transmission on two-wire circuits with special modems that split the frequencies and thus achieve full duplex transmission.

14. Increase the speed of transmission. The faster the speed of transmission by the modem, the more cost-effective are the data communications. Since error rates may increase with speed, you may need more error detection and correction facilities.

15. Utilize a reverse channel capability for control signals (supervisory) and to keep the carrier wave up in both directions.

16. Consider the following special controls on dial-up modems when the data communication network allows incoming dial-up connections: change the telephone numbers at regular intervals; keep the telephone numbers confidential; remove the telephone numbers from the modems in the computer operations area; require that each "dial-up terminal" have an electronic identification circuit chip to transmit its unique identification to the front end communication processor; do not allow automatic call receipt and connection (always have a person intercept the call and make a verbal identification); have the central site call the various terminals that will be allowed connection to the system; utilize dial-out only where an incoming dialed call triggers an automatic dial-back to the caller (in this way the central system controls those telephone numbers to which it will allow connection).

17. Ensure that there is adequate physical security at remote sites.

18. Ensure that the front ends, concentrators, modems, and the like can handle the automatic answering and automatic outward dialing of calls. This would increase efficiency and accuracy when it is preprogrammed into the system.

19. Review the techniques for testing used to validate the hardware operation to ensure integrity. Testing, including that of personnel, should uncover departures from the specified operation.

MULTIPLEXER, CONCENTRATOR, SWITCH

These three hardware devices enable the data communication network to operate in the most efficient manner. The *multiplexer* is a device that combines, in one data stream, several simultaneous data signals from independent stations. The *concentrator* performs the same functions as a multiplexer except it is intelligent and therefore can perform some of the functions of a front end communication processor. A *switch* is a device that allows the interconnection between any two circuits (lines) connected to the switch. There might be two distinct types of switch: a switch that performs message switching between stations (terminals) might be located within the data communication network facilities that are owned and operated by the organization; a circuit or line switching switch that interconnects various circuits might be located at (and owned by) the telephone company central office. For example, organizations perform message switching and the telephone company performs circuit switching.

1. Ensure that the system can switch messages destined for a down station/ terminal to an alternate station/terminal.

2. Determine whether the system can perform message switching to transmit messages between stations/terminals.

3. In order to avoid lost messages in a message-switching system, provide a store and forward capability. This is where a message destined for a busy station is stored at the central switch and then forwarded at a later time when the station is no longer busy.

4. Review the message- or transaction-logging capabilities to reduce lost messages, provide for an audit trail, restrict messages, prohibit illegal messages, and the like. These messages might be logged at the remote station (intelligent terminal), they might be logged at a remote concentrator/remote front end processor, or they might be logged at the central front end communication processor/central computer.

5. Transmit messages promptly to reduce risk of loss.

6. Identify each message by the individual user's password, the terminal, and the individual message sequence number.

7. Acknowledge the successful or unsuccessful receipt of all messages.

8. Utilize physical security controls throughout the data communication network. This includes the use of locks, guards, badges, sensors, alarms, and administrative measures to protect the physical facilities, data communication networks, and related data communication equipment. These safeguards are required for access monitoring and control to protect data communication equipment and software from damage by accident, fire, or environmental hazard, either intentional or unintentional.

9. For data communication equipment, check the manufacturer's mean time between failures (MTBF) in order to ensure that the data communication equipment has the largest MTBF.

10. Review the maintenance contract and mean time to fix (MTTF) for all data communication equipment. Maintenance should be both fast and availa-

ble. Determine from where the maintenance is dispatched, and if tests can be made from a remote site (for example, in many cases modems have remote loopback capabilities).

11. Ensure that the front ends, concentrators, modems, and the like handle the automatic answering and automatic outward dialing of calls. This would increase the efficiency and accuracy when it is preprogrammed into the system.

12. Ensure that all inbound and outbound messages are logged by the central processor, the front end, or remote concentrator in order to ensure against lost messages, to keep track of message sequence numbers (identify illegal messages), and to use for system restart should the entire system crash.

13. Ensure that there is adequate physical security at remote sites, especially for terminals, concentrators, multiplexers, and front end communication processors.

14. Determine whether the multiplexer/concentrator/remote front end hardware has redundant logic and backup power supplies with automatic fallback capabilities in case the hardware fails. This will increase the uptime of the many stations/terminals that might be connected to this equipment.

15. Consider uninterruptible power supplies at large multiplex/concentrator type remote sites.

16. Consider multiplex/concentrator equipment that has diagnostic lights, diagnostic capabilities, and the like.

17. If a concentrator is being used, is it performing some of the controls that usually are performed by the front end communication processor, therefore increasing the efficiency and correctness of data transmission?

18. Consider having concentrators and front ends perform two levels of editing. In the first level, the front end may add items to a message, reroute the message, or rearrange the data for further transmission. It also may check a message address for accuracy and perform parity checks. In the second level of editing, the concentrator or front end is programmed to perform specific edits of the different transactions that enter the system. This editing is an application system type of editing and deals with message content rather than form. It is specific to each application program being executed.

19. Have the concentrators, front ends, and central computers handle the message priority system, if one exists. A priority system is set up to permit a higher line utilization to certain areas of the network or to ensure that certain transactions are handled before other transactions of lesser importance.

20. Ensure that the front ends and concentrators can perform miscellaneous functions such as triggering remote alarms if certain parameters are exceeded, performing multiplexing operations internally, signaling abnormal occurrences to the central computer, slowing up input/output messages when the central computer is overburdened because of heavy traffic, and the like.

21. Ensure that the concentrators and front ends can validate electronic terminal identification.

22. See that messages are checked for valid destination address.

23. Ensure adequate error detection and control capabilities. These might include echo-checking, where a message is transmitted to a remote site and the remote site echoes the message back for verification; it might include forward error correction, where special hardware boxes automatically correct some errors upon receipt of the message; or it might include detection with retransmission. Detection with retransmission is the most common and cost-effective form of error detection and correction. This may include identification of errors by reviewing the parity bit or utilizing a special code to identify errors in individual characters during transmission. A more prevalent form is to utilize a polynomial (mathematical algorithm) to detect errors in message blocks. Whichever way is used, when a message error is detected, it is retransmitted until it is received correctly.

24. Safe store all messages. All transactions/messages should be protected in case of a disastrous situation such as power failure.

25. Ensure that there are adequate backup facilities (local and remote) to back up key pieces of hardware and communication circuits/lines.

26. Consider backup power capabilities for large facilities such as the central site and various remote concentrators.

27. Review the physical security (local and remote) for circuits/lines (especially the local loop), hardware/software, physical facilities, storage media, and the like.

28. Ensure that all sensitive communication programs and data are stored in protected areas of memory or disk storage.

29. Ensure that there is a policy for the use of test equipment. Modern test equipment may offer a new vulnerability to the organization. This test equipment is connected easily to communication lines, and all messages can be read in clear "English language." Test equipment should not be used for monitoring lines "for fun"; it should be locked up (key lock or locked hood) when it is not in use and after normal working hours when it is not

needed for testing and debugging. Programs written for programmable test equipment should be kept locked up and out of the hands of those who do not need these programs.

30. Review the operational procedures, for example, the administrative regulations, policies, and day-to-day activities supporting the security/safeguards of the remote site. These procedures may include:

    - Specifying the objectives of the EDP security for an organization, especially as they relate to data communications.
    - Planning for contingencies of security "events," including recording of all exception conditions and activities.
    - Assuring management that other safeguards are implemented, maintained, and audited, including background checks, security clearances and hiring of people with adequate security-oriented characteristics; separation of duties; mandatory vacations.
    - Developing effective safeguards for deterring, detecting, preventing, and correcting undesirable security events.
    - Looking for the existence of current administrative regulations, security plans, contingency plans, risk analysis, and personnel understanding of management objectives, and then reviewing the adequacy and timeliness of the specified procedures in satisfying these requirements.

31. Review the preventive maintenance and scheduled diagnostic testing such as cleaning, replacement, and inspection of equipment to evaluate its accuracy, reliability, and integrity. This may include schedules for testing and repair, adequate testing of software program changes submitted by the vendor, inventories of replacement parts (circuit boards), past maintenance records, and the like.

32. Review the organization's fault isolation/diagnostics, including the techniques used to ascertain the integrity of the various hardware/software components comprising the total data communication entity. These techniques are used to audit, review, and control the total data communication environment and to isolate the offending elements either on a periodic basis or upon detection of a failure. These techniques may include diagnostic software routines, electrical loopback, test message generation, administrative and personnel procedures, and the like.

33. Review the techniques used for testing to validate the hardware and software operation to ensure integrity. Testing, including that of personnel, should reveal departures from the specified operation.

34. Review error recording to reduce lost messages. All errors in transmission of messages in the system should be logged. This log should include the type of error, the time and date, the terminal, the circuit, the terminal operator, and the number of times the message was retransmitted before it was received correctly.

COMMUNICATION CIRCUITS (LINES)

The common carrier facilities are used as links (a link is the interconnection of any two stations/terminals) to interconnect the organization's stations/terminals. These communication circuits include, not to the exclusion of others, satellite facilities, public switched dial-up facilities, point-to-point private lines, multiplexed lines, multipoint or loop configured private lines, WATS service, and many others.

1. When feasible, use digital data transmission, because it has a lower error rate than analog data transmission.

2. Review the maintenance contract and mean time to fix (MTTF) for all data communication equipment. Maintenance should be both fast and available. Determine from where the maintenance is dispatched. Also determine if tests can be made from a remote site (for example, in many cases modems have remote loopback capabilities).

3. Consider backing up key circuits/lines. This circuit backup may take the form of a second leased line, modems that have the ability to go to the public dial-up network when a leased line fails, or manual procedures where the remote stations can transmit verbal messages using the public dial-up network.

4. Utilize four-wire circuits in a pseudo-full duplex transmission mode. In other words, keep the carrier wave up in each direction on alternate pairs of wires in order to reduce turnaround time and gain efficiency during transmission.

5. If needed, use full duplex transmission on two-wire circuits with special modems that split the frequencies and thus achieve full duplex transmission.

6. Utilize a reverse channel capability for control signals (supervisory) and to keep the carrier wave up in both directions.

7. Physically trace out and, as best as possible, secure the local loop communication circuits/lines within the organization or facility. After these lines leave the facility and enter the public domain, they cannot be physically secured.

8. Consider conditioning the voice grade circuits in order to reduce the number of errors during transmission. This may be unnecessary with the newer microprocessor-based modems that perform automatic equalization and balancing.

9. Within an organizational facility, fiber optics (laser) communication circuits can be used to totally preclude the possibility of wiretapping.

10. Ensure that there are adequate recovery facilities and/or capabilities for a loss of various communication circuit/lines.

11. Consider installing the capabilities to fall back to the public dial network from a leased-line configuration.

12. When utilizing multidrop or loop circuits, review the uptime problems. These types of configurations are more cost-effective than point-to-point configurations, but when there is a circuit failure close to the central site, all terminals/stations downline are disconnected.

13. Review the physical security (local and remote) for circuits/lines (especially the local loop), hardware/software, physical facilities, storage media, and the like.

14. With regard to data security, consider encrypting all messages transmitted.

15. Lock up telephone equipment rooms and install alarms on the doors of those telephone equipment rooms that contain the basic data communication circuits.

16. Do not put communication lines through the public switchboard unless it is a new electronic switchboard (ESS) and the intent is to gain verbal identification of incoming dial-up communication calls.

17. Consider packet switching networks which use alternate routes for different packets of information from the same message. This would offer a form of security in case someone were intercepting messages.

18. Ensure that there is a policy for the use of test equipment. Modern test equipment may offer a new vulnerability to the organization. This test equipment is connected easily to communication lines, and all messages can be read in clear "English language." Test equipment should not be used for monitoring lines "for fun"; it should be locked up (key lock or locked hood) when it is not in use and after normal working hours when it is not

needed for testing and debugging. Programs written for programmable test equipment should be kept locked up and out of the hands of those who do not need these programs.

19. Review the operational procedures, for example, the administrative regulations, policies, and day-to-day activities supporting the security/safeguards of the data communication network. These procedures may include:

    - Specifying the objectives of the EDP security for an organization, especially as they relate to data communications.

    - Planning for contingencies of security "events," including recording of all exception conditions and activities.

    - Assuring management that other safeguards are implemented, maintained, and audited, including background checks, security clearances and hiring of people with adequate security-oriented characteristics; separation of duties; mandatory vacations.

    - Developing effective safeguards for deterring, detecting, preventing, and correcting undesirable security events.

    - Reviewing the cost effectiveness of the system and the related benefits such as better efficiency, improved reliability, and economy.

    - Looking for the existence of current administrative regulations, security plans, contingency plans, risk analysis, and personnel understanding of management objectives, and then reviewing the adequacy and timeliness of the specified procedures in satisfying these requirements.

ERROR HANDLING, DATA COMMUNICATION

Error handling refers to the methodologies and controls for handling errors at a remote distributed site or at the centralized computer site. This also may involve the error-handling procedures of a distributed data processing system at the distributed site. It must ensure that when errors are discovered, they are corrected promptly and reentered into the system for processing.

1. In order to avoid lost messages in a message-switching system, provide a store and forward capability. This is where a message destined for a busy station is stored at the central switch and then forwarded at a later time when the station is no longer busy.

2. Review the message- or transaction-logging capabilities to reduce lost messages, provide for an audit trail, restrict messages, prohibit illegal mes-

sages, and the like. These messages might be logged at the remote station (intelligent terminal), they might be logged at a remote concentrator/remote front end processor, or they might be logged at the central front end communication processor/central computer.

3. Ensure that all inbound and outbound messages are logged by the central processor, the front end, or remote concentrator in order to ensure against lost messages, to keep track of message sequence numbers (identify illegal messages), and to use for system restart should the entire system crash.

4. See that the front end collects message traffic statistics and performs correlations of traffic density and circuit availability. These analyses are mandatory for the effective management of a large data communication network. Some of the items included in a traffic density report might be the number of messages handled per hour or per day on each link of the network, the number of errors encountered per hour or per day, the number of errors encountered per program or per program module, the terminals or stations that appear to have a higher than average error record, and the like.

5. Ensure adequate error detection and control capabilities. These might include echo-checking, where a message is transmitted to a remote site and the remote site echoes the message back for verification, or it might include forward error correction, where special hardware boxes automatically correct some errors upon receipt of the message, or it might include detection with retransmission. Detection with retransmission is the most common and cost-effective form of error detection and correction. This may include identification of errors by reviewing the parity bit or utilizing a special code to identify errors in individual characters during transmission. A more prevalent form is to utilize a polynomial (mathematical algorithm) to detect errors in message blocks. Whichever way is used, when a message error is detected, it is retransmitted until it is received correctly.

6. When reviewing error detection in transmission, first determine whatever error rate can be tolerated, then determine the extent and pattern of errors on the communication links used by the organization. Finally, review the error detection and correction methodologies in use and determine if they are adequate for the application systems utilizing the data communication network. In other words, a purely administrative message network (no critical financial data) would not require error detection and correction capabilities equal to those for a network that transmits critical financial data.

7. Identify all default options in the software and their impact if they do not operate properly.

8. Review the organization's fault isolation/diagnostics, including the techniques used to ascertain the integrity of the various hardware/software components comprising the total data communication entity. These techniques are used to audit, review, and control the total data communication environment and to isolate the offending elements either on a periodic basis or upon detection of a failure. These techniques may include diagnostic software routines, electrical loopback, test message generation, administrative and personnel procedures, and the like.

9. Review the training and education of employees with regard to the data communication network. Employees must be trained adequately in this area because of the high technical competence required for data communication networks.

10. Ensure that there is adequate documentation, including a precise description of programs, hardware, system configurations, and procedures intended to assist in the prevention of problems, identification of problems, and recovery from problems. The documentation should be sufficiently detailed to assist in reconstructing the system from its parts.

11. Review error recording to reduce lost messages. All errors in transmission of messages in the system should be logged. This log should include the type of error, the time and date, the terminal, the circuit, the terminal operator, and the number of times the message was retransmitted before it was received correctly.

12. Review the error correction procedures. A user's manual should specify a cross reference of error messages to the appropriate error code generated by the system. These messages help the user interpret the error that has occurred and suggest the corrective action to be taken. Ensure that the errors are in fact corrected and the correct data reentered into the system.

LOCAL LOOP (LINES)

The local loop is the communication facility between the customer's premises and the telephone company's central office or the central office of any other special common carrier. It is assumed to be metallic pairs of wires.

1. When feasible, use digital data transmission, because it has a lower error rate than analog data transmission.

2. Physically trace out and, as best as possible, secure the local loop communication circuits/lines within the organization or facility. After these lines

leave the facility and enter the public domain, they cannot be physically secured.

3. With regard to data security, consider encrypting all messages transmitted.

4. Ensure that there are adequate recovery facilities and/or capabilities for a software failure, loss of key pieces of hardware, and loss of various communication circuit/lines.

5. Ensure that there are adequate backup facilities (local and remote) to back up key communication circuits/lines.

6. Review the physical security (local and remote) for circuits/lines (especially the local loop), hardware/software, physical facilities, storage media, and the like.

7. Lock up telephone equipment rooms and install alarms on the doors of those telephone equipment rooms that contain the basic data communication circuits.

8. Do not put communication lines through the public switchboard unless it is a new electronic switchboard (ESS) and the intent is to gain verbal identification of incoming dial-up data communication calls.

9. Review the organization's fault isolation/diagnostics, including the techniques used to ascertain the integrity of the various hardware/software components comprising the total communication entity. These techniques are used to audit, review, and control the total data communication environment and to isolate the offending elements either on a periodic basis or upon detection of a failure. These techniques may include diagnostic software routines, electrical loopback, test message generation, administrative and personnel procedures, and the like.

10. Review the techniques for testing used to validate the hardware and software operation to ensure integrity. Testing, including that of personnel, should reveal departures from the specified operation.

11. Ensure that there is a policy for the use of test equipment. Modern day test equipment may offer a new vulnerability to the organization. This test equipment is easily connected to communication lines, and all messages can be read in clear "English language." Test equipment should not be used for monitoring lines "for fun"; it should be locked up (key lock or locked hood) when it is not in use and after normal working hours when it is not needed for testing and debugging. Programs written for programmable test equipment should be kept locked up and out of the hands of those who do not need them.

DATA ENTRY AND VALIDATION, DATA COMMUNICATION

Data entry and validation refers to the validation of data, either at the time of transmission or during transmission. The validation may take place at a remote site with an intelligent terminal, at the central site's front end communication processor, or at a distributed intelligence site's concentrator or remote front end communication processor.

1. Identify each message by the individual user's password, the terminal, and the individual message sequence number.

2. Consider the following special controls on dial-up modems when the data communication network allows incoming dial-up connections: change the telephone numbers at regular intervals; keep the telephone numbers confidential; remove the telephone numbers from the modems in the computer operations area; require that each "dial-up terminal" have an electronic identification circuit chip to transmit its unique identification to the front end communication processor; do not allow automatic call receipt and connection (always have a person intercept the call and make a verbal identification); have the central site call the various terminals that will be allowed connection to the system; utilize dial-out only where an incoming dialed call triggers an automatic dial-back to the caller (in this way the central system controls those telephone numbers to which it will allow connection).

3. See that each inbound and outbound message is serial numbered as well as time and date stamped at the time of logging.

4. Consider having concentrators and front ends perform two levels of editing. In the first level the front end may add items to a message, reroute the message, or rearrange the data for further transmission. It also may check a message address for accuracy and perform parity checks. In the second level of editing, the concentrator or front end is programmed to perform specific edits of the different transactions that enter the system. This editing is an application system type of editing and deals with message content rather than form and is specific to each application program being executed.

5. Ensure that the concentrators and front ends can validate electronic terminal identification.

6. See that messages are checked for valid destination address.

7. Ensure adequate error detection and control capabilities. These might include echo-checking, where a message is transmitted to a remote site and

the remote site echoes the message back for verification; or it might include forward error correction, where special hardware boxes can automatically correct some errors upon receipt of the message; or it might include detection with retransmission. Detection with retransmission is the most common and cost-effective form of error detection and correction. This may include identification of errors by reviewing the parity bit or by utilizing a special code to identify errors in individual characters during transmission. A more prevalent form is to utilize a polynomial (mathematical algorithm) to detect errors in message blocks. Whichever way is used, when a message error is detected, it is retransmitted until it is received correctly.

8. When reviewing error detection in transmission, first determine whatever error rate can be tolerated, then determine the extent and pattern of errors on the communication links used by the organization, and then review the error detection and correction methodologies in use and determine if they are adequate for the application systems utilizing the data communication network. In other words, a purely administrative message network (no critical financial data) would not require error detection and correction capabilities equal to those for a network that transmits critical financial data.

9. Ensure that there is an audit trail logging facility to assist in the reconstruction of data files and the reconstruction of transactions from the various stations. There should be the capability to trace back to the terminal end user.

10. Provide some tables for checking for access by terminals, people, database, and programs. These tables should be in protected areas of memory.

11. Protect against concurrent file updates. If the data management software does not provide this protection, the data communication software should.

12. Ensure that the system supports password protection (multilevel password protection).

13. Review the techniques used for testing to validate the hardware and software operation to ensure integrity. Testing, including that of personnel, should reveal departures from the specified operation.

ERRORS AND OMISSIONS, DATA COMMUNICATION

The following controls relate to the accidental or intentional transmission of data that are in error, including the accidental or intentional loss of data or omission of data that should have been entered or transmitted via the on-line system. This type of threat includes, but is not limited to, inaccurate data, incomplete data, malfunctioning hardware, and the like.

1. Ensure that the system can switch messages destined for a down station/terminal to an alternate station/terminal.

2. Determine whether the system can perform message switching to transmit messages between stations/terminals.

3. In order to avoid lost messages in a message-switching system, provide a store and forward capability. This is where a message destined for a busy station is stored at the central switch and then forwarded at a later time when the station is no longer busy.

4. Review the message- or transaction-logging capabilities to reduce lost messages, provide for an audit trail, restrict messages, prohibit illegal messages, and the like. These messages might be logged at the remote station (intelligent terminal), they might be logged at a remote concentrator/remote front end processor, or they might be logged at the central front end communication processor/central computer.

5. Transmit messages promptly to reduce risk of loss.

6. Acknowledge the successful or unsuccessful receipt of all messages.

7. When feasible, use digital data transmission, because it has a lower error rate than analog data transmission.

8. Consider modems with automatic equalization (built-in microprocessors for circuit equalization and balancing) in order to compensate for amplitude and phase distortions on the line. This will reduce the number of errors in transmission and may decrease the need for conditioned lines.

9. Consider conditioning the voice grade circuits in order to reduce the number of errors during transmission. This may be unnecessary with the newer microprocessor-based modems that perform automatic equalization and balancing.

10. Ensure that all inbound and outbound messages are logged by the central processor, the front end, or remote concentrator in order to ensure against lost messages, to keep track of message sequence numbers (identify illegal messages), and to use for system restart should the entire system crash.

11. See that each inbound and outbound message is serial numbered as well as time and date stamped at the time of logging.

12. Consider having concentrators and front ends perform two levels of editing. In the first level, the front end may add items to a message, reroute the

message, or rearrange the data for further transmission. It also may check a message address for accuracy and perform parity checks. In the second level of editing, the concentrator or front end is programmed to perform specific edits of the different transactions that enter the system. This editing is an application system type of editing and deals with message content rather than form and is specific to each application program being executed.

13. Have the concentrators, front ends, and central computers handle the message priority system, if one exists. A priority system is set up to permit a higher line utilization to certain areas of the network or to ensure that certain types of transactions are handled before other transactions of lesser importance.

14. See that the front end collects message traffic statistics and performs correlations of traffic density and circuit availability. These analyses are mandatory for the effective management of a large data communication network. Some of the items included in a traffic density report might be the number of messages handled per hour or per day on each link of the network, the number of errors encountered per hour or per day, the number of errors encountered per program or per program module, the terminals or stations that appear to have a higher than average error record, and the like.

15. Ensure that the front ends and concentrators can perform miscellaneous functions such as triggering remote alarms if certain parameters are exceeded, performing multiplexing operations internally, signaling abnormal occurrences to the central computer, slowing up input/output messages when the central computer is overburdened because of heavy traffic, and the like.

16. Ensure that there is a message intercept function for inoperable terminals or invalid terminal addresses.

17. See that messages are checked for valid destination address.

18. Ensure adequate error detection and control capabilities. These might include echo-checking, where a message is transmitted to a remote site and the remote site echoes the message back for verification; it might include forward error correction, where special hardware boxes can automatically correct some errors upon receipt of the message; or it might include detection with retransmission. Detection with retransmission is the most common and cost-effective form of error detection and correction. This may include identification of errors by reviewing the parity bit or utilizing a special code to identify errors in individual characters during transmission.

A more prevalent form is to utilize a polynomial (mathematical algorithm) to detect errors in message blocks. Whichever way is used, when a message error is detected, it is retransmitted until it is received correctly.

19. When reviewing error detection in transmission, first determine whatever error rate can be tolerated; then determine the extent and pattern of errors on the communication links used by the organization. Finally, review the error detection and correction methodologies in use and determine if they are adequate for the application systems utilizing the data communication network. In other words, a purely administrative message network (no critical financial data) would not require error detection and correction capabilities equal to those for a network that transmits critical financial data.

20. Ensure that there is an audit trail logging facility to assist in the reconstruction of data files and the reconstruction of transactions from the various stations. There should be the capability to trace back to the terminal end user.

21. With regard to the efficiency of modems, review to see if they have multiple-speed switches so the transmission rate can be lowered when the line error rates are high.

22. Utilize four-wire circuits in a pseudo-full duplex transmission mode. In other words, keep the carrier wave up in each direction on alternate pairs of wires in order to reduce turnaround time and gain efficiency during transmission.

RESTART AND RECOVERY, DATA COMMUNICATION

Should the data communication network fail, it must have restart and recovery capabilities. In other words, how does the software operate in a failure mode? How long does it take to recover from a failure? This restart and recovery threat also includes backup for key portions of the data communication network and the contingency planning for backup, should there be a failure at any point in the data communication network.

1. Consider using modems that have either manual or remote actuated loopback switches for fault isolation to ensure the prompt identification of malfunctioning equipment. These are extremely important in order to increase the uptime and to identify faults.

2. Use front panel lights on modems to indicate if the circuit/line is function-

ing properly (carrier signal is up). This may not be a viable alternative with organizations that have hundreds of modems.

3. Consider a modem with alternate voice capabilities for quick troubleshooting between the central site and a major remote site.

4. Consider placing unused backup modems in critical areas of the data communication network.

5. Consider using modems that have an automatic or semi-automatic dial backup capability in case the leased line fails.

6. Ensure that all inbound and outbound messages are logged by the central processor, the front end, or remote concentrator in order to ensure against lost messages, to keep track of message sequence numbers (identify illegal messages), and to use for system restart should the entire system crash.

7. Ensure that there are adequate restart and recovery software routines to recover from items such as a trapped machine check, where instead of bringing down the entire data communication system, a quick recovery can be made and only the one transaction need be retransmitted.

8. Ensure that there are adequate restart and recovery procedures to effect both a warm start and a cold start. In other words, a data communication system should never completely fail so the user has to perform a cold start (start up as if it is a new day, all message counters cleared). The system should go into a warm start procedure, where only parts of the system are disabled and recovery can be made while the system is operating in a degraded mode.

9. Ensure that there is an audit trail logging facility to assist in the reconstruction of data files and the reconstruction of transactions from the various stations. There should be the capability to trace back to the terminal end user.

10. Identify all default options in the software and their impact if they do not operate properly.

11. Ensure that there are adequate recovery facilities and/or capabilities for a software failure, loss of key pieces of hardware, and loss of various communication circuit/lines.

12. Ensure that there are adequate backup facilities (local and remote) to back up key pieces of hardware and communication circuits/lines.

13. Consider backup power capabilities for large facilities such as the central site and various remote concentrators.

---

14. Consider installing the capabilities to fall back to the public dial network from a leased-line configuration.

15. Review the physical security (local and remote) for circuits/lines (especially for local loop), hardware/software, physical facilities, storage media, and the like.

16. Review the training and education of employees with regard to the data communication network. Employees must be trained adequately in this area because of the high technical competence required for data communication networks.

17. Ensure that there is adequate documentation, including a precise description of programs, hardware, system configurations, and procedures intended to assist in the prevention of problems, identification of problems, and recovery from problems. The documentation should be sufficiently detailed to assist in reconstructing the system from its parts.

MESSAGE LOSS OR CHANGE, DATA COMMUNICATION

The following controls relate to the loss of messages as they are transmitted through the data communication system, or the accidental/intentional changing of messages during their transmission.

1. Ensure that the system can switch messages destined for a down station/terminal to an alternate station/terminal.

2. Determine whether the system can perform message switching to transmit messages between stations/terminals.

3. In order to avoid lost messages in a message-switching system, provide a store and forward capability. This is where a message destined for a busy station is stored at the central switch and then forwarded at a later time when the station is no longer busy.

4. Review the message- or transaction-logging capabilities to reduce lost messages, provide for an audit trail, restrict messages, prohibit illegal messages, and the like. These messages might be logged at the remote station (intelligent terminal), they might be logged at a remote concentrator/remote front end processor, or they might be logged at the central front end communication processor/central computer.

5. Transmit messages promptly to reduce risk of loss.

6. Acknowledge the successful or unsuccessful receipt of all messages.

7. Stringently control circuit test equipment. This specifically covers network monitoring equipment with which to read the data going over the communication lines. It is even more important to stringently control the programmable network monitoring equipment. This equipment is necessary, although the personnel authorized to use it should be limited. The uses to which it should be put should clearly be stated. It should be locked up in some fashion when it is not in use (this locking up might involve removing it from the area, installing a keylock switch in place of the on/off switch, or covering it with a lockable metal hood so it is unusable).

8. Review error recording to reduce lost messages. All errors in transmission of messages in the system should be logged and this log should include the type of error, the time and date, the terminal, the circuit, the terminal operator, and the number of times the message was retransmitted before it was correctly received.

9. Within an organizational facility, fiber optics (laser) communication circuits can be used to totally preclude the possibility of wiretapping.

10. Consider logging inbound and outbound messages at the remote site.

11. Ensure that all inbound and outbound messages are logged by the central processor, the front end, or remote concentrator in order to ensure against lost messages, to keep track of message sequence numbers (identify illegal messages), and to use for system restart should the entire system crash.

12. See that each inbound and outbound message is serial numbered as well as time and date stamped at the time of logging.

13. Consider having concentrators and front end perform editing. The front end may check a message address for accuracy and perform parity checks.

14. Ensure adequate error detection and control capabilities. These might include echo-checking, where a message is transmitted to a remote site and the remote site echoes the message back for verification, or it might include forward error correction, where special hardware boxes can automatically correct some errors upon receipt of the message, or it might include detection with retransmission. Detection with retransmission is the most common and cost-effective form of error detection and correction. This may include identification of errors by reviewing the parity bit or utilizing a special code to identify errors in individual characters during transmission. A more prevalent form is to utilize a polynomial (mathematical algorithm) to detect errors in message blocks. Whichever way is used, when a message error is detected, it is retransmitted until it is received correctly.

15. Ensure that there is an audit trail logging facility to assist in the reconstruction of data files and the reconstruction of transactions from the various stations. There should be the capability to trace back to the terminal end user.

16. Safe store all messages. All transactions/messages should be protected in case of a disastrous situation such as power failure.

PEOPLE CONTROLS, DATA COMMUNICATION

Individuals are responsible for managing, operating, and maintaining the data communication network and its equipment; writing software programs for the data communication system; and working at the remote stations/terminals.

1. Identify each message by the individual user's password, the terminal, and the individual message sequence number.

2. Utilize physical security controls throughout the data communication network. This includes the use of locks, guards, badges, sensors, alarms, and administrative measures to protect the physical facilities, data communication networks, and related data communication equipment. These safeguards are required for access monitoring and control to protect data communication equipment and software from damage by accident, fire, or environmental hazard, either intentional or unintentional.

3. Consider the following special controls on dial-up modems when the data communication network allows incoming dial-up connections: change the telephone numbers at regular intervals; keep the telephone numbers confidential; remove the telephone numbers from the modems in the computer operations area; require that each "dial-up terminal" have an electronic identification circuit chip to transmit its unique identification to the front end communication processor; do not allow automatic call receipt and connection (always have a person intercept the call and make a verbal identification); have the central site call the various terminals that will be allowed connection to the system; utilize dial-out only where an incoming dialed call triggers an automatic dial-back to the caller (in this way the central system controls those telephone numbers to which it will allow connection).

4. Ensure that there is adequate physical security at remote sites, especially for terminals, concentrators, multiplexers, and front end communication processors.

5. See that each inbound and outbound message is serial numbered, as well as time and date stamped, at the time of logging.

6. Provide some tables for checking for access by terminals, people, database, and programs. These tables should be in protected areas of memory.

7. For personnel who work in critical or sensitive areas, consider enforcing the following policies: insist that they take at lease five consecutive days of vacation per year, check with their previous employers, perform an annual credit check, and have them sign hiring agreements stating that they will not sell programs, etc.

8. Review the operational procedures, for example, the administrative regulations, policies, and day-to-day activities supporting the security/safeguards of the data communication network. These procedures may include:

    • Specifying the objectives of the EDP security for an organization, especially as they relate to data communications.

    • Planning for contingencies of security "events," including recording of all exception conditions and activities.

    • Assuring management that other safeguards are implemented, maintained, and audited, including background checks, security clearances and hiring of people with adequate security-oriented characteristics; separation of duties; mandatory vacations.

    • Developing effective safeguards for deterring, detecting, preventing, and correcting undesirable security events.

    • Reviewing the cost effectiveness of the system and the related benefits such as better efficiency, improved reliability, and economy.

    • Looking for the existence of current administrative regulations, security plans, contingency plans, risk analysis, and personnel understanding of management objectives, and then reviewing the adequacy and timeliness of the specified procedures in satisfying these.

9. Review the communication system's console log that shows "network supervisor terminal commands" such as: disable or enable a line or station for input or output, alternately route traffic from one station to another, change the order and/or frequency of line or terminal service (polling, calling, dialing), and the like.

10. Ensure that there is a policy for the use of test equipment. Modern test equipment may offer a new vulnerability to the organization. This test equipment is connected easily to communication lines, and all messages can be read in clear "English language." Test equipment should not be used

for monitoring lines "for fun"; it should be locked up (key lock or locked hood) when it is not in use and after normal working hours when it is not needed for testing and debugging. Programs written for programmable test equipment should be kept locked up and out of the hands of those who do not need these programs.

11. Review the training and education of employees with regard to the data communication network. Employees must be trained adequately in this area because of the high technical competence required for data communication networks.

12. Review the error correction procedures. A user's manual should specify a cross-reference of error messages to the appropriate error code generated by the system. These messages help the user interpret the error that has occurred and suggests the corrective action to be taken. Ensure that errors are, in fact, corrected and that the corrected data are reentered into the system.

13. Provide personnel with an education program in security matters so the responsibility for security is firmly fixed and clearly understood by those having such responsibilities.

FRONT END COMMUNICATION PROCESSOR

This hardware device interconnects all the data communication circuits (lines) to the central computer or distributed computer. It performs a subset of the following functions: error detection and correction, logging, message switching, store and forward, statistical data gathering, polling/addressing, insertion/deletion of line control codes, and the like.

1. Ensure that the system can switch messages destined for a down station/terminal to an alternate station/terminal.

2. Determine whether the system can perform message switching to transmit messages between stations/terminals.

3. In order to avoid lost messages in a message-switching system, provide a store and forward capability. This is where a message destined for a busy station is stored at the central switch and then forwarded at a later time when the station is no longer busy.

4. Review the message- or transaction-logging capabilities to reduce lost messages, provide for an audit trail, restrict messages, prohibit illegal mes-

sages, and the like. These messages might be logged at the remote station (intelligent terminal), they might be logged at a remote concentrator/remote front end processor, or they might be logged at the central front end communication processor/central computer.

5. Transmit messages promptly to reduce risk of loss.

6. Identify each message by the individual user's password, the terminal, and the individual message sequence number.

7. Acknowledge the successful or unsuccessful receipt of all messages.

8. For data communication equipment, check the manufacturer's mean time between failures (MTBF) in order to ensure that the data communication equipment has the largest MTBF.

9. Review the maintenance contract and mean time to fix (MTTF) for all data communication equipment. Maintenance should be both fast and available. Determine from where the maintenance is dispatched and whether tests can be made from a remote site (for example, in many cases modems have remote loopback capabilities).

10. Ensure that there is adequate physical security at remote sites, especially, for terminals, concentrators, multiplexers, and front end communication processors.

11. Determine whether the multiplexer/concentrator/remote front end hardware has redundant logic and backup power supplies with automatic fallback capabilities in case the hardware fails. This will increase the uptime of the many stations/terminals that might be connected to this equipment.

12. If a concentrator is being used, is it performing some of the controls that are usually performed by the front end communication processor and therefore increasing the efficiency and correctness of data transmissions?

13. See if the polling configuration list can be changed during the day in order to exclude or include specific terminals. This would allow the positive exclusion of a terminal, as well as allowing various terminals to come on-line and off-line during the working day.

14. Can the front ends, concentrators, modems, and the like handle the automatic answering and automatic outward dialing of calls? This would increase the efficiency and accuracy when it is preprogrammed into the system.

15. Ensure that all inbound and outbound messages are logged by the central processor, the front end, or remote concentrator in order to ensure against lost messages, to keep track of message sequence numbers (identify illegal messages), and to use for system restart should the entire system crash.

16. See that each inbound and outbound message is serial numbered, as well as time and date stamped, at the time of logging.

17. Ensure that there is a time-out facility so the system does not get hung up trying to poll/address a station. Also, if a particular station "times out" four or five consecutive times, it should be removed from the network configuration polling list so time is not wasted on this station (improves communication efficiency).

18. Consider having concentrators and front ends perform two levels of editing. In the first level, the front end may add items to a message, reroute the message, or rearrange the data for further transmission. It also may check a message address for accuracy and perform parity checks. In the second level of editing, the concentrator or front end is programmed to perform specific edits of the different transactions that enter the system. This editing is an application system type of editing and deals with message content rather than form. It is specific to each application program being executed.

19. Have the concentrators, front ends, and central computers handle the message priority system, if one exists. A priority system is set up to permit a higher line utilization to certain areas of the network or to ensure that certain transactions are handled before other transactions of lesser importance.

20. See that the front end collects message traffic statistics and performs correlations of traffic density and circuit availability. These analyses are mandatory for the effective management of a large data communication network. Some of the items included in a traffic density report might be the number of messages handled per hour or per day on each link of the network, the number of errors encountered per hour or per day, the number of errors encountered per program or per program module, the terminals or stations that appear to have a higher than average error record, and the like.

21. Ensure that the front ends and concentrators can perform miscellaneous functions such as triggering remote alarms if certain parameters are exceeded, performing multiplexing operations internally, signaling abnormal occurrences to the central computer, slowing up input/output messages when the central computer is overburdened because of heavy traffic, and the like.

22. Ensure that the concentrators and front ends can validate electronic terminal identification.

23. Ensure that there is a message intercept function for inoperable terminals or invalid terminal addresses.

24. See that messages are checked for valid destination address.

25. Ensure adequate error detection and control capabilities. These might include echo-checking, where a message is transmitted to a remote site and the remote site echoes the message back for verification; it might include forward error correction, where special hardware boxes can correct automatically some errors upon receipt of the message; or it might include detection with retransmission. Detection with retransmission is the most common and cost-effective form of error detection and correction. This may include identification of errors by reviewing the parity bit or utilizing a special code to identify errors in individual characters during transmission. A more prevalent form is to utilize a polynomial (mathematical algorithm) to detect errors in message blocks. Whichever way is used, when a message error is detected, it is retransmitted until it is received correctly.

26. When reviewing error detection in transmission, first determine whatever error rate can be tolerated, and then determine the extent and pattern of errors in the communication links used by the organization. Finally, review the error detection and correction methodologies in use to determine if they are adequate for the application system utilizing the data communication network. In other words, a purely administrative message network (no critical financial data) would not require error detection and correction capabilities equal to those for a network that transmits critical financial data.

27. Ensure that there are adequate restart and recovery software routines to recover from items such as a trapped machine check, where instead of bringing down the entire data communication system, a quick recovery can be made and only the one transaction need be retransmitted.

28. Ensure that there are adequate restart and recovery procedures to effect both a warm start and a cold start. In other words, a data communication system should never fail completely so the user has to perform a cold start (start up as if it is a new day, all message counters cleared). The system should go into a warm start procedure, where only parts of the system are disabled and recovery can be made while the system is operating on a degraded mode.

29. Ensure that the documentation of the front end software is comprehensive.

30. For entering sensitive or critical systems commands, restrict these commands to one master input terminal and ensure strict physical custody over this terminal. In other words, restrict the personnel who can use this terminal.

31. Ensure that there are adequate recovery facilities and/or capabilities for loss of key pieces of hardware, and loss of various communication circuit/lines.

32. Ensure that there are adequate backup facilities (local and remote) to back up key pieces of hardware and communication circuits/lines.

33. Consider backup power capabilities for large facilities such as the central site and various remote concentrators.

34. Review the preventive maintenance and scheduled diagnostic testing such as cleaning, replacement, and inspection of equipment to evaluate its accuracy, reliability, and integrity. This may include schedules for testing and repair, adequate testing of software program changes submitted by the vendor, inventories of replacement parts (circuit boards), past maintenance records, and the like.

35. Review the organization's fault isolation/diagnostics, including the techniques used to ascertain the integrity of the various hardware/software components comprising the total data communication entity. These techniques are used to audit, review, and control the total data communication environment and to isolate the offending elements either on a periodic basis or upon detection of a failure. These techniques may include diagnostic software routines, electrical loopback, test message generation, administrative and personnel procedures, and the like.

36. Review error recording to reduce lost messages. All errors in transmission of messages in the system should be logged. This log should include the type of error, the time and date, the terminal, the circuit, the terminal operator, and the number of times the message was retransmitted before it was received correctly.

RELIABILITY (UPTIME), DATA COMMUNICATION

The reliability of the data communication network and its uptime includes the organization's ability to keep the data communication network operating, the mean time between failures (MTBF) at a minimum, and the ability to minimize the time to repair equipment when it malfunctions. Low reliability/breakdown of hardware, reliability of software, and the maintenance of these two items are the chief threats here.

1. Ensure that the system can switch messages destined for a down station/terminal to an alternate station/terminal.

2. Consider using modems that have either manual or remote actuated loopback switches for fault isolation to ensure the prompt identification of malfunctioning equipment. These are extremely important in order to increase the uptime and to identify faults.

3. Use front panel lights on modems to indicate if the circuit/line is functioning properly (carrier signal is up). This may not be a viable alternative with organizations that have hundreds of modems.

4. Consider a modem with alternate voice capabilities for quick troubleshooting between the central site and a major remote site.

5. For data communication equipment, check the manufacturer's mean time between failures (MTBF) in order to ensure that the data communication equipment has the largest MTBF.

6. Consider placing unused backup modems in critical areas of the data communication network.

7. Consider using modems that have an automatic or semi-automatic dial backup capability in case the leased line fails.

8. Review the maintenance contract and mean time to fix (MTTF) for all data communication equipment. Maintenance should be both fast and available. Determine from where the maintenance is dispatched, and determine if tests can be made from a remote site (for example, in many cases modems have remote loopback capabilities).

9. Increase data transmission efficiency. The faster the modem synchronization time, the lower will be the turnaround time and thus more throughput to the system.

10. Consider modems with automatic equalization (built-in microprocessors for circuit equalization and balancing) in order to compensate for amplitude and phase distortions on the line. These will reduce the number of errors in transmission and may decrease the need for conditioned lines.

11. Utilize four-wire circuits in a pseudo-full duplex transmission mode. In other words, keep the carrier wave up in each direction on alternate pairs of wires in order to reduce turnaround time and gain efficiency during transmission.

12. If needed, use full duplex transmission on two-wire circuits with special modems that split the frequencies and thus achieve full duplex transmission.

13. Utilize a reverse channel capability for control signals (supervisory) and to keep the carrier wave up in both directions.

14. Consider conditioning the voice grade circuits in order to reduce the number of errors during transmission. This may be unnecessary with the newer microprocessor-based modems that perform automatic equalization and balancing.

15. Use four-wire circuits in such a fashion that there is little or no turnaround time. This can be done by using two wires in each direction and keeping the carrier signal up.

16. Determine whether the multiplexer/concentrator/remote front end hardware has redundant logic and backup power supplies with automatic fallback capabilities in case the hardware fails. This will increase the uptime of the many stations/terminals that might be connected to this equipment.

17. Consider uninterruptible power supplies at large multiplexer/concentrator type remote sites.

18. Consider multiplexer/concentrator equipment that has diagnostic lights, diagnostic capabilities, and the like.

19. If a concentrator is being used, is it performing some of the controls that are usually performed by the front end communication processor and therefore increasing the efficiency and correctness of data transmissions?

20. Can the front ends, concentrators, modems, and the like handle the automatic answering and automatic outward handling of calls? This would increase efficiency and accuracy when it is preprogrammed into the system.

21. For efficiency, ensure that the central system can address either a group of terminals (group address), several terminals at a time (multiple address), or one terminal at a time (single address), or can send a broadcast message simultaneously to all stations/terminals in the system.

22. Ensure that there is a time-out facility so the system does not get hung up trying to poll/address a station. Also, if a particular station "times out" four or five consecutive times, it should be removed from the network configuration polling list so time is not wasted on this station (improves communication efficiency).

23. See that the front end collects message traffic statistics and performs correlations of traffic density and circuit availability. These analyses are mandatory for the effective management of a large data communication network. Some of the items included in a traffic density report might be the number of messages handled per hour or per day on each link of the network, the number of errors encountered per hour or per day, the number of errors encountered per program or per program module, the terminals or stations that appear to have a higher than average error record, and the like.

24. Ensure that the front ends and concentrators can perform miscellaneous functions such as triggering remote alarms if certain parameters are exceeded, performing multiplexing operations internally, signaling abnormal

occurrences to the central computer, slowing up input/output messages when the central computer is overburdened because of heavy traffic, and the like.

25. Ensure that there are adequate restart and recovery software routines to recover from items such as a trapped machine check, where instead of bringing down the entire data communication system, a quick recovery can be made and only the one transaction need be retransmitted.

26. Ensure that there are adequate restart and recovery procedures to effect both a warm start and a cold start. In other words, a data communication system should never completely fail so the user has to perform a cold start (start up as if it is a new day, all message counters cleared). The system should go into a warm start procedure, where only parts of the system are disabled and recovery can be made while the system is operating in a degraded mode.

27. Make available a systems trace capability to assist in locating problems.

28. Ensure that the documentation of the system software is comprehensive.

29. Provide adequate maintenance for the software programs.

30. Identify all default options in the software and their impact if they do not operate properly.

31. Ensure that there are adequate recovery facilities and/or capabilities for a software failure, loss of key pieces of hardware, and loss of various communication circuit/lines.

32. Ensure that there are adequate backup facilities (local and remote) to back up key pieces of hardware and communication circuits/lines.

33. Consider backup power capabilities for large facilities such as the central site and various remote concentrators.

34. Consider installing the capabilities to fall back to the public dial network from a leased-line configuration.

35. When utilizing multidrop or loop circuits, review the uptime problems. These types of configurations are more cost-effective than point-to-point configurations, but when there is a circuit failure close to the central site, all terminals/stations downline are disconnected.

36. Review the physical security (local and remote) for circuits/lines (especially the local loop), hardware/software, physical facilities, storage media, and the like.

37. Review the preventive maintenance and scheduled diagnostic testing such as cleaning, replacement, and inspection of equipment to evaluate its accu-

racy, reliability, and integrity. This may include schedules for testing and repair, adequate testing of software program changes submitted by the vendor, inventories of replacement parts (circuit boards), past maintenance records, and the like.

38. Determine whether there is a central site for reporting all problems encountered in the data communication network. This usually results in faster repair time.

39. Review the organization's fault isolation/diagnostics, including the techniques used to ascertain the integrity of the various hardware/software components comprising the total data communication entity. These techniques are used to audit, review, and control the total data communication environment and to isolate the offending elements either on a periodic basis or upon detection of a failure. These techniques may include diagnostic software routines, electrical loopback, test message generation, administrative and personnel procedures, and the like.

40. Review the training and education of employees with regard to the data communication network. Employees must be trained adequately in this area because of the high technical competence required for data communication networks.

41. Ensure that there is adequte documentation, including a precise description of programs, hardware, system configurations, and procedures intended to assist in the prevention of problems, identification of problems, and recovery from problems. The documentation should be sufficiently detailed to assist in reconstructing the system from its parts.

42. Review the techniques for testing used to validate the hardware and software operation to ensure integrity. Testing, including that of personnel, should reveal departures from the specified operation.

APPENDIX 3

AUDIT AND CONTROL OF COMMUNICATION NETWORKS

This paper (publication 500–57) is a follow-up of the first National Bureau of Standards (NBS) invitational workshop on audit and evaluation of computer security. The earlier paper was published in NBS special publication 500–19 (part X).

In order to better understand what is meant by a teleprocessing environment, Figure A3–1 was developed to show examples of the alternative teleprocessing network configurations that might be available. These networks are among those that might be faced when conducting a security review in today's teleprocessing environment. It should be noted that there might be combinations of networks, where for example a multidrop configuration might have a local loop at each of the drops. Also, where this figure depicts "transmission lines" the audit and control expert reviewing the network might find various transmission media, such as satellite circuits, microwave transmission, fiber optics, or copper wire pairs.

Source: Computer Science & Technology: Audit and Evaluation of Computer Security II: System Vulnerabilities and Controls, ed. Zella G. Ruthberg. Proceedings of the NBS Invitational Workshop, Miami Beach, Fla., Nov. 28–30, 1978. National Bureau of Standards Special Publication SP 500–57.

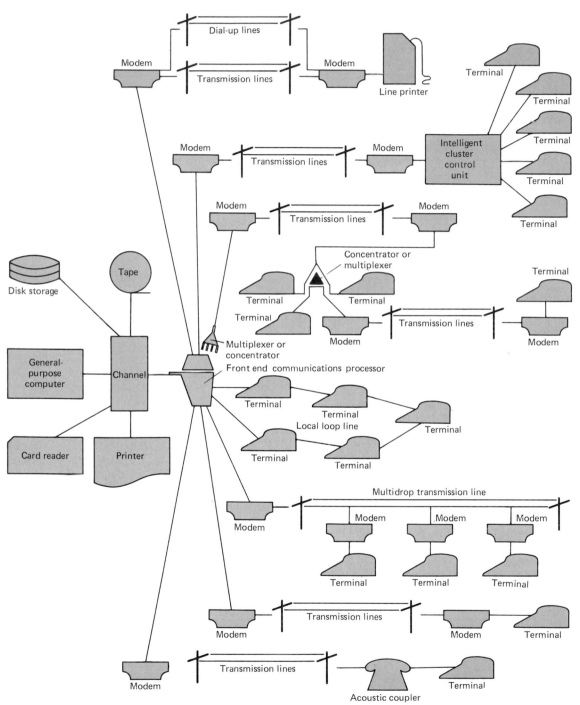

FIGURE A3–1 Network configurations.

DEFINITION OF THE
COMMUNICATION COMPONENT SECURITY AUDIT

For the purpose of this paper a computer security audit is defined as an independent evaluation of the controls employed to ensure the accuracy and reliability of the data maintained on or generated by a teleprocessing network, the appropriate protection of the organization's information assets (including hardware, software, and data) from all significant anticipated threats or hazards, and the operational reliability and performance assurance of all components of the automated data processing system.

With regard to the communication component, all modes of data transmission and associated equipment should be considered. Specific vulnerabilities should be identified along with appropriate safeguards, such as interception of microwave transmission, with encryption serving as the countering control.

THE CONTROL MATRIX

This paper presents a matrix that relates the various vulnerabilities to the specific controls that might be available to mitigate them (see Figure A3–2, The Control Matrix). The vulnerabilities are listed across the top of the matrix and are defined in a later section of this paper. The controls are listed down the left vertical axis of the matrix and are also defined in a later section of this paper. Within the cells of the matrix there is either an X or an O whenever the control is an appropriate countermeasure to a specific vulnerability. An X indicates a primary control that can be used to mitigate the specific vulnerability; an O indicates a secondary control that might be useful in mitigating the specific vulnerability. To apply the matrix, first identify the vulnerability that may be present in your teleprocessing network. Next, proceed down the column of the specific vulnerability and identify whether the controls in the left vertical column are applicable.

The control matrix can be used in two other ways to assist the auditor. The first is to determine the exposures that will be faced by the organization whenever one of the vulnerabilities does, in fact, occur. These exposures are listed at the bottom of the matrix, below each vulnerability column. For example, if the vulnerability "Message Lost" occurred, then the organization would be subjected to exposures A, E, F, and G. These exposures are defined in the list of exposures that follows.

The second use to which the matrix can be put is to specifically identify the various components of the network where the controls might be most effectively located. To do this, the auditor would choose a specific control such as "Sequence Number Checking" and follow across that row to the right-hand side of the ma-

	Vulnerabilities									
Controls	Message Lost	Misrouting	Message Alteration	Disruption	Disaster	Disclosure (privacy)	Message Insertion	Theft (physical)	Duplicate Message	Components Where Controls Are Located
Sequence number checking	X	0					X		X	9–10, 17
Sending and receiving identification	0	X				0				9–10, 17
Transaction journal	X	0	X	X	X		X		X	9–11, 17
Positive acknowledgment	X	X	X	X			X		X	9–11, 17
Time and date stamp	X								X	9–11, 17
Periodic message reconciliation	·X		X	0			X		X	10, 17
Checksum on message address		X								9–11, 17
Error detection code			X	0						9–12, 17
Error correction code			X	0						9–12, 17
Key redundancy code			X	0						9–12, 17
Echoplexing	X	0	X	0			0		0	10
Error logging	0	0	X	X					0	9–10
Backup equipment and facilities				X	X					1, 7–11, 17
Physical security	0	0	0	X	X	X	0	X	0	1–17
Recovery procedures	0		0	X	X				0	1–17
Communication policy	0	0	0	0	0	0	0	0	0	1–17
Life support system				X	X					4, 8–11
Device disconnection detection	0							X		7, 9–11
Built-in device address							X	0		9–11, 17
Encryption	X	X	X			X	X		X	7–11, 17
Unlisted phone number (dial-up)							X	X		9–11
Low error rate facilities	X		X	X						1–3, 7
Software controls and testing	X	X	X	X			X	X	X	9–11, 17
Documentation	0	0	0	0	0	0	0	0	0	1–16
Emanation control						X				1–4, 6–11, 13–17
Training and education	0	0	0	0	0	0	0	0	0	1–17
Exposures	A,E, F,G	A,D E,F, I	A,D E,G, H	C-I	C-I	F,G, I	B,D, E,G, H	C,E, G,H,	D,E, G,I	

FIGURE A3–2 The control matrix.

LIST OF EXPOSURES

A.   Erroneous record keeping
B.   Unacceptable accounting
C.   Business interruption
D.   Erroneous management decisions
E.   Fraud
F.   Statutory sanctions
G.   Excessive costs/deficient revenues
H.   Loss or destruction of assets
I.   Competitive disadvantage

The above items A through I represent the various exposures that the organization faces whenever some sort of a vulnerability (threat or concern) takes place. In other words, the result of a threat might be one of these exposures.

---

trix, where there are numbers, in this case, 9–10, 17. These numbers indicate those specific components of a data communication network where the controls might be located. These 17 components are defined at the end of this report.

INTERRELATIONS OF SECURITY CONTROLS

The auditor should recognize that the security controls shown in the matrix have complex interrelations in solving certain security problems. There are no linear equations that show how these controls add to or subtract from one another. The security controls required in a worst case analysis of an intentional assault on a communication system constitute a highly structured set of interrelationships.

For example, encryption is a valuable security control in a communication system. It is not, however, a complete solution in and of itself. The security objectives of a communication system can only be satisfied when encryption is used in conjunction with several other controls. In particular, sequence numbers must be used to detect attempts to add, delete, or replay messages by a technically competent penetrator. A cryptographic error detection code must be used to detect alteration of messages. Encryption key management must be performed to ensure authentication of communicating devices.

In addition, message reconciliation must be performed during and at the end of every session to ensure that all messages transmitted have been received. Emanation controls prevent the loss of encryption keys and plaintext messages through undesirable electronic phenomena.

These constitute the necessary set of nondiscretionary controls required for secure communication. In addition, certain discretionary, human-oriented controls are required to support the encryption system. Physical security must prevent theft or unauthorized use of a device containing a valid encryption key. Maintenance and testing must ensure the correct operation of the controls. Documentation must explain how the controls must be used. Finally, the user must be educated and trained in the use of these controls.

DEFINITION OF THE VULNERABILITIES

The following list defines the vulnerabilities that are listed across the top of the control matrix. These vulnerabilities could be interpreted as the concerns or threats to which a data communication network might be subjected.

- **Message Lost** Refers to a message that never reaches its intended destination.
- **Misrouting** Is said to occur in a message-switching network when a message intended for a destination, e.g., Node A is sent to another destination, Node B.
- **Message Alteration** Refers to unauthorized (accidental or intentional) modification of an authentic message.
- **Disruption** A temporary or intermittent service outage affecting one or more of the network components which may result in one or more of the following consequences: denial of service, misrouting, message alteration, messages lost, duplicate message, etc.
- **Disaster** An interruption resulting in denial of service for an extended period of time as the result of an accident, natural catastrophe, or sabotage. The distinction between a disaster and a disruption is based upon the length of service outage and upon the permanence of the damage to the affected components.
- **Disclosure (Privacy)** Unauthorized access to any data is disclosure. If the data are personally identifiable to an individual or legal person, then the unauthorized disclosure is a privacy violation.
- **Message Insertion** The addition of an extraneous unauthorized message at any component in the network. This vulnerability is never accidental and does not include duplicate messages.
- **Theft (Physical)** Physical theft refers to unauthorized removal of any hardware component.
- **Duplicate Message** The insertion or processing of multiple copies of an otherwise authorized message. This can occur accidentally or intentionally.

DEFINITIONS OF THE CONTROLS

The following list defines each of the controls listed down the left verticle axis of the control matrix.

- **Sequence Number Checking** A method where all messages contain an integral sequence number for each level of the communication system. Verification techniques must detect duplicate and missing numbers, reject duplicates, and report missing messages.

- **Sending and Receiving Identification** A method where sufficient information is contained in the message to uniquely identify both the sender and receiver of a message.

- **Transaction Journal** A method of capturing sufficient system and message level data to establish an adequate audit trail or to have an actual copy of each and every transaction transmitted in the network.

- **Positive Acknowledgment** A method where the receipt of each message is positively confirmed back to the sender.

- **Time and Date Stamp** An automatic procedure whereby each message contains time and date information for each major processing node.

- **Periodic Message Reconciliation** System facilities to verify completeness of processing by periodically providing summary information to reconcile number of messages, dollar values, control totals, etc., both sent and received.

- **Checksum on Message Address** A procedure that verifies the message address using hashing or other summing type of totals.

- **Error Detection Code** A method of inserting extra (redundant) bits of information to permit detection and correction of errors at the receiving equipment without retransmission of the original message.

- **Error Correction Code** A method of inserting extra (redundant) bits of information to permit detection and correction of errors at the receiving equipment without retransmission of the original message.

- **Key Redundancy Code** The insertion of duplicate information in key fields of the message stream (such as dollar amounts, description identifiers, quantities, etc.) which can be compared at the receiving equipment for correctness.

- **Echoplexing** A verification procedure by which each character received by the receiving station equipment is transmitted back to the originating equipment.

- **Error Logging** A software program that records error messages, by line, terminal, and also type and frequency. This recording is to measure the degree of reliability and performance of the communication system. Statistical analysis and management reports are required for evaluation and corrective action to minimize error rates.

- **Backup Equipment and Facilities** Duplicate or alternate equipment (power, air conditioning, etc.), software, and procedures to be invoked whenever a major outage occurs with the primary system. Also a physical facility located away from the primary site and capable of supporting the original primary site telecommunication function at an acceptable operational level.

- **Physical Security** The ability to have proper physical security over the data communication facilities, software, and all other aspects of the teleprocessing network. This includes restrictive access controls over personnel, adequate fire protection, backup electrical equipment, and any other aspects of physical security with regard to maintaining the integrity of the data communication network.

- **Recovery Procedures** A set of written procedures that clearly defines responsibilities and procedures for operational programming and supervisory personnel to allow for the orderly recovery of the system to operational status or to recover from excessive error rates.

- **Communication Policy** A statement of agency or corporate policy regarding design, use, and maintenance of communication components including security objectives and penalties for not achieving these objectives.

- **Life Support System** Equipment, techniques, and procedures that will eliminate or minimize damages caused by disasters, occurrences such as fire, power failures, flood, environmental changes, etc.

- **Device Disconnection Detection** The use of electrical control signals or other mechanisms to detect physical disconnection of communication system components.

- **Built-in Device Address** The imbedding of a device address or identifier via hardware or software mechanisms in communication system components.

- **Encryption** The transformation of data (cleartext) to an unintelligible form (ciphertext) through the use of an algorithm under the control of a key such as the federal Data Encryption Standard (DES) (FIPS Pub. 46).

- **Unlisted Phone Number (Dial-Up)** The acquisition and use of unlisted telephone numbers for the communication system component that can be accessed via dial-up lines.

- **Low Error Rate Facilities** The selection and use of data transmission facilities with characteristically low error rates such as conditioned lines or digital transmission lines.

- **Software Controls and Testing**   The procedures employed in development, installation, and maintenance of software in communication system components to ensure the correctness, integrity, and availability of the software.
- **Documentation**   The generation, revision, and maintenance of manuals dealing with appropriate design, maintenance, and operational aspects of the communication system.
- **Emanation Control**   The use of shielding and associated techniques to suppress electromagnetic, acoustic, and radio frequency emanations from communication system components.
- **Training and Education**   The development, presentation, and periodic review of educational materials dealing with correct operation and maintenance of the communication system.

GENERAL DEFINITIONS OF COMPONENTS

The following list of items enumerates and defines the components of a data communication network. In some cases the item listed may be a characteristic of data transmission rather than an actual component.

1. **Circuits**   A circuit can be a single communication facility or a combination of different types of communication facilities such as:
   - **Satellite**   A facility that uses ultra-high-frequency signaling relayed through a device orbiting the earth.
   - **Microwave**   A facility that uses high-frequency signaling which passes through terrestrial relay points.
   - **Fiber optics**   A facility that transmits signals through the use of optical media utilizing a fiberglass-like cable.
   - **Wire**   A facility that transmits through a metallic conductor. This facility may utilize long-distance copper wire pairs, coaxial cable, or the copper wire local loop between a user premises and the telephone company's switching office.
2. **Analog Transmission**   Transmission of a continuously variable signal which has an almost infinite number of states (an example of an analog signal is a sine wave).
3. **Digital Transmission**   Transmission of a discretely variable signal such as discrete voltage levels (an example is signaling which is composed of either a positive or a negative voltage).

4. **Carrier Switch/Facility** A communication facility supplied by a commercial vendor of telecommunication services that provides for the interconnection of transmission devices (an example would be the telephone company's switching office or the Telenet Packet switches).

5. **Configurations** These are the methods of connecting communication devices. There are many examples of communication configurations, some of which were shown in Figure A3–1. Examples of these configurations might be as follows:

   • **Dedicated/private leased lines** These circuits are always available to the customer for transmission and generally are used with on-line real-time-systems.

   • **Dial/switched circuits** A circuit connection which is established by dialing a telephone or establishing a physical or logical connection before data can be transmitted.

   • **Point-to-point circuits** This method provides a communication path between two points. It can be a dial-up or a dedicated circuit.

   • **Multidrop circuits** This method allows for the sharing of a communication facility. It is similar to a party line telephone call because several input/output terminals share the same line. Only one terminal can be transmitting on the line at a time.

   • **Local cable** This method of connecting communication devices consists of a privately owned cable or wire interconnecting many terminals with the computer system.

6. **Packet Switching (Value Added Networks—VAN) System** A type of data communication technique that allows for messages to be divided or segmented into packets and routed dynamically through a network to the final destination point.

7. **Interface Unit** The device that connects a data transmitting (terminal) or receiving unit to the transmission facility. An example of this would be a modem, a digital service unit, or a device that converts voltage signaling to light signaling.

8. **Multiplexer** A device that combines several independent data streams into one data stream at a higher signaling speed for transmission to a similar device that separates the high-speed signal into the original independent data streams. Note: Some of the multiplexers are software-driven and are similar to concentrators; however, most of them are nonintelligent hard-wired devices.

9. **Concentrator** A programmable device that will perform the same func-

tion as a multiplexer with added functions such as data storage (buffering), message error checking, data flow control, polling, etc.

10. **Front End Communication Processor**   A programmable device that interfaces a communication network to a host computer. Some of the functions that can be performed by a front end are polling, code and speed conversion, error detection and correction, store and forward functions, format checking, data flow control, network statistics gathering, message authentication, communication routing and control, and the like.

11. **Message Switch**   A privately owned programmable device that accepts messages from many users, stores them, and at some time after receiving them transmits them to their intended destination. This device generally receives messages at slow speeds over dial-up lines.

12. **Protocols**   Software or hardware rules that facilitate the transmission between devices. Some protocols provide for error control.

13. **Test Equipment (Technical Control Facility)**   A combination of equipment that facilitates the physical monitoring, diagnostics, and restoration of communication systems should they fail. They can contain circuit patching, spare equipment, and alternate switches, and they might involve message text monitoring or quantitative measure equipment.

14. **Audio Response Unit**   A unit that accepts analog, audio voice, or digital signals and converts them to digital computer signaling or can also convert digital signals from a computer into human understandable voice signals.

15. **Auto Answering**   A device that automatically answers a telephone and establishes a connection between data communication devices.

16. **Auto Dialing Unit**   A device that accepts computer signals and automatically dials the telephone number of a remote communication device.

17. **Terminals**   An input/output device that is used to enter messages into the system and/or receive messages from the system.

RISK ANALYSIS FOR DATA COMMUNICATION NETWORKS

The matrix concept that is used for conducting a risk analysis was first introduced in Chapter 8 in the section "Matrix of Controls." Read that section before proceeding further.

The risk analysis methodology shows how to develop average annual loss figures for all of the potential threats that might occur. Various estimates will have to be made; therefore, some of these dollar figures will be subject to question. Estimating various "dollar loss figures" and "probabilities of occurrence" are hazards of any risk analysis methodology.

When possible, have three to five of the most knowledgable people assist in making these estimates (Delphi group).

There are four steps that must be followed in this risk analysis methodology. These are presented below.

STEP ONE: DEVELOP THE EXPECTED LOSS MATRIX

First identify all of the threats to the data communication network and all of the component parts that make up the network. A *threat* is an adverse occurrence that could cause a dollar loss, malfunction, or other unwanted situation in the data communication network. A *component* is one of the individual parts that, when joined together, make up the data communication network.

Threats and components were discussed in Chapter 8. Figure A4–1 shows the empty matrix with the threats listed horizontally across the top and the individual component parts listed vertically down the left axis.

STEP TWO: ESTIMATE THE EXPECTED LOSSES

The next task is to estimate the expected loss that would occur for each threat/component cell of the matrix. For example, starting with the cell in the upper left hand corner of Figure A4–1 try to estimate the expected loss that would occur if messages were lost or accidentally changed at the terminal. In Figure A4–2, notice that a dollar loss *per single occurrence* has been entered in the lower half of the cell all the way down column 1 (below the threat "Message Loss or Change").

The method used most often to obtain such estimates is to assemble a Delphi Team. This small group of three to five experts arrives at a consensus as to their estimate of the dollar loss per single occurrence with regard to each individual threat and its corresponding component. This estimate is then recorded in each individual cell.

What a Delphi Team usually does is discuss and understand the threat "Message Loss or Change." Then they try to allocate the loss per occurrence (each time a message is lost or changed) among each of the various components. The Delphi Team may address each cell (threat to component) individually. Alternatively, they may look at the overall threat and ask themselves what the dollar loss per occurrence might be should this threat occur in relation to each component down the column. The net result is the Delphi Team's best estimate as to the individual per-occurrence loss. Each of these dollar values is placed in the lower left corner of the cell as shown in Figure A4–2.

Figure A4–3 is an aid to making this estimate. Because of the difficulty of arriving at a specific dollar estimate, it is better to agree that a certain loss might fall within a range. The midpoint of that range is then used as the best estimate of the dollar loss per single occurrence. This is the method that was used to determine loss figures for the threat "Message Loss or Change" in column 1 of the matrix in Figure A4–2.

Let us assume that the logic of the Delphi Team was the following. If a message is lost at a terminal, in a front end processor, in the host computer, or in the software, it will be discovered very quickly because of existing control techniques. Therefore, we will have a very low dollar loss per occurrence. If the message is lost on the communication circuits, it can disappear for some period of time because controls are lacking in the current system. Therefore, this type of message

THREATS / COMPONENTS	Message Loss or Change	Errors and Omissions	Network Unavailable	Lack of Recovery Capability		
Terminals						
Modems						
Communication circuits						
Front end processor						
Host computer						
Protocols/ software						

FIGURE A4–1 Empty matrix with threats/components.

THREATS COMPONENTS	Message Loss or Change	Errors and Omissions	Network Unavailable	Lack of Recovery Capability			
Terminals	$50.						
Modems	0						
Communication circuits	$550.						
Front end processor	$50.						
Host computer	$50.						
Protocols/software	$50.						

FIGURE A4–2 Enter dollar loss per single occurrence.

Mid-point	Range
$0	No loss
$50	Less than $100
$550	Between $100 & $1,000
$3,000	Between $1,000 & $5,000
$7,500	Between $5,000 & $10,000
$30,000	Between $10,000 & $50,000
$75,000	Between $50,000 & $100,000
$175,000	Between $100,000 & $250,000
$375,000	Between $250,000 & $500,000
$750,000	Between $500,000 & $1,000,000
$1,500,000	Between $1,000,000 & $2,000,000
$3,500,000	Between $2,000,000 & $5,000,000
$7,500,000	Between $5,000,000 & $10,000,000
Use Actual Estimated Loss	Greater than $10,000,000

FIGURE A4–3 Expected loss ranges.

loss might result in a larger dollar loss. Finally, the Delphi Team agreed that a message cannot be lost in a modem; it can be lost in the terminal or on the communication circuits, but not in the modem. Figure A4–4 shows the dollar loss per occurrence estimates for the entire matrix of four threats and six components.

STEP THREE: ESTIMATE HOW OFTEN EACH THREAT IS LIKELY TO OCCUR

Figure A4–5 is the annual loss multiplier table that can be used to estimate the occurrence rate (not a true probability) for each threat in relation to its corresponding component.

The same Delphi Team looks at the threat "Message Loss or Change" to estimate how often this threat might occur. Using the annual loss multiplier table (Figure A4–5) their choices range, in a continuous fashion, from never to several times a day (25 times).

Looking now at Figure A4–6, notice that their *best estimates* were: messages might be lost at the terminals at the rate of one each day (365 was taken from Figure A4–5), at modems only one each year, at the communication circuits once each day, and at the front end/host computer/software, once each month. The numbers 365, 1, 365, 12, 12, 12 that go down column 1 of Figure A4–6 were chosen from Figure A4–5 by the Delphi Team.

THREATS / COMPONENTS	Message Loss or Change	Errors and Omissions	Network Unavailable	Lack of Recovery Capability				
Terminals	$50.	$50.	$550.	$550.				
Modems	0	0	$550.	$550.				
Communication circuits	$550.	$50.	$550.	$550.				
Front end processor	$50.	$50.	$3,000.	$3,000.				
Host computer	$50.	$50.	$7,500.	$7,500.				
Protocols/ software	$50.	$550.	$7,500.	$7,500.				

FIGURE A4–4 Dollar losses per single occurrence.

Time Value Perception	Basis of Computation	Annual Loss Multiplier
Never	—	Zero
Once each 300 years	1/300	0.0033
Once each 100 years	1/100	0.01
Once each 50 years	1/50	0.02
Once each 10 years	1/10	0.10
Once each 3 years	1/3	0.33
Once each year	1/1	1.0
Once each 4 months	12/4	3.0
Once each month	12/1	12.0
Once each week	52/1	52.0
Once each day	365/1	365.0
Several times a day (10 times)	365/0.1	3,650.0
Several times a day (25 times)	365/0.04	9,125.0

FIGURE A4–5 Annual loss multiplier.

Figure A4–7 shows the entire matrix completed with all of the individual dollar losses per single occurrence recorded in the lower left half of the cells and the individual occurrence rates (annual loss multipliers) recorded in the upper right half of the cells. This is a "working matrix." The dollar loss and occurrence rate for each threat to component interface has been identified. All that remains to be done is to multiply these two numbers.

STEP FOUR: CALCULATE/COPY THE AVERAGE ANNUAL LOSS ONTO THE RISK ANALYSIS ANNUAL LOSS MATRIX

The last task is to multiply the dollar loss per occurrence by the occurrence rate (annual loss multiplier) in order to obtain the "average annual loss" for each cell. Figure A4–8 shows this calculation copied onto the average annual loss matrix.

In summary, the matrix now shows the average annual dollar loss for all of the threat-to-component interfaces (individual cells) as well as for individual threats and individual components.

- The total at the bottom of each column is the average annual loss associated with each individual threat.
- The total in the rightmost column is the average annual loss for each individual component of the system.

COMPONENTS \ THREATS	Message Loss or Change	Errors and Omissions	Network Unavailable	Lack of Recovery Capability			
Terminals	365 / $50.						
Modems	1 / 0						
Communication circuits	365 / $550.						
Front end processor	12 / $50.						
Host computer	12 / $50.						
Protocols/ software	12 / $50.						

FIGURE A4–6 Enter occurrence rates for each threat/component.

THREATS / COMPONENTS	Message Loss or Change	Errors and Omissions	Network Unavailable	Lack of Recovery Capability			
Terminals	365 / $50.	3,650 / $50.	52 / $550.	0.33 / $550.			
Modems	1 / 0	1 / 0	12 / $550.	0.33 / $550.			
Communication circuits	365 / $550.	3 / $50.	3 / $550.	0.33 / $550.			
Front end processor	12 / $50.	12 / $50.	1 / $3,000.	0.33 / $3,000.			
Host computer	12 / $50.	12 / $50.	1 / $7,500.	0.33 / $7,500.			
Protocols/ software	12 / $50.	12 / $550.	1 / $7,500.	0.33 / $7,500.			

FIGURE A4–7 Matrix of estimated dollar losses/occurrence rates.

THREATS COMPONENTS	Message Loss or Change	Errors and Omissions	Network Unavailable	Lack of Recovery Capability	TOTALS
Terminals	$18,250	$182,500	$28,600	$182	$229,532
Modems	0	0	$6,600	$182	$6,782
Communication circuits	$200,750	$150	$1,650	$182	$202,732
Front end processor	$600	$600	$3,000	$990	$5,190
Host computer	$600	$600	$7,500	$2,475	$11,175
Protocols/ software	$600	$6,600	$7,500	$2,475	$17,175
TOTALS	$220,800	$190,450	$54,850	$6,486	$472,586

FIGURE A4–8 Average annual loss matrix.

- The dollar value in each cell is the average annual loss for each threat-to-component relationship.
- The total in the lower right corner ($472,586) is the average annual loss for the entire data communication network.

In order to refine or make constructive changes to the matrix, the Delphi Team can examine each average annual loss in each cell to determine how realistic the figure is in light of their current knowledge. For example, the interface of the cell for "Message Loss or Change" and "Communications Circuits" shows an average annual loss of $200,750. If the network uses microwave-based circuits maintained by a major telephone company, this estimate would be very high. On the other hand, if it uses tropospheric scatter circuits that are maintained by four small independent telephone companies among the various islands in the South Pacific, this might not be too high a figure. As this demonstrates, the matrix approach permits an iterative process to refine earlier calculations and possibly modify some of the estimated dollar losses per occurrence and estimated occurrence rates.

CIRCUIT COSTS (TARIFFS)

VOICE GRADE CIRCUIT (SERIES 2000/3000)

The cost of a voice grade private circuit is comprised of the mileage charges and a station terminal charge (line termination at each end of the circuit). The monthly charges are determined by calculating the mileage between the two cities, multiplying the mileage by the appropriate rate, and then adding in a station terminal charge at each end. The station terminal is the "plug" into which the modem is connected. This cost calculation is comprised of three steps:

- Using the vertical and horizontal coordinates, calculate the line mileage (this formula was shown in Figure 10–24).
- Determine whether the cities being connected are a category A rate center, category B rate center, or a combination of the two (Figure A5–1).
- Calculate the mileage charges and the station terminal charges from the rate tables.

The alphabetical list of cities in Figure A5–1 are the category A (large cities as determined by AT&T) rate centers and the vertical and horizontal coordinates are included for each. Using this list, you can determine both the mileage between cities and the category A rate centers. For very accurate work, AT&T Tariff Number 264 (which lists over 19,000 rate centers) should be used.

	V H		V H
Abilene, Tex.	8698 – 4513	Bridgeport, Conn.	4841 – 1360
Akron, Ohio	5637 – 2472	Bristow, Okla.	7799 – 4216
Albany, Ga.	7649 – 1817	Brockton, Mass.	4465 – 1205
Albany, N.Y.	4639 – 1629	Buffalo, N.Y.	5075 – 2326
Albuquerque, N.Mex.	8549 – 5887	Buffalo Peace	
Alexandria, La.	8409 – 3168	Bridge, N.Y.	5074 – 2334
Allentown, Pa.	5166 – 1585	Burlington, Iowa	6449 – 3829
Altoona, Pa.	5460 – 1972	Burlington, Vt.	4270 – 1808
Amarillo, Tex.	8266 – 5076	Calais, Me.	3561 – 1208
Anaheim, Calif.	9250 – 7810	Cambridge, Mass.	4425 – 1258
Anniston, Ala.	7406 – 2304	Camden, N.J.	5249 – 1453
Antonia, Mo.	6880 – 3507	Canton, Ohio	5676 – 2419
Apollo, Tex.	8958 – 3482	Cape Girardeau, Mo.	7013 – 3251
Appleton, Wis.	5589 – 3776	Carson City, Nev.	8139 – 8306
Asheville, N.C.	6749 – 2001	Casper, Wyo.	6918 – 6297
Atlanta, Ga.	7260 – 2083	Casselton, N.Dak.	5633 – 5241
Atlantic City, N.J.	5284 – 1284	Cedar Rapids, Iowa	6261 – 4021
Augusta, Ga.	7089 – 1674	Centralia, Ill.	6744 – 3311
Austin, Tex.	9005 – 3996	Champaign–Urbana, Ill.	6371 – 3336
Baker, Calif.	8888 – 7537	Charleston, S.C.	7021 – 1281
Bakersfield, Calif.	8947 – 8060	Charleston, W. Va.	6152 – 2174
Baltimore, Md.	5510 – 1575	Charlotte, N.C.	6657 – 1698
Baton Rouge, La.	8476 – 2874	Chattanooga, Tenn.	7098 – 2366
Beaumont, Tex.	8777 – 3344	Cheshire, Conn.	4755 – 1366
Beckley, W. Va.	6218 – 2043	Chesterfield, Mass.	4595 – 1478
Benton Ridge, Ohio	5847 – 2784	Cheyenne, Wyo.	7203 – 5958
Berlin, N.J.	5257 – 1408	Chicago, Ill.	5986 – 3426
Bethia, Va.	5957 – 1491	Chico, Calif.	8057 – 8668
Billings, Mont.	6391 – 6790	Chipley, Fla.	7927 – 1958
Biloxi, Miss.	8296 – 2481	Cincinnati, Ohio	6263 – 2679
Binghampton, N.Y.	4943 – 1837	Clarksville, Tenn.	6988 – 2837
Birmingham, Ala.	7518 – 2446	Clarksburg, W. Va.	5865 – 2095
Bismarck, N.Dak.	5840 – 5736	Clearwater, Fla.	8203 – 1206
Blacksburg, Va.	6247 – 1867	Cleveland, Ohio	5574 – 2543
Bloomington, Ind.	6417 – 2984	Cocoa, Fla.	7925 – 0903
Blue Ridge Summit, Pa.	5518 – 1746	Collinsville, Ill.	6781 – 3455
Boise, Idaho	7096 – 7869	Colorado Springs, Colo.	7679 – 5813
Boone, Iowa	6394 – 4355	Columbia, S.C.	6901 – 1589
Boston, Mass.	4422 – 1249	Columbus, Ga.	7556 – 2045
Brewton, Ala.	8001 – 2244	Columbus, Miss.	7657 – 2704

	V	H		V	H
Columbus, Ohio	5972	2555	Fort Lauderdale, Fla.	8282	0557
Concord, N.H.	4326	1426	Fort Morgan, Colo.	7335	5739
Conyers, Ga.	7243	2016	Fort Myers, Fla.	8359	0904
Corpus Christi, Tex.	9475	3739	Fort Pierce, Fla.	8054	0737
Crestview, Fla.	8025	2128	Fort Walton Beach, Fla.	8097	2097
Crosby, N. Dak.	5495	6199	Fort Wayne, Ind.	5942	2982
Dallas, Tex.	8436	4034	Fort Worth, Tex.	8479	4122
Danville, Ky.	6558	2561	Framingham, Mass.	4472	1284
Davenport, Iowa	6273	3817	Frankfort, Ky.	6462	2634
Dayton, Ohio	6113	2705	Fresno, Calif.	8669	8239
Daytona Beach, Fla.	7791	1052	Gainesville, Fla.	7838	1310
Decatur, Ala.	7324	2585	Gastonia, N.C.	6683	1754
De Kalb, Ill.	6061	3591	Glenwood Springs, Colo.	7651	6263
Delta, Utah	7900	7114	Grand Forks, N.Dak.	5420	5300
Denver, Colo.	7501	5899	Grand Island, Nebr.	6901	4936
Des Moines, Iowa	6471	4275	Grand Junction, Colo.	7804	6438
Detroit, Mich.	5536	2828	Grand Rapids, Mich.	5628	3261
Dickinson, N.Dak.	5922	6024	Greeley, Colo.	7345	5895
Dodge City, Kans.	7640	4958	Green Bay, Wis.	5512	3747
Dodgeville, Wis.	5963	3890	Greensboro, N.C.	6400	1638
Dover, Del.	5429	1408	Greenville, Miss.	7888	3126
Duluth, Minn.	5352	4530	Greenville, N.C.	6250	1226
Eau Claire, Wis.	5698	4261	Greenville, S.C.	6873	1894
El Paso, Tex.	9231	5655	Greenwood, Miss.	7798	2993
Ennis, Tex.	8514	3970	Gulfport, Miss.	8317	2511
Eureka, Calif.	7856	9075	Hackensack, N.J.	4976	1432
Evansville, Ind.	6729	3019	Harlingen, Tex.	9820	3663
Fairmont, W. Va.	5808	2091	Harrisburg, Pa.	5363	1733
Fairview, Kans.	6956	4443	Hartford, Conn.	4687	1373
Fall River, Mass.	4543	1170	Hattiesburg, Miss.	8152	2636
Fargo, N.Dak.	5615	5182	Hayward, Calif.	8513	8660
Fayetteville, Ark.	7600	3872	Helena, Mont.	6336	7348
Fayetteville, N.C.	6501	1385	Herndon, Va.	5644	1640
Findlay, Ohio	5828	2766	Hinsdale, Ill.	6023	3461
Fitzgerald, Ga.	7539	1684	Hot Springs, Ark.	7827	3554
Flagstaff, Ariz.	8746	6760	Houghton, Mich.	5052	4088
Flint, Mich.	5461	2993	Houston, Tex.	8938	3536
Florence, S.C.	6744	1417	Huntington, N.Y.	4918	1349
Forrest City, Ark.	7555	3232	Huntington, W. Va.	6212	2299
Fort Collins, Colo.	7331	5965			

	V H		V H
Huntsville, Ala.	7267 – 2535	Lodi, Calif.	8397 – 8532
Huron, S.Dak.	6201 – 5183	Longview, Tex.	8348 – 3660
Indianapolis, Ind.	6272 – 2992	Logan, Utah	7367 – 7102
Iowa City, Iowa	6313 – 3972	Los Angeles, Calif.	9213 – 7878
Iron Mountain, Mich.	5266 – 3890	Louisville, Ky.	6529 – 2772
Jackson, Mich.	5663 – 3009	Lubbock, Tex.	8598 – 4962
Jackson, Miss.	8035 – 2880	Lynchburg, Va.	6093 – 1703
Jackson, Tenn.	7282 – 2976	Lyons, Nebr.	6584 – 4732
Jacksonville, Fla.	7649 – 1276	Macon, Ga.	7364 – 1865
Jasper, Ala.	7497 – 2553	Madison, Wis.	5887 – 3796
Johnson City, Tenn.	6595 – 2050	Madisonville, Ky.	6845 – 2942
Joliet, Ill.	6088 – 3454	Manchester, N.H.	4354 – 1388
Joplin, Mo.	7421 – 4015	Manhattan, Kans.	7143 – 4520
Julian, Calif.	9374 – 7544	Marion, Ill.	6882 – 3202
Kalamazoo, Mich.	5749 – 3177	Mattoon, Ill.	6502 – 3291
Kansas City, Kans.	7028 – 4212	McComb, Miss.	8262 – 2823
Kansas City, Mo.	7027 – 4203	Medford, Oreg.	7503 – 8892
Kennewick, Wash.	6595 – 8391	Memphis, Tenn.	7471 – 3125
Key West, Fla.	8745 – 0668	Meridian, Miss.	7899 – 2639
Kingsport, Tenn.	6570 – 2107	Miami, Fla.	8351 – 0527
Klamath Falls, Oreg.	7510 – 8711	Midland, Tex.	8934 – 4888
Knoxville, Tenn.	6801 – 2251	Milwaukee, Wis.	5788 – 3589
La Crosse, Wis.	5874 – 4133	Minneapolis, Minn.	5781 – 4525
Lafayette, La.	8587 – 2996	Mobile, Ala.	8167 – 2367
Lake Charles, La.	8679 – 3202	Mojave, Calif.	8993 – 7899
Lake City, Fla.	7768 – 1419	Monroe, La.	8148 – 3218
Lamar, Colo.	7720 – 5403	Montgomery, Ala.	7692 – 2247
Lansing, Mich.	5584 – 3081	Mooers Forks, N.Y.	4215 – 1929
La Plata, Md.	5684 – 1528	Morgantown, W. Va.	5764 – 2083
Laredo, Tex.	9681 – 4099	Morristown, N.J.	5035 – 1478
Laredo, Tex.	9683 – 4098	Morristown, Tenn.	6699 – 2183
Las Cruces, N.Mex.	9132 – 5742	Muncie, Ind.	6130 – 2925
Las Vegas, Nev.	8665 – 7411	Muskogee, Okla.	7746 – 4042
Laurel, Miss.	8066 – 2645	Nashua, N.H.	4394 – 1356
Laurinburg, N.C.	6610 – 1437	Nashville, Tenn.	7010 – 2710
Lawrence, Mass.	4373 – 1311	Nassau, N.Y.	4961 – 1355
Leesburg, Va.	5634 – 1685	Neche, N.D.	5230 – 5456
Little Rock, Ark.	7721 – 3451	Newark, Ill.	6123 – 3527
Littleton, Mass.	4432 – 1327	Newark, N.J.	5015 – 1430
Locust, N.C.	6613 – 1640	New Bern, N.C.	6307 – 1119

	V	H		V	H
New Brunswick, N.J.	5085 – 1434		Potsdam, N.Y.	4404 – 2054	
New Haven, Conn.	4792 – 1342		Pottstown, Pa.	5246 – 1563	
New London, Conn.	4700 – 1242		Poughkeepsie, N.Y.	4821 – 1526	
New Market, Md.	5558 – 1676		Prescott, Ariz.	8917 – 6872	
New Orleans, La.	8483 – 2638		Providence, R.I.	4550 – 1219	
Newport News, Va.	5908 – 1260		Provo, Utah	7680 – 7006	
New York City, N.Y.	4997 – 1406		Racine, Wis.	5837 – 3535	
Norfolk, Va.	5918 – 1223		Raleigh, N.C.	6344 – 1436	
North Bend, Nebr.	6698 – 4739		Reading, Pa.	5258 – 1612	
North Bend, Wash.	6354 – 8815		Red Oak, Iowa	6691 – 4465	
North Brook, Ill.	5954 – 3479		Redwood City, Calif.	8556 – 8682	
Oakland, Calif.	8486 – 8695		Reno, Nev.	8064 – 8323	
Ocala, Fla.	7909 – 1227		Richmond, Va.	5906 – 1472	
Ogden, Utah	7480 – 7100		Roanoke, Va.	6196 – 1801	
Oklahoma City, Okla.	7947 – 4373		Rochester, N.Y.	4913 – 2195	
Omaha, Nebr.	6687 – 4595		Rockford, Ill.	6022 – 3675	
Orangeburg, S.C.	6980 – 1502		Rock Island, Ill.	6276 – 3816	
Orlando, Fla.	7954 – 1031		Rocky Mount, N.C.	6232 – 1329	
Panama City, Fla.	8057 – 1914		Rosendale, N.Y.	4813 – 1564	
Parkersburg, W. Va.	5976 – 2268		Roswell, N.Mex.	8787 – 5413	
Pendelton, Oreg.	6707 – 8326		Sacramento, Calif.	8304 – 8580	
Pensacola, Fla.	8147 – 2200		Saginaw, Mich.	5404 – 3074	
Peoria, Ill.	6362 – 3592		Salina, Kans.	7275 – 4656	
Petersburg, Va.	5961 – 1429		Salinas, Calif.	8722 – 8560	
Petoskey, Mich.	5120 – 3425		Salt Lake City, Utah	7576 – 7065	
Philadelphia, Pa.	5251 – 1458		San Angelo, Tex.	8944 – 4563	
Philadelphia, Pa.	5257 – 1501		San Antonio, Tex.	9225 – 4062	
Philadelphia, Pa.	5222 – 1493		San Bernardino, Calif.	9172 – 7710	
Phoenix, Ariz.	9135 – 6748		San Diego, Calif.	9468 – 7629	
Pine Bluff, Ark.	7803 – 3358		San Francisco, Calif.	8492 – 8719	
Pittsburgh, Pa.	5621 – 2185		San Jose, Calif.	8583 – 8619	
Plano, Ill.	6096 – 3534		San Luis Obispo, Calif.	9005 – 8349	
Plymouth, Mich.	5562 – 2891		Santa Fe, N.Mex.	8389 – 5804	
Pocatello, Idaho	7146 – 7250		Santa Rosa, Calif.	8354 – 8787	
Polk City, Fla.	8067 – 1067		Sarasota, Fla.	8295 – 1094	
Pontiac, Mich.	5498 – 2895		Scranton, Pa.	5042 – 1715	
Port Angeles, Wash.	6206 – 9061		Searcy, Ark.	7581 – 3407	
Port Huron, Mich.	5367 – 2813		Seattle, Wash.	6336 – 8896	
Portland, Me.	4121 – 1334		Seguin, Tex.	9161 – 3981	
Portland, Oreg.	6799 – 8914		Shreveport, La.	8272 – 3495	

	V	H		V	H
Sidney, Nebr.	7112 – 5671		Tulsa, Okla.	7707 – 4173	
Sikeston, Mo.	7099 – 3221		Tupelo, Miss.	7535 – 2825	
Sioux City, Iowa	6468 – 4768		Twin Falls, Idaho	7275 – 7557	
Sioux Falls, S.Dak.	6279 – 4900		Ukiah, Calif.	8206 – 8885	
Socorro, N.Mex.	8774 – 5867		Van Nuys, Calif.	9197 – 7919	
South Bend, Ind.	5918 – 3206		Waco, Tex.	8706 – 3993	
Spartanburg, S.C.	6811 – 1833		Wadena, Minn.	5606 – 4915	
Spokane, Wash.	6247 – 8180		Waldorf, Md.	5659 – 1531	
Springfield, Ill.	6539 – 3513		Warrenton, Va.	5728 – 1667	
Springfield, Mass.	4620 – 1408		Washington, D.C.	5622 – 1583	
Springfield, Mo.	7310 – 3836		Washington, D.C.	5603 – 1598	
Stamford, Conn.	4897 – 1388		Washington, D.C.	5632 – 1590	
Stevens Point, Wis.	5622 – 3964		Waterloo, Iowa	6208 – 4167	
Stockton, Cal.	8435 – 8530		Waycross, Ga.	7550 – 1485	
St. Joseph, Mo.	6913 – 4301		Westchester, N.Y.	4921 – 1416	
St. Louis, Mo.	6807 – 3482		West Glendive, Mont.	5963 – 6322	
St. Paul, Minn.	5776 – 4498		West Palm Beach, Fla.	8166 – 0607	
St. Petersburg, Fla.	8224 – 1159		West Sweetgrass, Mont.	5829 – 7475	
Succasunna, N.J.	5038 – 1508		Wheeling, W. Va.	5755 – 2241	
Sunnyvale, Calif.	8576 – 8643		White River Jct., Vt.	4327 – 1585	
Superstition–Apache Junction, Ariz.	9123 – 6669		Wichita, Kans.	7489 – 4520	
Sweetwater, Tex.	8737 – 4632		Williamsport, Pa.	5200 – 1873	
Syracuse, N.Y.	4798 – 1990		Williamstown, Ky.	6353 – 2636	
Tallahassee, Fla.	7877 – 1716		Wilmington, Del.	5326 – 1485	
Tampa, Fla.	8173 – 1147		Winchester, Ky.	6441 – 2509	
Terre Haute, Ind.	6428 – 3145		Winston-Salem, N.C.	6440 – 1710	
Thomasville, Ga.	7773 – 1709		Winter Garden, Fla.	7970 – 1069	
Toledo, Ohio	5704 – 2820		Winter Haven, Fla.	8084 – 1034	
Topeka, Kans.	7110 – 4369		Woodstock, Ill.	5964 – 3587	
Traverse City, Mich.	5284 – 3447		Worcester, Mass.	4513 – 1330	
Trenton, N.J.	5164 – 1440		Wyoming Switch, Minn.	5686 – 4521	
Troy, Ala.	7771 – 2136		Youngstown, Ohio	5557 – 2353	
Troy, N.Y.	4616 – 1633		Yuma, Ariz.	9385 – 7171	
Tucson, Ariz.	9345 – 6485				
Tully, N.Y.	4838 – 1953				

FIGURE A5–1 Category A rate centers including vertical and horizontal coordinates (V&H).

For estimation purposes, this list of category A rate centers is quite adequate. Actually, in a real situation the final cost is determined by the common carrier. By using the list of category A rate centers, however, you can estimate circuit costs within 1 percent of the final figure.

The specific procedure is to:

- Calculate the mileages using the vertical and horizontal coordinates or an air atlas mileage table. An air atlas mileage table will have to be used for any cities other than category A rate centers because we have listed only category A rate centers here.

- Utilize the cost schedules in Figure A5 2 to calculate the monthly charges. Remember that these charges are levied against the organization on a monthly basis.

Use Schedule I in Figure A5–2 when the two cities that are being connected go from a category A rate center to a category A rate center. Schedule II is used when the communication circuit goes between a category A rate center and a category B rate center (category B rate centers are any city that is not listed on the category A rate cities shown in Figure A5–1). Schedule III is used when the two cities that are being connected are both category B rate centers (neither city is listed in the list of category A rate centers shown in Figure A5–1).

Remember to add a station terminal charge (shown in Figure A5–2) for each terminal connection. For example, on a point-to-point circuit there might be a station terminal charge for each end of this circuit. On a multidrop circuit there might be a station terminal charge at the front end communication processor and

| | Schedule | | |
Mileage	I	II	III
First mile	$73.56	$75.00	$76.43
Next 14 miles, each (2–15)	2.59	4.77	6.35
Next 10 miles, each (16–25)	2.16	4.47	5.48
Next 15 miles, each (26–40)	1.62	2.89	4.03
Next 20 miles, each (41–60)	1.62	1.95	3.03
Next 20 miles, each (61–80)	1.62	1.95	2.31
Next 20 miles, each (81–100)	1.62	1.95	1.95
Next 900 miles, each (101–1000)	.94	.94	.97
Each additional mile (1001 and over)	.58	.58	.58

Station Terminal Charge:
$36.00 per circuit termination (full or half duplex) per month

FIGURE A5–2 Voice grade price schedule (interstate).

Type C	
• Point-to-point without switching	$30.00/month
• Two point with switching	$44.00/month
• Multipoint without switching	$44.00/month
Type D	
• Point-to-point only	$21.00/month
• Multidrop (3-point only)	$71.00/month

FIGURE A5–3 Conditioning charges (per month).

several more station terminal charges (one for each dropoff) as the line goes out to the various terminal locations.

If a high-speed modem requires conditioning of a voice grade circuit, a monthly cost will have to be added (as shown in Figure A5–3), depending upon the type of conditioning that is ordered. There are various subclasses within Type C and Type D conditioning. In Figure A5–3 we show only a single average cost for Type C and Type D. These are adequate for estimation purposes.

DIGITAL CIRCUIT

For estimating the costs of a digital circuit, our presentation utilizes the American Telephone and Telegraph Dataphone Digital Service pricing (Dataphone Digital Service is a registered trademark of the American Telephone and Telegraph Company). This digital service is available on a point-to-point or multipoint basis in either full duplex or half duplex. To calculate the cost of a digital circuit, the cost of four different elements has to be accumulated. Each is a monthly charge.

- Digital station terminals (the plug into which the terminal/modem is attached)
- Central office data access line (this is the local loop mileage between your site and the telephone company central office)
- Mileage charges (IXC)
- Modem (this is the data service unit)

Figure A5–4 is a listing of these four cost elements.

First, determine the number of digital station terminals that are needed (one for each dropoff or terminal). Second, determine the central office data access line charges. Third, calculate the line mileage costs between cities by using the vertical and horizontal coordinates or an air atlas. These costs depend upon the speed

1. Digital Station Terminals

	2400 BPS	4800 BPS	9600 BPS	56,000 BPS	1.544 MBPS
(A)	$36.00	$36.00	$47.00	$180.00	(C)
(B)	100.00	208.00	383.00	908.00	(C)

(A) When connecting to AT&T circuits.

(B) When connecting to another common carrier's circuits.

2. Central Office Data Access Line.

	2400 BPS	4800 BPS	9600 BPS	56,000 BPS	1.544 MBPS
	$64.00	$172.00	$336.00	$728.00	$721.00

This is the local loop mileage.

	+ (C)	+ (C)	+ (C)	+ (C)	+ $87./mile

(C) Not yet tariffed.

3. Mileage Charges

	2400, 4800 9600 BPS	56,000 BPS	1.544 MBPS
First mile	$73.56	$367.80	$2,884.00
Next 14 miles (2–15)	2.59	12.95	92.30
Next 10 miles (16–25)	2.16	10.80	92.30
Next 75 miles (26–100)	1.62	8.10	92.30
Next 100 miles (101–200)	.94	4.70	92.30
Next 300 miles (201–500)	.94	4.70	72.11
Next 500 miles (501–1000)	.94	4.70	57.70
Each additional mile Over 1000	.58	2.90	57.70

4. Modem (Data Service Unit)

2400, 4800, 9600 56,000 BPS	56,000 BPS
$45.00/Month	$576.00/Month

FIGURE A5–4 Digital circuit price schedule.

of transmission, *not* whether they are category A/B rate centers. Finally, add in the cost of the digital modems at each line termination. When comparing the cost of digital circuits with that of regular voice grade (analog) circuits, be sure to include modem costs. Modems cost much less for digital than they do for analog circuits.

SATELLITE CIRCUIT

Voice grade equivalent point-to-point communication circuits can be furnished via a satellite connection. The cost of these circuits are made up of the satellite channel charge, the local access channel (local loop), and the station termination charge. Channel conditioning also can be bought.

Figure A5–5 lists the charges for a satellite circuit. In order to calculate the cost of a satellite channel, determine the appropriate circuit charge (notice that satel-

---

1. **Satellite Channel Charge:** Four-wire voice grade circuits, monthly charge

ATL	CHI	DAL	HOU	LAS	NYC	PHL	SFC	WDC	
—	B	B	B	F	B	—	F	B	Atlanta
	—	B	B	C	B	—	C	—	Chicago
		—	—	D	D	—	D	D	Dallas/Fort Worth
			—	D	D	—	D	D	Houston
				—	F	F	—	F	Los Angeles
					—	—	F	—	New York City
						—	F	—	Philadelphia
							—	F	San Francisco
								—	Washington D.C.

Monthly Charge				
A	B	C	D	F
—	$690	$765	$765	$965

2. **Local Loop Access Channels:** Per circuit end

Monthly $37

3. **Station Termination:** Per circuit terminated at a customer location

	Monthly
With echo suppressor	$28
With echo canceller	58

4. **Channel Conditioning:** Monthly charge per voice grade circuit

Type C2—$28   Type C4—$58

FIGURE A5–5 Satellite circuit price schedule.

lite channels are available only between major cities), add the local loop access channel cost (one is required at each end of the circuit) and the monthly cost of the station termination (the plug at each dropoff). If channel conditioning is needed, this will be an additional monthly charge.

DIAL-UP CIRCUIT

For dial-up communication circuits, as well as leased circuits, it should be recognized that there are two distinctly different rate schedules. One is for *interstate* calls (calls between states) and the other is for *intrastate* calls (calls within the state). Figure A5–6 shows an interstate dial-up rate table. If the cost of calls wholly within the home state needs to be calculated, contact the local telephone company to request an intrastate dial-up call rate table.

U.S. Interstate (Continental U.S.)							
	Sun.	Mon.	Tues.	Wed.	Thurs.	Fri.	Sat.
8 A.M.–5 P.M.	Night	Day					
5 P.M.–11 P.M.	Evening Rate: 60% of Day Rate						
11 P.M.–8 A.M.	Night Rate: 40% of Day Rate						

Day Rate

Mileage	First Minute	Each Additional Minute
1-10	$.32	$.16
11-22	.40	.22
23-55	.48	.28
56-124	.57	.37
125-292	.58	.39
293-430	.59	.42
431-925	.62	.43
926-1910	.64	.44
1911-3000	.74	.49
3001-4250	.76	.51
4251-5750	.79	.53

FIGURE A5–6 Dial-up price schedule.

To calculate the cost of a dial-up call, determine the mileage between the calling party's location and the called party's location. This mileage figure leads to the correct portion of the rate table. Now only the number of minutes of the duration of the call need be determined. By multiplying the appropriate rate (see Figure A5–6) by the number of minutes, the cost of the call is calculated. As shown, evening and night rate calls cost less.

WIDEBAND CIRCUIT (SERIES 8000)

The Series 8000 wideband data transmission service is for high-speed data transmission of 50,000 bits per second or greater. The circuit that is supplied is a 48,000 hertz bandwidth circuit. It can be delivered as a single 48,000 hertz circuit or subdivided into 12 voice grade channels of 4000 hertz each. These channels must begin at one point and end at one point.

This service has two basic monthly cost categories. First, calculate the interexchange channel (IXC) mileage cost between telephone company central offices. Second, calculate the cost of a service terminal (one at each end of the circuit). Figure A5–7 shows these costs.

Mileage Costs (per month)		
	Miles	$/Mile/Month
	1–250	23.40
	251–500	16.45
	501 and Over	11.70
Service Terminals (per month)		684.00

FIGURE A5–7 Wideband (Series 8000) price schedule.

PACKET SWITCHING NETWORK (PUBLIC)

A packet switched network has three components to its cost. These are usage, network access ports, and network interface equipment. The usage is easy to calculate because it depends upon how many packets are transmitted and the rate charged per 1000 packets.

The *network access port* is the line that connects your premises to the packet switched service. It is made up of either a dial-up access charge or a dedicated circuit access charge.

The *network interface equipment* is the hardware that interfaces the terminals

1. **Usage**
   Per 1,000 packets where a packet is a 128-character block:
   Interstate (U.S.) $1.55/1,000 packets*

   *Volume discounts and off-hour usage discounts are available.

2. **Network Access Port**
   - Dial access (300–1200 bps) $6.90 per hour
   - Includes access line to local telephone company exchange
   - Dedicated (includes access circuit and modems)

1–100 Mile Circuit	Cost/Month
300–1200 bps	$550.00
1800 bps	$700.00
2400–4800 bps	$925.00
9600 bps	$1,400.00
Each additional 100 miles of circuit	$125.00

3. **Network Interface Equipment**

	Cost/Month
Basic unit	$800.00
4 asynchronous ports (9600 bps)	200.00
8 asynchronous ports (9600 bps)	240.00
4 synchronous ports (9600 bps)	240.00
8 HDLC protocol ports (9600 bps)	480.00
1 synchronous port (56,000 bps)	240.00

FIGURE A5–8 Packet switching price schedule.

to the packet switched network. In effect, it is the front end packet switching node (SN). Figure A5–8 lists the figures that make up the three cost components of a packet switched service.

SELECTED HARDWARE COSTS

Figure A5–9 shows some costs for selected pieces of hardware. These costs represent averages and therefore might not reflect the correct cost of a specific piece of hardware. For this reason, it is recommended that you consult reference material (Auerbach or Datapro) or a specific hardware manufacturer to determine the spe-

Modem (300–2400 bps)	$30 per month
Modem (4800 bps)	$125 per month
Modem (9600 bps)	$175 per month
Digital modems were included in Figure A5-4.	
Video terminal	$160 per month
Front end communication processor (depending upon features)	$2,000 per month
Multiplexers	$50–250 per month

FIGURE A5–9 Selected hardware costs.

cific cost of hardware under consideration. Also, these reference materials provide a better picture of the availability of specific types of hardware. For software, the various software vendors must be contacted to determine their purchase/lease cost for specific packages.

APPENDIX 6

EVALUATING TELEPROCESSING MONITORS[1]

Once data processing management decides to implement a teleprocessing package, organization becomes a key issue. The project's success or failure is dependent on this factor. With proper planning and coordination, the system can usually be implemented within predetermined budgetary and time constraints. Without proper organization, the project is destined for failure, because of missed deadlines, cost overruns, and overlooked users whose requirements were never considered.

The basic organizational framework suggested for this task is based on a committee structure. Traditionally, two levels of committees (middle and upper management), consisting of three types, are required to ensure an effective implementation. Depending on the size and structure of the organization, the high-level management representatives may vary in seniority. In some cases, even the company president may become involved in the decision to purchase the monitor:

- **DP Steering Committee** Ideally chaired by a vice-president (or equivalent), this body is enacted to manage the implementation of the system. To be effective, it must include high-level management representatives from the user, data processing, and finance departments. This committee directs sev-

[1]This section is reprinted with permission of SDA Products, 71 Fifth Avenue, New York, NY 10003.

eral ad hoc committees and reports directly on implementation progress to a senior management-level executive. The first responsibility of the Steering Committee is the preparation of a charter to be approved by top management.

- **Standing Subcommittees** These groups, usually already established within the organization's data processing department, are responsible for such functions as system performance evaluation and long-range planning. Since implementation of a teleprocessing monitor has a significant impact on both of these areas, these subcommittees should interface with the DP Steering Committee and ad hoc committees, as required.

- **Ad Hoc Committees** These are temporary working groups organized specifically for special projects, such as implementing the teleprocessing monitor. Although there are three separate functions to be performed, usually by three distinct committees or teams, it is desirable to include the same systems personnel familiar with the hardware configuration, systems software, and applications on every team to ensure better continuity. This organization is facilitated by the fact that the functions are performed sequentially.

Before considering these functions individually, it is important to note that implementing a teleprocessing monitor requires a high level of technical competence, not only in systems software but also in hardware and applications. Selecting a specific teleprocessing monitor can greatly influence the choice of a hardware and network configuration because monitors are often limited in the types of hardware they support. A fixed network or host mainframe configuration can severely limit the choice of a teleprocessing package. The performance of the system ultimately depends on the right combination of software, terminals, memory size, and mainframe. Consequently, a successful implementation must involve the coordination and cooperation of highly competent individuals knowledgable in each of these areas. Thus, the members of the following teams should be carefully chosen.

The ad hoc committees can be categorized as follows:

- A technical design team to gather users' requirements and prepare functional specifications. The team should be involved in systems planning and should have a thorough knowledge of the existing and proposed hardware configuration. Additional members from the data processing department include a systems programmer and applications programmer. One or more representatives from the user department, ideally at the management level, are included to round out the team.

- A selection team to evaluate various vendor proposals against the stated functional specifications. In addition to the systems personnel, the team should include a representative from the finance department. Questions concerning costs and benefits can become significant budgetary concerns when a teleprocessing package is being selected. Thus, it is advantageous to include during the initial evaluation phase someone experienced in budgetary planning. It is also important to identify and estimate all initial and ongoing costs because some are commonly overlooked.

- An implementation team to ensure that the system is installed properly, verified in accordance with a predefined benchmark testing program, and accepted. The same basic personnel who acted as the technical design team should be appointed to carry out this task.

THE CHARTER

The successful implementation of a teleprocessing monitor involves a sizeable commitment of time, effort, and money. To utilize and protect this investment to its fullest, the highest corporate management should sanction the entire effort. The most effective means to accomplish this is to have the company president (or equivalent) sign a written charter that clearly defines the planned implementation.

The charter itself need not be lengthy, but, as a minimum, it should accomplish three objectives. First, it must identify clearly, concisely, and as quantitatively as possible the objectives of the implementation. Monetary savings, for instance, are usually an important consideration and should be delineated where applicable. This estimate will be inaccurate at this stage, but it should not be overly optimistic.

Providing better service to customers is usually another critical factor; anticipated benefits should be expressed in quantitative terms, such as percentage reduction in turnaround time. In conjunction with this, the charter should establish the overall implementation plan, including a schedule and procedures for implementing short-term and long-range goals. A budgetary estimate is also prepared on the basis of estimated manpower and computer resources required. There is a complete list of evaluation criteria in the section, Prepare a Plan in Chapter 10.

Second, the charter should detail the specific interface between the project management (high-level committee) and the company president and should include reporting requirements. Generally, a report is prepared after the evaluation of proposals to recommend the selection of a specific package. Often, a formal presentation is made in conjunction with submission of this report.

After the system has been installed and accepted, a final report covering the highlights of the implementation effort should be prepared for top management.

Third, the charter must define clearly the organizational framework on which this effort is based. It is here that a committee structure is authorized to manage and conduct the project.

COLLECTING USER NEEDS

The basis for selection of a teleprocessing monitor is to choose that product with the greatest efficiency and least cost that most effectively satisfies the present and future needs of the user community. The end result of this detailing of user needs will be a feature list. The list will include user needs weighted so as to point out the relative importance of each item.

The following list outlines the categories for user need studies (Chapter 10, Network Design Fundamentals, should be reviewed).

- **Device Support**   The nature of the applications will give an indication of certain device requirements, such as formatted CRTs, printers, remote CPUs, etc. The network processing distribution and traffic volume will further define device characteristics, such as whether they are local, switched, or leased multipoint, and whether they are dedicated or shared with other applications such as a text editor.

- **Implementation Approach**   The skill level and orientation of the application programmers is a key element; monitors differ widely in their program interface architecture. Some are simple, others complex; monitors also differ in the extent to which the user directly programs functions or alternately can utilize automatic system services and utilities. Certain program interface requirements dictate well-trained, senior-level personnel. The unavailability of system utilities may add many work-months to application implementation time. The potential programming language must be defined, as well as the user's willingness to accept "dialects" or "extensions" to programming languages as required in some systems.

- **Transaction Data**   Information must be gathered on the characteristics of message traffic: total volume, peak volume, message sizes, distribution across various program functions, response time requirements, etc. Also, data must be obtained on probable message processing program characteristics: number and sizes of programs, program I/O activity, amount of processing per message, etc.

- **Data File Requirements**   The need to support certain file structures and to interface with database management systems must be defined.

- **Data Integrity Requirements**   If the system will do on-line updating, the cost effect of a data error situation must be considered. Determine the likelihood of any errors being detected through means external to the on-line system, such as balancing. The more costly the errors, and the less likely their external detection, the more comprehensive the required monitor scheme. Also consider data recovery requirements in the event of system failure or program error.

• **Security and Accounting Requirements** Present and future security needs may require control levels on a terminal, transaction, file and/or user basis, depending on the nature of certain transactions and on the sensitivity of accessed data. As a correlation, an accounting and measurement or statistics scheme may also be required to monitor and report on system usage, etc.

EVALUATION LIST FOR TELEPROCESSING MONITORS

The input obtained in the preceeding requirements analysis will be used to weight the rankings of the feature/performance evaluation list. The list consists of detailed elements falling into fourteen categories (each category is discussed thoroughly in the rest of this section):

- Communications
- Queue Management
- Program Management
- File/Database Management
- Task/Resource Control
- Restart/Recovery
- Utility Features
- Program/System Isolation
- Statistics/Accounting
- Operating Environment and General Information
- Implementation
- Performance
- Cost
- Vendor

COMMUNICATIONS

Transaction monitors are often termed communications monitors since they function in an on-line environment. The communications support contained in a transaction monitor falls into two broad categories: logical device management and physical device control.

Device control requirements vary considerably within the monitor, ranging from minimal to extensive. If the transaction monitor is interfaced to a comprehensive telecommunication access program such as IBM's TCAM or VTAM, or is connected to a network management minicomputer, then the monitor is relieved of nearly all physical control responsibilities. These external facilities will provide for polling, dialing, error detection, error correction, and other such activities. The TP monitor must have an appropriate interface to these facilities. If external control is utilized, then the monitor evaluation must focus on the quality of the interface provided. Care should be made to detect if any of the following communication interface weaknesses exist:

- *A simple product interface with the true support requirements hidden elsewhere.* For example, in TCAM the required support can vary from a few lines of user-supplied code to literally thousands of lines within the TCAM Message Control Program, all as a function of the level of TCAM support within the monitor. The same situation can arise with minicomputer front ends when extensive specialized coding is required.
- *A limited support interface.* For example, any TP monitor can have a simple VTAM interface developed in a short time; however, to use all the SNA facilities properly requires extensive support within the monitor.
- *The external interface is in lieu of other interfaces.* For example, you cannot mix BTAM and TCAM, nor BTAM and VTAM.
- *The external interface results in loss of standard logical communications facilities, such as terminal level security, etc.*

If you are to utilize direct control, such as BTAM, rather than an external system, these are the important items for consideration:

- What and how many terminals are supported and in what manner (leased, switched, etc.)?
- What user coding, if any, is required?
- Extent of control such as control terminal commands for starting/stopping of devices, lines, and polling, line time-out and switched line disconnect control, automatic error recovery, message rerouting, etc.
- Network statistics covering utilization and error situations.

The key element in communications support within the transaction processing monitor is the logical management of devices. The terminal's network traffic is integrally interrelated with message processing by the application programs. This relationship and incumbent support requirements exist irrespective of the physical device connection. The important aspects of device management are:

- **Message Processing**   Teleprocessing messages are unwieldy for processing in application programs because of the variable formatting and awkward control character requirements. The monitor should provide comprehensive message editing and formatting (mapping) services to isolate applications from this complexity. All supported terminals should appear identical to the application to permit device independence. Preferably, CRT formatting specifications can be automatically produced, based upon designing model formats from the devices themselves.

- **Single/Multiple Inputs**   The monitor should permit the devices to enter multiple messages when logical responses are not required, as in data collection operations. But an update transaction should cause logical rejection of a second message if the first message has not been acknowledged with a program output. This allows interactive conversational processing and high-speed data entry.

- **Security**   This should be related to a specific device transaction, and/or account code, so that an operator must not only be knowledgable of the requisite security code but also have physical access to a device and be authorized to enter specific transactions.

- **CRT Control**   On CRT devices the timing of the outputs must be consistent with operator utilization: a second output should not be sent while an operator is constructing an input. The monitor should be aware of and appropriately manage this CRT processing.

- **Screen Browsing**   An interactive page browsing technique is most appropriate for lengthy, multiscreen outputs. With this facility, the monitor saves the multiple screens and permits the CRT operator to request forward and backward paging as well as specific page number retrievals.

- **Message Switching/Broadcasting**   This should be an automatic monitor function and be linked with the CRT processing to ensure that message switching does not overlay screens in process.

- **Transaction Code**   A transaction code can be specified in every input but can be more effective if these techniques can be available: operator sign-on to transaction code, provided in the screen format, or program controlled.

The communications support should be table-driven and modular so that only specifically required options are implemented.

QUEUE MANAGEMENT

Queues are a vital part of a transaction processing system. The queue may contain messages, message streams, and/or data, depending on their usage. Generally, these queues fall into three broad categories:

- **Destination Queues** These are static queues of predefined message destinations: all of the programs and all of the terminals in the system. The purpose of these queues is to permit message processing to function asynchronously from network servicing. The telecommunications functions are relatively slow and costly. Therefore, network service should receive the highest priority. The destination queues permit messages to enter and leave the monitor at network speeds, despite any momentary backup in message processing capabilities. These queues can be either core, disk, or both. To minimize disk space, it is desirable to support queues which are blocked, spanned, and reusable. Queues whose disk block size must be maximum message size waste considerable space if messages are variable length. Optimally, the queues can be configured so that for input the normal message processing volumes can be accommodated in core, and any excess rolled to disk. For output queues only the next message need stay in core with any excess on disk. The queues should self-prime so that loading from disk to core occurs as the core frees up; without priming, transmission is slowed for disk retrieval.

- **Transient Data Queues** These are scratch pad areas used for temporary message storage, conversational processing save areas, working storage areas, system-wide table information storage, and other such transient uses. A transient queue element is a single string of data created on demand by a program and assigned an identity that can be used for subsequent retrieval. Optimally, these transient queues would have variable-length keys, and the program could specify at creation whether an element should be core- or disk-resident, based upon immediacy needs. If it is core-resident, the monitor could page out these areas upon excessive demand in a low-core situation.

- **Dynamic Data Queues** These are similar to the transient queues but contain multiple strings of data, such as multiple segments of a message, multiple related messages, multiple extracted data file records, etc. These queues should be variable-length keyed items. However, their nature makes disk residency a natural. Ideally, these queues could be interpartitional to allow passing messages and/or data between batch and on-line, or across multiple on-line systems. This dynamic data queuing facility will be useful for creating and managing lengthy printed reports, batching transactions for off-line processing, saving selected data file records for reprocessing, etc.

In all queuing situations, the monitor should provide for restart/recovery considerations. Since queues are such an integral monitor function, any recovery should be automatic without requiring added user coding or special considerations when application programs are written.

PROGRAM MANAGEMENT

The program management function is the heart of the transaction processing system. Effectively, the TP monitor is a mini operating system that schedules, executes, and supervises user applications in response to the real-time receipt of messages. The ideal monitor performs this function in an efficient manner and provides a simplified, transparent environment for the user application program.

The program scheduling function can be a critical element in the TP monitor. In a low-volume system, user programs can be demand-scheduled, i.e., executed in direct response to messages. However, in an overload situation, demand scheduling causes serious slowdowns when it exhausts the available real storage.

In a higher-volume system, a more intelligent program scheduling function can result in higher throughputs and lower response times. The ideal scheduling technique attempts to optimize storage since this is the recognized scarce resource. This high-volume scheduling technique operates cognizant of all the queued input and uses an algorithm to determine message scheduling. The algorithm includes message priority and program resource demand. This scheduling technique processes messages in a more effective sequence than the order of their random or burst arrival and processes parallel messages through a single reentrant user program. The scheduling priority should operate with a relative, round-robin technique. High-priority messages receive preferential, but not absolute system access.

An orderly and effective user program environment is a major goal for the transaction monitor. The program interface techniques and operating environment have a critical effect on application programming productivity. Programmer productivity has two aspects: required skill levels and programmer time per application. Time per program is to a great extent a function of any preprogrammed utility support and intrinsic monitor features which may greatly reduce coding requirements.

The factors affecting time and skills for user programming are a function of these system attributes:

- **Language Support**   It is important that the application programmers can write in their native language. Therefore, the monitor should support COBOL, PL/1, Assembler Language, and FORTRAN, as appropriate. In addition to the primary programming language, consider that a requirement may arise to support an application in another language. A COBOL-oriented insurance company often finds PL/1 or FORTRAN convenient for an actuarial application. Any limitation in the use of the language or the compilers should be carefully noted. If the monitor provides its own language or dialect, then this clear departure from the standard batch approach must be carefully evaluated.

• **Service Request Processing** Monitors differ widely in the techniques by which application programs access monitor services. The simplest approach is the use of standard CALL statements. Another approach has the application return to the monitor with status settings to indicate the desired service. More involved interfaces have macros or commands where a special structure is used to invoke services. Finally, the user program may be written in a special language that includes implicit monitor functions.

Again, the greater the disparity between the batch and the on-line approaches, the greater the likelihood of increased training, higher required skill levels, more costly maintenance, and lower productivity. Additionally, some interface approaches entail application program manipulation of system labels. This increased integration of application code and the monitor tables also increases training and skill and may lower productivity.

• **Reentrancy** To ensure fast response time in high volumes, the multithreaded monitor may not be enough; the applications themselves must have a multithread capacity, or be reentrant. PL/1 can be readily supported reentrantly. Assembler Language is easier to write reentrantly if special monitor macros are provided. COBOL reentrancy requires extensive monitor support if it is to be transparent to the code. Under some monitors, pseudoreentrancy can be achieved if the application itself performs such chores as obtaining and managing dynamic areas. Clearly, this programmer-involved approach to reentrancy affects training, skill, and productivity.

FILE/DATABASE MANAGEMENT

Timesharing systems are typified by many users doing different things with different data; transaction processing systems have many users performing the same functions on the same data files. Therefore, the transaction processing systems generally provide centralized file control. The purpose of centralized file control is data integrity. This feature prevents data anomalies caused by parallel message processing against the same structures and provides recovery following failures. The monitor file services must also be designed for storage and CPU efficiency.

To protect files from conflicting simultaneous updates, the monitor must reserve data being updated. This exclusive control can be on the file or the record level. Some monitors may restrict the exclusive control to only one record at a time, while others may permit the updating of a series of records, as is the typical requirement in many applications. If the file is to be simultaneously updated by both batch and on-line, as would be required in any 24-hour running system, this exclusive control must span all systems.

The centralized file control can facilitate restart/recovery by capturing before and after images of updated records onto scratch pad areas for immediate recovery or onto log files for total system crash recovery.

The centralized nature of file processing also readily permits gathering extensive file processing statistics. This feature facilitates usage-demand basis file record buffering to minimize storage usage. File-related control terminal commands should include file open and close commands.

Most of the preceding file-related features are automatically provided for a database file by a commercial DBMS linked to the monitor. It is relatively trivial to provide a simplistic Data Base/Data Communications (DB/DC) link, but an integrated support is highly complex. Comprehensive DB/DC support requires the implementation of the following activities:

- **Connection Establishment** This is where the bridge between the DB and the comprehensive DC system is built These DB/DC systems should function as separate operating system tasks. The database system will issue real or implied operating system wait state functions, which would destroy DC performance if it functioned under the same task as the DBMS. At the more sophisticated level, the DB and DC systems will operate as separate jobs within the operating system. The advantage of this multijob approach is that the DBMS can be shared by both on-line and batch. Further, being a separate job, the DBMS can be operational for longer and differing periods of time than the DC system. The DB/DC link is normally effected through an intertask or interregion Event Control Block (ECB) channel. (Two pairs of ECBs are shared by the DB/DC system, a pair for communication in each direction. One ECB in a pair is used by the sender of a command function to indicate the desire for an action by the recipient; the other ECB in the pair is used to acknowledge to the sender the receipt of the command.) At the time of the establishment of connection, a technique should also be implemented to detect the abnormal failure of the DB or DC system by the other system. If the DB system is to operate as a subtask to the DC system, it is normally attached by the DC system at this time.

- **Command Processing** To process a DBMS command issued by the on-line application, the monitor must intercept the request and pass it to the DBMS. This is normally accomplished by a routine operation under the monitor whose entry point is the same name as that in the DBMS. Therefore, when an application requests DBMS services, it will access this interface routine. The interface routine passes the command to the DBMS while allowing the monitor to perform other activities during command servicing.

- **Contingency Processing** The integrated DB/DC relationship should provide for these error conditions:

  Application failure When an on-line application program fails, any outstanding DBMS activities should be canceled, and the updating activities by this failing program should be reversed or canceled, if the DBMS software so provides.

Monitor failure   When the entire TP system fails, then all outstanding on-line application program activity should be canceled and provisions should be made for an orderly restoration of processing.

DBMS failure   If the DBMS fails, then the monitor can take two effective alternate paths (preferably at user option): either suspend database application programs and continue processing non-DB messages, or suspend all processing and prepare for an orderly recovery

• **Checkpointing**   The difficulties in recovery of on-line updating of files or databases stem from the parallel message processing activities of the monitor. In order to satisfy response time requirements successfully, the monitor is concurrently processing many messages, and these messages may all be updating files and databases. If a high-volume monitor recovers only those messages in process at failure time, it may ignore important intermessage relationships and thereby create data anomalies. The periodic establishment of common DB/DC system equilibrium points provides an effective point for recovery; these equilibrium points are commonly called checkpoints although they are not related to operating system checkpoints.

A checkpoint is a time instance where all updating activities are queue scheduled. The most effective checkpoint methodology would be as follows:

A time lapse or message count causes the monitor to begin checkpoint activities.

The monitor requests the DBMS to prepare (but not yet take) a checkpoint.

The DBMS brings concurrent batch programs into a checkpoint-ready state, queue-scheduled (quiesces) and then returns to the teleprocessing monitor.

The monitor quiesces the applications by delaying new update scheduling until after the checkpoint.

A time-out value ensures that this message quiescing is not excessively lengthy as a result of one long-running message. If such a situation is encountered, the checkpoint is canceled, and a new time interval or message count will select the next checkpoint attempt.

If the on-line programs can be quiesced (use count of zero) in a reasonable time period, then the DB is requested to take a checkpoint (buffer write out, etc.).

The monitor then takes an identical checkpoint by writing a special record to the message log file; update message processing immediately resumes.

The total time for checkpointing should be about one-half second. Since all nonupdate program monitor activity continues during checkpointing, end users should see little if any change in system performance.

- **Recovery** The recovery scheme must marry message recovery with data recovery. The DBMS will be recovered by data reversal to checkpoint. The monitor must ensure that if the updates caused by a message have been reversed, these messages be reexecuted. This is a synchronized message/database recovery.

TASK/RESOURCE CONTROL

The standard operating system services are designed primarily to support batch functions. Therefore, the higher the volume, the more the monitor must directly provide more efficient replacements for operating system services. The efficient, high-volume monitor must provide its own task and storage management. In high-volume environments, the monitor appears as a single task to the operating system, and it owns a large block of storage. The monitor suballocates CPU resources from its single task to all the candidate work activities (threads). CPU time and storage allocation control is called multithreading. This multithreading should be priority-driven (The user should be aware that this priority effect is significant only if the monitor becomes CPU-bound since priorities affect scarce resources only. Scheduling priority in conjunction with dispatching priority is nearly always more effective than dispatching priority alone.)

In medium-volume systems, the monitor may have several operating system tasks established at initialization time and then use these tasks to execute programs. This is called multitasking.

RESTART/RECOVERY

There are two different error situations confronting the transaction monitor: those that can be accommodated on-line and those which cause system termination and require a restart.

Following an error, a well-designed monitor can dynamically respond to the error conditions while continuing normal operation.

The monitor should trap data file errors, teleprocessing errors, and certain monitor specification errors—for example, queue full conditions, low storage situations, etc. Appropriate corrective action should be taken for each of these situations. The control terminal operator should be notified of the error situation and be provided corrective action commands where such would be meaningful and valid.

If an application program fails, the following should be performed:

- The control terminal should be notified of the error.

- An analysis of the error situation should be performed to determine if system integrity has been compromised, and if so, a graceful shutdown should be attempted. If not, the originating terminal for the failed message should be notified.
- An indicative dump with appropriate additional diagnostics should be obtained. Provisions should be available to display the dump on-line or immediately print it, even though the teleprocessing system is not terminating (many spooling systems do not start printing until job completion).
- A table indicator should instruct the monitor whether this program should be suspended for further messages following an error or remain a currently available program. The control terminal operator should have an override capability to this decision.
- All resources owned by this failing program should be freed.
- If possible, the monitor should dynamically reverse any file or database updated so far by this program (DBMS technology may limit this capability).
- Provisions should exist to permit a new program version to be dynamically invoked, if and when available, to replace this erroneous program.

A total system failure could result from power loss, machine problems, operating system problems, etc. In these cases, the monitor should provide for a rapid, warm restart/recovery for these activities:

- Message queues
- Programs in process at failure time
- Files
- Databases

If the monitor supports combined core and disk queuing, then a reconstruction technique is necessary, since core can be overlaid or unavailable following failures. If disk queues are involved, the monitor should have provisions for recovery of disk file destruction (headcrash, etc.).

A logging or journaling facility can provide information on those programs in process at failure time that will require restarting. The monitor should especially note messages that are restarted, so that if they are again in a failure situation, the monitor can protect itself from a potential recursive error loop situation.

The recovery of files and databases, as discussed previously, involves recording before/after file images and using those values to do file reversals, or if the file or its indexes are destroyed, to rebuild it.

Logging is the critical component for recovery. The accuracy and timeliness of log data affect the level of recovery. For this reason, the monitor should provide for synchronized logging, wherein associated activities are temporarily suspended until the log data are physically written. This would ensure that a file update has not been made without first recording a before image of the record on the log. Since synchronization can be costly in terms of time and resources, it should be selectively available on a file, program, and terminal basis, as should logging itself.

UTILITY FEATURES

As part of implementing a monitor, many functions can be incorporated that are effectively on-line or batch application functions but are of a generalized nature. The availability of these features can lessen the user's programming requirements. The following are the more common available and valuable utility features:

- Message switching/broadcasting/copying
- Message mapping: editing, conversion, formatting
- Data entry/verification/batching
- On-line table-driven file/database inquiry
- On-line table-driven file/database update
- Query languages
- Multiscreen browsing
- System simulation/performance prediction
- On-line screen format generation and maintenance
- Performance/network monitoring and control
- System/terminal/file/program utilization and accounting reports

PROGRAM/SYSTEM ISOLATION

The on-line system is a vehicle that provides for the growing requirement for services over many years. This growth can be a major problem for it often has an impact on a system's reliability. Specifically, on-line systems have encountered reliability problems in the following situations:

- On-line testing
- New application implementation
- Maintenance modifications to existing applications

- Different users operating concurrently, such as different departments in a corporate data center or different clients in a service bureau

Each of the above situations introduces potential system instability. A costly solution is to provide multiple, separate systems for these situations; a separate system for testing gives different or new users their own system. A more sensible approach is for the monitor to provide for program or system isolation. These are some of the available approaches to isolation:

- **Instruction Simulation** This is particularly effective for testing since it interprets each instruction prior to execution to determine if a core overlay or other error or unauthorized situation would result and can provide an instruction trace during this operation.
- **Program Isolation** This occurs when programs are separated from one another by hardware, either as separate jobs in separate partitions or by alternating the protect key facilities within a region during task switching. Although effective, this is a costly technique. This architecture does not readily support reentrancy, a significant limitation in a high-volume system.
- **System Isolation** This is similar to the above, but an entire multithread reentrant application system is isolated. Although it theoretically provides a lower level of separation, it is often logically acceptable. For example, with system separation, a failure in one order-entry program may bring down the order-entry system, but never the personnel system.

STATISTICS/ACCOUNTING

The monitor should provide the following information:

- **General System-Wide Information**

  File usage
  Core usage
  Program usage
  Terminal usage
  Other tuning information relevant to isolating roadblocks
  On-line statistics/status displays

- **Detailed Performance Information**: Response time for each different message type (includes at least mean, maximum, and standard deviation)

- **Accounting Information (for chargeback billing)**: For every message, with summary capability:

  CPU time

  I/O activity

  Storage utilization

  Resource utilization

OPERATING ENVIRONMENT AND GENERAL INFORMATION

The required environment for the monitor should be carefully examined, and information recorded on the following aspects:

- Operating systems supported
- Operating system modifications required (carefully evaluate any of these in light of the future direction of microcoded operating system services)
- Storage requirements; fixed storage (generally higher if the monitor has many features); variable storage (a function of monitor scheduling and storage management techniques)
- Disk space requirements
- Availability of monitor program source code
- Language in which the monitor coding is written
- Reentrant capabilities of monitor, that is, link pack eligibility.
- Tape or disk logging
- Product education and materials
- Completeness and accuracy of product documentation
- Product User Group and published proceedings
- User-contributed program distribution
- User exits availability

IMPLEMENTATION

There are two aspects to implementation considerations: product and user. The product implementation typically consists of some form of a customization process, such as a system generation. The requirements for this process should be investigated, but all products are generally the same. The major impact of the implementation process is to bring the user up to the necessary skill and experience levels that are required to utilize the system effectively.

A major aspect of monitor user skills will be training. Generally, the more features within the monitor, the greater will be the training requirements. You should carefully evaluate this education, since training can be a major factor in implementing effective, efficient, and reliable systems. In a complex, high-volume product, there should be formal education rather than just informal system engineering sessions. To evaluate the education program, ask to review the student class material, which should be of high quality and distinct from product documentation.

PERFORMANCE

The normal technique for evaluating performance would be to talk to other users operating similar systems.

Since no other monitor environment is identical to yours, a beneficial feature would be a system modeling facility whereby on your own machine you could run a model of the system that would be developed. This model system would have generated application programs that would execute against test files and experience all of the real overhead of the actual monitor, including storage requirements, channel contention, device contention, CPU overhead, operating system overhead, and such. At the conclusion of the model, the statistics should include response time and chargeback billing data.

COST

The monitor software may very well be the least costly item in an on-line system, but it is one of the most important items. It may be very difficult and costly to replace a monitor, once installed. The cost of a teleprocessing monitor also includes user programming productivity and hardware efficiency. The availability of features in one monitor may considerably reduce the programming efforts relative to another product. While it is difficult to enumerate all the variables, the following areas should be addressed:

- Product price
- Installation and training
- Cost of developing necessary features not available in the monitor, or requiring user exit code

VENDOR

Transaction processing products are prone to errors and require continuous development to remain state-of-the-art. Therefore, the product vendor and support must be examined closely. Carefully examine these attributes:

- Number of users of the entire product line
- Number of users with your configuration
- Length of time the product has been formally available to customers
- Vendor support and tracking system for user-reported problems, whether product, user, or nonproduct errors
- Frequency, availability and ease of application of product modifications, enhancements, and error corrections
- Is the current vendor the product owner or an agent?
- What has been the new release/enhancement history over the years; has the product remained state-of-the-art?
- How rapidly did the vendor interface the product with new operating system versions and upgrades?
- Who provides the local support, and what are their qualifications for such a complex, specialized environment?
- What is the financial viability of the product's developer (consider using a financial rating service)?
- When checking references, correlate the results with the complexity of the product and its usage at the referenced site. Small users are often less demanding of efficiency and features and may give uncharacteristically high ratings.

The key to the vendor analysis is their track record in this particular product. The general size and reputation of the vendor is not necessarily indicative of support for a specific product. The largest of hardware vendors have dropped support for many of their software products.

INDEX

485